IFIP Advances in Information and Communication Technology

623

Editor-in-Chief

Kai Rannenberg, Goethe University Frankfurt, Germany

IFIP – The International Federation for Information Processing

IFIP was founded in 1960 under the auspices of UNESCO, following the first World Computer Congress held in Paris the previous year. A federation for societies working in information processing, IFIP's aim is two-fold: to support information processing in the countries of its members and to encourage technology transfer to developing nations. As its mission statement clearly states:

IFIP is the global non-profit federation of societies of ICT professionals that aims at achieving a worldwide professional and socially responsible development and application of information and communication technologies.

IFIP is a non-profit-making organization, run almost solely by 2500 volunteers. It operates through a number of technical committees and working groups, which organize events and publications. IFIP's events range from large international open conferences to working conferences and local seminars.

The flagship event is the IFIP World Computer Congress, at which both invited and contributed papers are presented. Contributed papers are rigorously refereed and the rejection rate is high.

As with the Congress, participation in the open conferences is open to all and papers may be invited or submitted. Again, submitted papers are stringently refereed.

The working conferences are structured differently. They are usually run by a working group and attendance is generally smaller and occasionally by invitation only. Their purpose is to create an atmosphere conducive to innovation and development. Refereeing is also rigorous and papers are subjected to extensive group discussion.

Publications arising from IFIP events vary. The papers presented at the IFIP World Computer Congress and at open conferences are published as conference proceedings, while the results of the working conferences are often published as collections of selected and edited papers.

IFIP distinguishes three types of institutional membership: Country Representative Members, Members at Large, and Associate Members. The type of organization that can apply for membership is a wide variety and includes national or international societies of individual computer scientists/ICT professionals, associations or federations of such societies, government institutions/government related organizations, national or international research institutes or consortia, universities, academies of sciences, companies, national or international associations or federations of companies.

More information about this series at http://www.springer.com/series/6102

Zhongzhi Shi · Mihir Chakraborty ·
Samarjit Kar (Eds.)

Intelligence Science III

4th IFIP TC 12 International Conference, ICIS 2020
Durgapur, India, February 24–27, 2021
Revised Selected Papers

 Springer

Editors
Zhongzhi Shi
Institute of Computing Technology
Chinese Academy of Sciences
Beijing, China

Mihir Chakraborty
Jadavpur University
Kolkata, India

Samarjit Kar
Department of Mathematics
National Institute of Technology Durgapur
Durgapur, India

ISSN 1868-4238   ISSN 1868-422X (electronic)
IFIP Advances in Information and Communication Technology
ISBN 978-3-030-74828-9   ISBN 978-3-030-74826-5 (eBook)
https://doi.org/10.1007/978-3-030-74826-5

This Springer imprint is published by the registered company Springer Nature Switzerland AG
The registered company address is: Gewerbestrasse 11, 6330 Cham, Switzerland

Preface

This volume comprises the proceedings of the 4th International Conference on Intelligence Science (ICIS 2020). Artificial intelligence research has made substantial progress in some special areas so far. However, deeper knowledge on the essence of intelligence is far from sufficient and, therefore, many state-of-the-art intelligent systems are still not able to compete with human intelligence. To advance the research in artificial intelligence, it is necessary to investigate intelligence, both artificial and natural, in an interdisciplinary context. The objective of this conference series is to bring together researchers from brain science, cognitive science, and artificial intelligence to explore the essence of intelligence and the related technologies. The conference provides a platform for discussing some of the key issues facing intelligence science.

For ICIS 2020, we received more than 42 papers, of which 28 were included in the conference program as regular papers. All of the papers submitted were reviewed by three referees. We are grateful for the dedicated work of both the authors and the referees, and we hope that these proceedings will continue to bear fruit over the years to come.

A conference such as this could not succeed without help from many individuals, who contributed their valuable time and expertise. We want to express our sincere gratitude to the Program Committee members and referees, who invested many hours for reviews and deliberations. They provided detailed and constructive review reports that significantly improved the papers included in the conference program.

We are very grateful for the sponsorship of the following organizations: the Chinese Association for Artificial Intelligence (CAAI), the International Federation for Information Processing Technical Committee on Artificial Intelligence (IFIP TC12), the China Chapter of the International Society for Information Studies, and the National Institute of Technology, Durgapur, India, supported by the Institute of Computing Technology, Chinese Academy of Sciences, China. Thanks go to Professor Samarjit Kar as Chair of the Organizing Committee. We specially thank Shi Wang and Zeqin Huang for carefully checking materials for the proceedings.

Finally, we hope that you find this volume inspiring and informative.

March 2021
Zhongzhi Shi
Mihir Chakraborty

Organization

Sponsors

Chinese Association for Artificial Intelligence (CAAI)
China Chapter under International Society for Information Studies

Organizer

National Institute of Technology, Durgapur, India

Support

IFIP Technical Committee 12

General Chairs

Anupam Basu	NIT Durgapur, India
Yixin Zhong	Beijing University of Posts and Telecommunications, China

Program Chairs

Zhongzhi Shi	Institute of Computing Technology, Chinese Academy of Sciences, China
Mihir Chakraborty	Jadavpur University, India

Program Committee

Goutam Chakraborty, India
Mihir Chakraborty, India
Zhihua Cui, China
Sujit Das, India
Debabrata Datta, India
Debi Prosad Dogra, India
Akio Doi, Japan
Jiali Feng, China
Samarjit Kar, India
Hongbo Li, China
Yujian Li, China
Peide Liu, China
Chenguang Lu, China

Xudong Luo, China
Jinwen Ma, China
Bijay B. Pal, India
Zhongfeng Qin, China
Partha Pratim Roy, India
Chuan Shi, China
Zhongzhi Shi, China
Arif Ahmed Sk, Norway
Andrzej Skowron, Poland
Tanmoy Som, India
Nenad Stefanovic, Serbia
Guoying Wang, China
Shi Wang, China
Pei Wang, USA
Min Xie, China
Yinsheng Zhang, China
Chuan Zhao, China
Yixin Zhong, China
Fuzhen Zhuang, China
Xiaohui Zou, China

Organization Committee

Chairs

Zhao Chang   University of Utah, USA
Samarjit Kar   NIT Durgapur, India

General Secretaries

Parag K. Guha Thakurta   NIT Durgapur, India
Bibhash Sen   NIT Durgapur, India

Local Arrangement Chairs

Durbadal Mandal   NIT Durgapur, India
Partha Bhowmik   NIT Durgapur, India

Finance Chairs

Sajal Mukhopadhyay   NIT Durgapur, India
Ashis Kumar Dhara   NIT Durgapur, India

Publication Chairs

Tandra Pal   NIT Durgapur, India
Chiranjib Koley   NIT Durgapur, India

Publicity Chairs

Sudit Mukhopadhyay	NIT Durgapur, India
Tanmay De	NIT Durgapur, India

International Liaison

Suchismita Roy	NIT Durgapur, India
Seema Sarkar Mondal	NIT Durgapur, India

Contents

Intelligent Robot

Medical Artificial Intelligence

Extended Abstract

Brain Cognition

Mind Modeling in Intelligence Science

Zhongzhi Shi[✉]

Key Laboratory of Intelligent Information Processing, Institute of Computing Technology,
Chinese Academy of Sciences, Beijing 100190, China
shizz@ict.ac.cn

Abstract. Intelligence Science is an interdisciplinary subject which dedicates to joint research on basic theory and technology of intelligence by brain science, cognitive science, artificial intelligence and others. Mind modeling is the core of intelligence science. Here mind means a series of cognitive abilities, which enable individuals to have consciousness, sense the outside world, think, make judgment, and remember things. The mind model consciousness and memory (CAM) is proposed by the Intelligence Science Laboratory. The CAM model is a framework for artificial general intelligence and will lead the development of a new generation of artificial intelligence. This paper will outline the age of intelligence, mind model CAM, brain computer integration.

Keywords: Intelligence science · Age of intelligence · Mind modeling · CAM · ABGP agent

1 The Age of Intelligence

In the summer of 1956, John McCarthy, a young assistant professor at Dartmouth University, Minsky at Harvard University, Shannon at Bell Labs and Rochester, the information research center of IBM, initiated the conference. They invited Newell and Simon from Carnegie Mellon University, Selfridge and Solomon from MIT, and Samuel and More from IBM. Their research majors include mathematics, psychology, neurophysiology, information theory and computer science. They are interdisciplinary and discuss the possibility of artificial intelligence from different perspectives. McCarthy first introduced the term of artificial intelligence (AI) in the proposal for the Dartmouth summer research project on artificial intelligence. He defined AI as "making a machine's response way like the intelligence on which a person acts". Dartmouth conference marks the official birth of artificial intelligence.

For more than 60 years, the heuristic search and non-monotonic reasoning put forward by AI scholars have enriched the methods of problem solving. The research of big data, deep learning, knowledge discovery and so on has promoted the development of intelligent systems and achieved practical benefits. The progress of pattern recognition has enabled the computer to have the ability of listening, speaking, reading and seeing to a certain extent.

© IFIP International Federation for Information Processing 2021
Published by Springer Nature Switzerland AG 2021
Z. Shi et al. (Eds.): ICIS 2020, IFIP AICT 623, pp. 3–12, 2021.
https://doi.org/10.1007/978-3-030-74826-5_1

On February 14–16, 2011, Watson, the IBM artificial intelligence system, defeated two "Ever Victorious generals" Jennings and Rutter in the famous American quiz show jeopardy. From March 9 to 15, 2016, AlphaGo of Google DeepMind adopted deep reinforcement learning and Monte Carlo search algorithm to beat the Korean go champion Li Shisha 4:1.

After a continuous dialogue with other Silicon Valley technology tycoons in 2015, Elon Musk decided to jointly create OpenAI, hoping to prevent the catastrophic impact of AI and promote AI to play a positive role. On December 12, 2015, OpenAI, a non-profit artificial intelligence project, was officially launched. OpenAI put forward GPT model in improving language understanding by generative pre training. GPT-2, announced in February 2019, is an unsupervised translational language model trained on 8 million documents with a total of 40 Gb of text.

In May 2020, OpenAI released GPT-3, which contains 175 billion parameters more two orders of magnitude parameters than GPT-2 with 1.5 billion parameters, which is a great improvement over GPT-2. GPT-3 has achieved strong performance on many NLP datasets, including translation, question answering and cloze tasks, as well as tasks requiring immediate reasoning or domain adaptation, such as using a new word in a sentence or performing 3-digit operations. GPT-3 can generate news article samples that are difficult for human evaluators to distinguish. For GPT-3, its biggest value is the ability of self-learning without supervision, and the performance improvement can be achieved by simply expanding the scale.

In 1972, Dr. Christian Anfinsen, the Nobel Prize winner in chemistry, said that theoretically, the amino acid sequence of a protein should be able to completely determine its 3D structure. This hypothesis has inspired 50 years of research on protein 3D structure prediction based on amino acid sequence. International protein structure prediction competition (CASP) selects the protein structures that have been analyzed experimentally, but not yet published, as targets, and let research teams around the world use their own computational methods to predict their structures. In this year's CASP, the AlphaFold system predicted a median GDT score of 92.4 for all protein targets. Even for the most difficult protein targets, AlphaFold had a median GDT score of 87.0. Among the nearly 100 protein targets tested, AlphaFold predicted two-thirds of the protein targets with the same predicted structure as the experimental results.

From above major events we can see that the arrival of the age of intelligence. Intelligence science is an interdisciplinary subject which is jointly studied by brain science, cognitive science, artificial intelligence and others. Intelligence science not only to conduct functional simulation of intelligent behavior, but also should explore on the mechanism to explore new theory of intelligence, new technologies. It has penetrated into all fields of society and aroused extensive concern and research.

2 Mind Modeling

Mind is all of human being spiritual activities, including emotion, will, feeling, perception, image, learning, memory, thought, intuition and so on. It is frequent to utilize modern scientific method to study the form, process and law of human irrational psychology and rational cognition's integration. The technology of establishing mind model is often called the mind modeling, with purpose of exploring and studying the mechanism of human's thought in some ways, especially the human information processing mechanism, which also provides the design of corresponding artificial intelligence system with new architecture and technology [1].

Mind problem is a very complex nonlinear problem, and mind world must be studied by modern scientific method. Intelligence science is researching on a psychological or mind process, but it is not a traditional psychological science. It must look for the evidence of neurobiology and brain science, so as to provide a certainty basis for the mind problems. The mind world is different from the world described by modern logic and mathematics: the latter is a world without contradiction, while the mind world is full of contradiction. Logic and mathematics can only use deductive reasoning and analytical method to understand and grasp possible world, while human's mind can master the world in many ways such as deduction, induction, analogy, analysis, synthesis, abstraction, generalization, association and intuition. So mind world is more complex than the latter. So, from the poor, non-contradictory, deductive and relatively simple possible world, how can we enter the infinite, contradictory, using multiple logic and cognitive approach, more complex mind world? This is one of the basic issues of intelligence science.

Generally, models are used to express how mind works and understand the working mechanism of mind. In 1980, Newell first proposed the standard of mind modeling [2]. In 1990, Newell described human mind as a set of functional constraints and proposed 13 criteria for mind [3]. In 2003, on the basis of Newell's 13 criteria, Anderson et al. Put forward Newell test [4] to judge the criteria to be met by human mind model and the conditions needed for better work. In 2013, the literature [5] analyzed the standards of mind models. In order to construct mind model better, the paper proposed the criteria of mind modeling:

- Behave Flexibly
- Adaptive Behavior
- Real Time
- Large-Scale Knowledge Base
- Dynamic Behavior
- Knowledge Integration
- Use Language
- Consciousness
- Learning
- Development
- Evolution
- Brain

So far, there are many mind modeling methods, including symbolic based modeling, neural network based on connectionism mind modeling, agent based mind modeling, math based mind modeling, computer simulation based mind modeling, hybrid mind modeling. Here we only introduce symbolic mind model, connectionist mind model and hybrid mind model.

2.1 Symbolic Mind Model

The 1975 ACM Turing Award was presented jointly to Allen Newell and Herbert A. Simon at the ACM Annual Conference in Minneapolis, October 20, 1976. They gave Ten Turing Lecture entitled "Computer science as empirical inquiry: symbols and search". At this lecture they presented a general scientific hypothesis—a law of qualitative structure for symbol systems: the Physical Symbol System Hypothesis [6]. A physical symbol system has the necessary and sufficient means for general intelligent action. By "necessary" it means that any system that exhibits general intelligence will prove upon analysis to be a physical symbol system. By "sufficient" it means that any physical symbol system of sufficient size can be organized further to exhibit general intelligence. By "general intelligent action" it wishes to indicate the same scope of intelligence as we see in human action: that in any real situation behavior appropriate to the ends of the system and adaptive to the demands of the environment can occur, within some limits of speed and complexity. The Physical Symbol System Hypothesis clearly is a law of qualitative structure. It specifies a general class of systems within which one will find those capable of intelligent action.

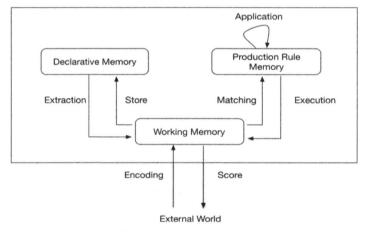

Fig. 1. Architecture of ACT

In 1983, Anderson elaborated the basic theory of ACT-R from every aspect of psychological processing activities in the book "cognitive structure". The general framework of ACT consists of three memory components: working memory, declarative memory and procedural memory in Fig. 1 [7].

- declarative memory, which is semantic network composed by interconnected concepts with different active intensities.
- procedural memory, a series of production rules of procedural memory.
- working memory, which contains the information that is currently active.

Declarative knowledge is represented by chunks. Chunk is similar to graph structure. Each chunk can encode a group of knowledge. Declarative knowledge is able to be reported, and not closely associated with the context. However, procedural knowledge usually cannot be expressed. It is automatically used and is targeted to be applied to specific context. It was tested that a variety of methods can be used to store information in declarative memory and then extract the stored information. The matching process is to find the correspondence between the information in working memory and conditions of production rules. The execution process is to put the actions generated by successfully matched production rules to the working memory. Matching all production activities before execution is also known as production applications. The last operation is done by the working memory. Then the rule execution is achieved. Through the "application process", the procedure knowledge can be applied to its own processing. By checking existing production rules, new production rules can be learned. To the greatest extent, Anderson explained skill acquisitions as knowledge compilation. It is the realization of transformation of declarative knowledge to procedural knowledge. Knowledge compilation has two sub-processes: the procedure programming and synthesis.

Procedure programming refers to the process of transforming declarative knowledge into procedural knowledge or production rules. Problem solvers initially solve the problem, such as mathematics or programming, according to knowledge obtained from books. In process of problem solving, the novice will combine weak problem solving methods, such as hill-climbing method or the method-goal analysis, to produce many sub-goals, and produce declarative knowledge. When repeatedly solve an event in the problem, some special declarative knowledge will be extracted repeatedly. At this moment, a new production rule is generated. While an application can learn new production rules indicates that according to ACT theory the process of learning is "learning from doing". Program knowledge can be described as a mode (IF part of production rule), and the action to be performed is described as an action (THEN part of production rule). This declarative knowledge to procedural knowledge transformation process, will also lead to reduction of processing on test speech. Related to this, the degree of automation on problem-solving will be increased.

In ACT-R, the Learning is realized by the growth and adjustment of micro knowledge units. This knowledge can be combined to produce complex cognitive processes. In the learning process, the environment plays an important role, because it established the structural problem object. This structure can assist chunk learning, and promote the formation of production rules. The importance of this step is that it re-emphasizes the importance of the essential characteristics in analytical environment for understanding human cognition. However, since the demise of behaviorism, cognitive revolution rise, this point was neglected.

2.2 Connectionist Mind Model

Adaptive Resonance Theory (ART) was proposed by S. Grossberg of Boston University in 1976 [8]. There are many versions of ART. ART1 is the earliest version. ART1 contains a master-slave algorithm with parallel architecture. It uses set operation in the activation and matching functions of the algorithm. It mainly deals with the problem of image recognition (i.e. black and white) with only 0 and 1. ART2 can process gray scale (i.e. analog value) input. ART3 has a multi-level search architecture, which integrates the functions of the first two structures and expands the two-layer neural network to any multi-layer neural network. Since ART3 incorporated the bio electrochemical reaction mechanism of neurons into the operation model of neurons, its function and ability were further expanded.

The basic architecture of ART1 is shown in Fig. 2. It consists of three parts: attention subsystem, adjustment subsystem and gain control. Attention subsystem and adjustment subsystem are complementary. ART model deals with familiar or unfamiliar events through the interaction between these two subsystems and control mechanism. In the attention subsystem, there are F_1 and F_2, which are composed of short-term memory units, namely STM-F_1 and STM-F_2. The connecting channel between F_1 and F_2 is long-term memory LTM. Gain control has two functions: one is used to distinguish bottom-up and top-down signals in F_1; the other is that F_2 can play a threshold role for signals from F_1 when input signals enter the system. The adjustment subsystem consists of A and STM reset wave channels.

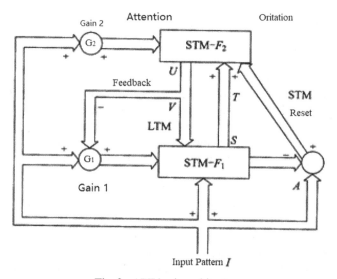

Fig. 2. ART basic architecture

In ART model, its working process adopts 2/3 rule. The so-called 2/3 rule is that in ART network, two of the three input signals must work to make neurons produce output signals. When double input from bottom up, among the three input signal sources of F_1, there is input signal I for input and input to F_1 generated after I passes gain control 1. Because these two input signals work, neurons in F_1 are activated and F_1 can generate signal output.

In 2008, Grossberg and Massimiliano put forward SMART (synchronous matching adaptive resonance theory) model [9], in which the brain coordinates a multi-stage thalamus and cortex learn process and stabilizes important information out of memory. SMART model shows bottom-up and top-down paths that work together and through coordinating the several processes of learning expectations, focus, resonance and synchronize to complete the above objectives. In particular, SMART model explains how to achieve needs about concentrating on learning through the brain subtle loop, especially the cell hierarchical organization in the new cortex loop.

The SMART model made different vibration frequencies associated with the spike timing-dependent plasticity (STDP) together. If the average incentive of presynaptic and postsynaptic cells is 10~20 ms, in other word, in STDP learning window, learning scenarios will be more easily to be restricted to the matching conditions. This model predicts STDP will further strengthen sync excitement in the regions of the related cortical and subcortical, and the long-term memory weights can be matched synchronous resonance to prevent or quickly reversed in fast learning rules. In the matched condition, the amplified γ vibration made the presynaptic excitement compress into a narrow time-domain window, which will help excitement to spread over the cortex hierarchical structure. The prediction is consistent with the rapid reduction of effects, postsynaptic excitatory of the observed lateral geniculate nucleus.

2.3 Hybrid Mind Model

In the mind activities, memory and consciousness play the most important role. Memory stores various important information and knowledge. Consciousness make human having the concept of self, according to the needs, preferences based goals, and do all kinds of cognitive activity according to memory information. Therefore, the main emphasis on mind model CAM are memory functions and consciousness functions [10]. Figure 3 shows the architecture of the mind model CAM, which includes 5 main modules: memory, consciousness, high-level cognitive functions, perception and effectors.

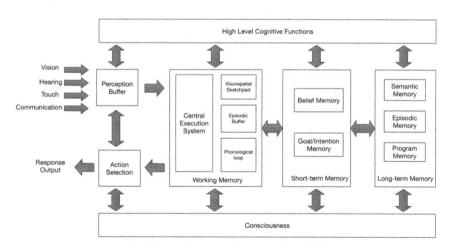

Fig. 3. Architecture of mind model CAM

3   Brain Computer Integration

Brain computer integration is a new form of intelligence, which is different from human intelligence and artificial intelligence. It is a new generation of intelligence science system which combines physics and biology. The brain computer integration intelligence is different from human intelligence and artificial intelligence in the following three aspects: (a) at the intelligent input, the idea of brain computer integration intelligence not only depends on the objective data collected by hardware sensors or the subjective information sensed by human facial features, but also combines the two effectively, and forms a new input mode with the prior knowledge of person; (b) in the stage of information processing, which is also an important stage of intelligence generation, a new way of understanding is constructed by integrating the cognitive mode of human beings with the computing ability of computer advantages. (c) at the output end of intelligence, the value effect of human beings in decision-making is added into the algorithm of gradual iteration of computer to match each other, forming an organic and probabilistic coordination optimize judgment. In the continuous adaptation of human computer integration, people will consciously think about inertia common sense behavior, and machines will also find the difference of value weight from the decision-making under different conditions of people.

Brain computer integration adopts a hierarchical architecture. Human beings analyze and perceive the external environment through the acquired perfect cognitive ability. The cognitive process can be divided into perception and behavior layer, decision-making layer, memory and intention layer, forming mental thinking. The machine perceives and analyzes the external environment through detecting data, and the cognitive process can be divided into awareness and actuator layer, planning layer, belief and motivation layer, forming formal thinking. The same architecture indicates that humans and machines can merge at the same level, and cause and effect relationships can also be generated at different levels. Figure 4 is the cognitive model of brain computer integration [11]. In

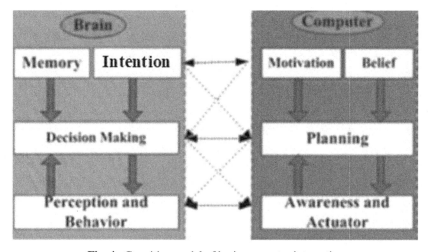

Fig. 4. Cognitive model of brain computer integration

this model the left part is simulated human brain in terms of CAM mind model; the right part is computer based on ABGP agent.

In brain computer integration, each brain and computer can be viewed as an agent playing special role and work together for a sharing goal. The agent cognitive model is illustrated in Fig. 5 [12], which agents deliberate the external perception and the internal mental state for decision-making. The model is represented as a 4-tuple: < Awareness, Belief, Goal, Plan > . Awareness has four basic characteristics:

(a) Awareness is knowledge about the state of a particular environment.
(b) Environments change over time, so awareness must be kept up to date.
(c) People maintain their awareness by interacting with the environment.
(d) Awareness is usually a secondary goal—that is, the overall goal is not simply. Awareness is described by the basic elements and relationships related to the agent's setting.

Belief can be viewed as the agent's knowledge about its environment and itself. Goal represents the concrete motivations that influence an agent's behaviors. Plan is used to achieve the agent's goals. Moreover, an important module motivation-driven intention is used to drive the collaboration of cyborg intelligent system.

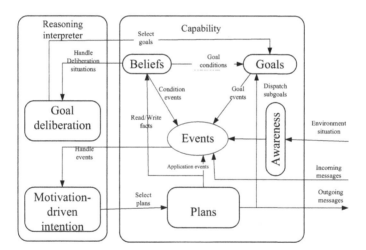

Fig. 5. Agent cognitive model

4 Conclusions

Mind modeling is the core of intelligence science. This paper described the symbolic mind model ACT-R, connectionist mind model SMART, hybrid mind model CAM. The mind model consciousness and memory (CAM) is proposed by the Intelligence Science Laboratory. The CAM model is a framework for artificial general intelligence and will lead the development of a new generation of artificial intelligence. This paper took brain computer integration as an application example of mind model CAM.

Intelligence science is an interdisciplinary subject which is jointly studied by brain science, cognitive science, artificial intelligence and others. Intelligence science not only to conduct functional simulation of intelligent behavior, but also should research on the mechanism to explore new theory and technologies for promoting the development of intelligence age.

Acknowledgements. This work is supported by the National Program on Key Basic Research Project (973) (No. 2013CB329502), National Natural Science Foundation of China (No. 61035003), National Science and Technology Support Program (2012BA107B02).

References

1. Shi, Z.Z.: Mind Computation. World Scientific, New Jersey (2017)
2. Newell, A.: Physical symbol systems. Cogn. Sci. **4**, 135–183 (1980)
3. Newell, A.: Unified Theories of Cognition. Harvard University Press, Cambridge Mass (1990)
4. Anderson, J.R., Lebiere, C.L.: The Newell test for a theory of cognition. Behav. Brain Sci. **26**, 587–637 (2003)
5. Shi, Z.Z.: Intelligence Science Is The Road To Human Level Artificial Intelligence. Keynotes Speaker, IJCAI WIS-2013 (2013)
6. Newell, A., Simon, H.A.: Computer science as empirical inquiry: symbols and search. In: Communications of the Association for Computing Machinery, ACM Turing Award Lecture, vol. 19, no. 3, pp. 113–126 (1976)
7. Anderson, J.R.: The Architecture of Cognition. Harvard University Press, Cambridge (1983)
8. Grossberg, S.: Adaptive pattern classification and universal recoding: I. parallel development and coding of neural detectors. Biol. Cybern. **23**, 121–134 (1976)
9. Grossberg, S., Versace, M.: Spikes, synchrony, and attentive learning by laminar thalamocortical circuits. Brain Res. **1218**, 278–312 (2008)
10. Shi, Z.Z.: On intelligence science. Int. J. Adv. Intell. **1**(1), 39–57 (2009)
11. Shi, Z.Z., Huang, Z.Q.: Cognitive model of brain-machine integration. In: Hammer, P., Agrawal, P., Goertzel, B., Iklé, M. (eds.) AGI 2019. LNCS (LNAI), vol. 11654, pp. 168–177. Springer, Cham (2019). https://doi.org/10.1007/978-3-030-27005-6_17
12. Shi, Z.Z., Yue, J.P., Ma, G., Yang, X.: CCF-Based awareness in agent model ABGP. In: Dillon, T. (ed.) IFIP AI 2015. IAICT, vol. 465, pp. 98–107. Springer, Cham (2015). https://doi.org/10.1007/978-3-319-25261-2_9

Some Discussions on Subjectivity of Machine and its Function

Chuyu Xiong[(✉)]

4 Merry Lane, Jericho, NY 11753, USA

Abstract. Many people think that, in order to improve the capabilities of a machine, the machine should have its initiative. Consequently, in order to make machine having its initiative, the consciousness of machine should be established. Here, we propose that instead of consciousness, the subjectivity of the machine should be considered. Thus we can avoid unnecessary arguments, and conduct fruitful discussions. We analyze the subjectivity of a machine and propose a working definition, and discuss 4 major aspects of subjectivity, namely, a priori and cogitating, active perception to outside, self awareness, and dynamic action. Dynamic action of a machine is crucially important and we suggest a possible way to approach it.

Keywords: Subjectivity of machine · A priori · Cogitating · Active perception · Self awareness · Dynamic action

1 Introduction

Currently, machines are able to do a lot of seemingly intelligent things, such as playing Go, recognizing human face, etc. On the other side, it also appears that machine has a lot of difficulties for more things, such as automatic driving, robot, and much more. We want to develop machines that can do these things. People have found that most difficulties for these tasks stem from this situation: machines lack initiative, consequently hard to adapt to complicated/novel situations. Contrast to this, human has conscious, so human can form a high degree of awareness of his environment and himself, and based on the awareness form a high degree of initiative. Thus, human indeed have much more abilities. Naturally, many researchers think that, in order to make machines become more capable, it is best to have some ways to equip machines with a kind of consciousness.

But, it is super difficult to go this direction. Consciousness of human is a major scientific challenge and whole scientific community are working hard to understand it. Yet, no satisfactory result has been achieved, and we do not expect

---

Great thanks for whole heart support of my wife. Thanks for Internet and research contents contributers to Internet.

Z. Shi et al. (Eds.): ICIS 2020, IFIP AICT 623, pp. 13–24, 2021.
https://doi.org/10.1007/978-3-030-74826-5_2

to see break through in foreseeable future. So far, people can only roughly argue about consciousness of machine. In fact, whether or not a machine can have consciousness is still an open question, and big controversial one.

We can step back and ask: do we must establish consciousness for machine first? Let's see some common explanations. According to wikiPedia[1]: "Consciousness, at its simplest, is sentience or awareness of internal or external existence." Still according to wikiPedia[2]: "subjectivity is: Something being a subject, broadly meaning an entity that has agency, meaning that it acts upon or wields power over some other entity (an object)." These explanations give us an useful hint.

The hint is: From perception to action, there are a huge amount of processes. We believe consciousness is actively involving with these processes (which is the reason why consciousness is so powerful). However, is it possible that we can distinguish these processes into different layers? Further, can we concentrate on the layers more close to action? We believe the answer is YES.

To help to explain this thought, we consider one example. When a car driver is approaching a stop sign, seems a lot of processes happen inside him: first, his perception channels activated, and excites several parts inside his brain, then these parts mutually communicates, all such communications form his awareness to the constant changing situation, and form his understanding to situation, then based on such understanding, his action is determined and executed. Consciousness involves with all these processes. So, people think that in order to have correct actions suiting to constantly changing situations, consciousness is necessary. However, when we look carefully, this example actually shows this point: there are layers, the most top layer is consciousness, then down to initiative, then down to subjectivity, which directly connects to action. This part, i.e. after consciousness activated and before executing action, is very interesting. This part is subjectivity.

Thus, we would like to propose: for the purpose to enhance capability of a machine, we can consider subjectivity of the machine, instead of consciousness of the machine. At least for now.

Of course, subjectivity of machine is not clear by this word itself. We do not have an appropriate and workable definition yet. However, it is already apparent that to discuss subjectivity is a much easier comparing to discuss consciousness. We are going to discuss subjectivity of machine in this article.

We will try to describe subjectivity of machine as reasonable as possible. The subjectivity, of course, is something happens inside the machine. Should we treat everything inside the machine as the subjectivity of the machine? There is no need to do that, and that is not right. Since our purpose is to enhance machine's capability, we can focus on those things that happen inside a machine and have important influences on the machine's capability. So, we ask what are these things?

---

[1] https://en.wikipedia.org/wiki/Subjectivity.
[2] https://en.wikipedia.org/wiki/Consciousness.

There are 4 kinds of important things: 1) a priori and cogitating; 2) active perception to the outside of the machine; 3) awareness of self; 4) dynamic action. They are 4 major aspects of subjectivity of machine. We can see that these 4 aspects can be studied in details and by solid technological approach, not like consciousness of machine that can not be studied technically now. We emphasize again, unlike the consciousness of machine, the 4 aspects can be clearly defined and understood. We can measure them, can build engineering model for them, and can develop scientific theory to study them. These 4 aspects form the working definition of subjectivity of machine.

Another point of view to see is: For human, subjectivity is after consciousness and before action. We can see that the 4 aspects fit quite well into this understanding. Such fitting supports the working definition.

Moreover, we can see that even if the machine has consciousness, when the consciousness works, it must work through the 4 aspects of subjectivity. That is to say, we need to establish the subjectivity of machine anyway.

Now, we can say that using subjectivity of machine, we can avoid unnecessary arguments and conduct productive research and development. However, as far as we know, there is no systematic research work on the subjectivity of machine. We have been working on universal learning machine for several years [2,4,6]. In the research process, we have consistent interest in the subjectivity of machine and its role in learning. We have thought over this topic for long. These thoughts are foundation of this article.

In this article we will not discuss how to realize the 4 aspects of subjectivity in a machine, which will be left to future and more in-depth researches. Following, in Sect. 2, we will talk more details of each aspect of subjectivity; in Sect. 3, we will see the functions of each aspect; in Sect. 4, we will discuss dynamic action, which is the most important aspect.

2 Major Aspects of Subjectivity of Machine

There are 2 ways to describe subjectivity of machine: One is behaviorism approach, another is mechanism approach. Behaviorism approach is to see the behavior of machine from outside, and use the behavior to understand the subjectivity of machine, while mechanism approach is to see how these aspects of subjectivity are formed and executed inside machine. Ideally, it is best to have both approaches and make them working together. However, due to the hardness and novelty of topic, we will consider the behaviorism approach more. In this way, we can at least achieve some good understanding. If possible, we will also try to see mechanism to these aspects.

However, there is one question that we have to address: Machines are controlled by programs, and programs are written and installed into machine by us. How can programs form subjectivity of machine? We believe that this will become clear as we examine 4 major aspects carefully.

2.1   A Priori and Internal Cogitating

A priori is the very essential part of machine. If the machine can learn, it must have a priori, otherwise, machine is impossible to learn. The learning is to adjust and modify according to external and internal information. In order to do such adjusting and modification, a machine must have some internal structures. At least a learning framework is needed, so that the external information can be used and internal information can be generated, and learning can be conducted through this framework [2,3]. Moreover, machine could have more, such as some knowledge that are part of what desired to learn. All these things together form a priori.

Look at the behavior of the machine. If the behavior is that response immediately follows stimulus by a fixed procedure, no matter how complicated and comprehensive the response procedure is, we think, the subjectivity is very small, even zero. However, in contrast, if we see such behavior that after stimulus, the machine cogitates before response, we think that the subjectivity is much higher.

So, if we see a machine can learn, and/or, it cogitates before action, then we say, machine must have a priori, and/or it must have internal cogitating. Such machine has higher subjectivity.

This aspect of subjectivity, i.e. a priori and cogitating, can be formed by programs inside machine. For learning, we have layout the basic principles [2,4]. For cogitating, it can be thought as a reasoning system (be simple or very complicated), and a decision is made after weighing concurrent outside and internal situations, not by pure fixed procedure. Most likely, these are realized by multiple programs working together.

2.2   Active Perception

Usually machines are equipped with some sensors to obtain information from the outside world. Many machines are like this: according to a preset program, a fixed sensor is used to get information according to a fixed channel and in a fixed range, and then put them as input. The key is to collect information within a preset range by a preset program. For this case, we think that the subjectivity of the machine is low or even zero, although the machine can obtain external information.

In such case, the capabilities of the machine are indeed limited. To improve the capabilities of the machine, it needs to actively perceive outside situation. If we see the behavior of machine having such characteristics, we think the subjectivity of the machine is higher. Active perception to outside can be done by programs working together.

2.3   Self Awareness

The machine's perception to self can be done by some sensors, just as the machine's perception to outside. However, many machines lack awareness of self, especially the awareness to their internal information processing. If the machine has awareness of self (no matter how small), it is a step to higher subjectivity. This is different from the perception to outside, for which, only active perception is considered as higher subjectivity. For awareness of self, no active is required. Of course, if the awareness of self is done actively, the subjectivity is even higher.

If we can observe a machine's behavior showing awareness of self, we know the machine has higher subjectivity. Awareness to self can be done by preset program. This is not a problem at all. The degree of awareness of self can be used as measure of subjectivity.

One point we should directly mention here: Awareness to self is indeed related to consciousness. But, awareness to self is well defined and we can fully understand it. This fact shows that considering subjectivity has advantages than considering consciousness.

2.4 Dynamic Action

What really is dynamic action of machine? Perhaps, we can first see what is dynamic action of people. When we are facing certain task, due to situation, we can act this way or that way, and do not have clear criteria to determine what to do. However, the possible ways are not equal. The consequences of actions are different (could significantly different). But, we do not know the consequences. So we can only judge actions after action is done. Facing with such situation, what do we do? Actually, facing with uncertainty, we try our best to act and often we indeed choose the best out from unknowns. This is dynamic action of people.

The dynamic action of machine should be understood in the same sense. That is, when a machine is facing uncertainty, how will it do? If from outside we can observe the behavior of machine to try its best, according to behaviorism approach, we can say the machine has dynamic action. Of course, "try its best" will be controversy. But, other than the possible controversy, the dynamic action can be clearly understood along this direction. Again, compare to consciousness, dynamic action is much easier to well define and understand clearly. This is advantage.

* * * * *

After discussed 4 major aspects of subjectivity of machine, we can come back to the question: Can programs in machine form subjectivity of machine? Now we can say, Yes. The program is put in by the designer of the machine. Therefore, the behaviors of the machine actually reflects the thoughts of the designer to some extent, and can even be regarded as the embodiment of the designer's subjectivity. If the machine is not a learning machine, then the subjectivity of the machine is completely determined by the designer's mind. If the machine is a learning machine, after learning, there is a distance between the machine's subjectivity and the designer's thinking, and this distance could be huge. In short, the subjectivity of the machine boils down to some properties formed by all programs in the machine. No doubt, subjectivity of machine need further researches. However, contrast to consciousness, subjectivity can be clearly defined and formed by technical steps. In fact, we have already briefly discussed how to form each aspects of subjectivity. For most important aspect, i.e. dynamic action, we will discuss more in Sect. 4.

3   Functions of Subjectivity of Machine

We will use some examples, Go game, image recognition, natural language processing and control systems, to see functions that the 4 aspects of subjectivity could play, which can us understand subjectivity better. These typical examples are chosen because they are easier to describe by us.

3.1   Go Game

The purpose of the Go program is to play Go game and win it. A Go program has two operating modes, one is learning and the other is competition. The two operating modes may be mixed, but distinguishing them can help our discussion. Let's talk about learning first. The learning is very crucial. At the beginning, the Go program only had a framework for learning, such as artificial neural networks and Monte Carlo searching trees. Relying on these, Go programs can gradually gain the ability to play chess through training. We can see that these neural networks and searching trees are a priori of the Go program, which has a decisive influence on the Go program. Without it, there is no Go program than can win human master.

During the game, the Go program uses the program established from learning. It will process input information. At this time, the program will form such cogitating: evaluate several possible moves, then pick the moves that the program thinks as the best (not necessarily with the highest evaluation value), and then output (i.e. play). If no move is satisfactory, the program can go back to evaluate more moves. The cogitating is indeed present here and is very essential. This shows the high ability of Go program.

The outside world of Go program is very simple: the board of game, which can be formalized as a 19×19-dimensional binary vector. Go program does not need any active perception to outside. Also, Go program needs no self awareness. But it might need dynamic action, which we will discuss later.

3.2   Image Recognition

An image recognition system also has two operating modes, one is training and another is working. During training, the image recognition system, as the Go program, relies on a priori for learning.

Some image recognition systems are so-called end-to-end. After this system is trained, it actually appears as a mathematical function (very complicated, though). An image (usually a bit array) is input and a vector (usually, probabilities of choices) is output. Therefore, such a system, there is no internal cogitating. Such a system shows very low subjectivity.

However, we could do much better. The amount of information input by these systems is much larger than that of the Go program (for example, compare 19×19 to $1024 \times 1024 \times 3$), and the complexity of the objects contained inside input is also much higher. Image recognition requires trade-off and take choices. For example, when a face recognition program processes input, in order to improve recognition efficiency, it is necessary to discard some of the insignificant information and focus on the key parts. But, which parts to focus? This is cogitating. A system with such ability has higher subjectivity.

Most of the current image recognition systems do not actively perceive. However, if an image recognition system wants to improve performance and improve recognition ability, active perception is required. For example, if you want to reduce energy consumption and save computing resources, you can use this approach: For 1024×1024 precision color images, first input low-precision images, such as 128×128 precision images, but then actively collect local high-precision images at the proper place. Since it is local, although with high precision, the amount of data is much smaller. In doing so, the efficiency will be greatly improved. In fact, the human eye works like this. Moreover, if in some special situations, such as defaced images, high-precision partial images of the defaced area and the surrounding area can be collected, the defacement can be overcome and the correct recognition can be achieved. Such an ability would be impossible to achieve without active perception. With active perception, the efficiency of the machine are improved, and subjectivity is higher.

Most of the existing image recognition programs have no need to aware themselves. But, for some situation, it can certainly do better with self-awareness. Suppose in an image recognition program, there is a competition between outcome A and outcome B. This situation is quite common. If there is no internal awareness, the image recognition program cannot know the specific situation of the competition, only knows the result, such as A is winning. What if the winning is only by a very small advantage? If program aware the situation, that is, only a very small advantage, program may have new initiatives, such as doing it again, re-entering data, etc. If this is possible, the efficacy will increase. And, subjectivity become higher.

Perhaps, most existing image recognition programs no need to have dynamic action. However, when the requirements to the programs are increased, such as increasing the recognition rate, reducing energy consumption, enabling it to work successfully under various difficult conditions, increasing the recognition speed, etc., it will needs dynamic action.

3.3 Natural Language Processing

A good natural language processing system should be online learning, that is, even at work, it is still in a learning mode, and the learned results are applied immediately to the current work. The ability of natural language processing is very dependent on a priori.

For active perception to outside, we can consider an automatic translation software (a special natural language processing). It could be so-called end-to-end software (many such programs indeed are), so it has no active perception, it throws everything into a huge artificial neural network and output of the network is the translation. The subjectivity in this case is quite low. However, to have active perception, translation software can do much better.

Besides to translation software, there are many kinds of natural language processing systems, such as speech recognition, press release writing, and so on. For some of them, active perception is absolutely necessary, such as, speech recognition in a noisy background.

A software for writing needs awareness of self, for example, it needs to recall what it writes before at some point, so it can write consistently later. If a natural language processing system can do dynamic action, it could perform much better.

3.4   Controlling System

There are many kinds of control systems, and many levels of complexity. We can use an automatic driving system as an example to discuss. Such system must be obtained from learning. So, a priori plays a central role in learning. In order to ensure high reliability and high security, this system cannot learn online, so after the training, the system is fixed. However, the internal cogitating of this system is very complicated. This is because the input of automatic driving are very complicated from many sources, and it is impossible to simply normalize. The system must select and sort the input information in real time, which is high level of active perception. It must also need to have a high level of self-awareness. Such system has a large knowledge base and will meet novel situations. System needs to act correctly when novel situation appears and such action must not conflict with existing knowledge as much as possible. This is dynamic action. Such a system inevitably shows extremely high subjectivity.

In traditional control system, feedback is integrated into the system without cogitating. So, feedback just appears as one term in control equation. This actually limits the ability of controlling system greatly. With cogitating and dynamic action, a controlling system will have a much stronger ability.

* * * * *

These examples demonstrate that a machine with subjectivity will have much better abilities. This is exactly what we want.

4   Further Discussions on Dynamic Action

From previous discussions, we can see that dynamic action is in the center. We know that a priori and cogitating can be completely realized by a preset program, but if a machine posses dynamic action, a priori and cogitating of the machine can be better. The active perception and the awareness of self can also be completely realized by a preset program. Yet, these aspects are very closely related to dynamic action. Therefore, dynamic action of machine is the most crucial aspect of subjectivity.

What really is dynamic action of a machine? How can a machine has dynamic action? The questions are not easy. We do not expect to discuss them without controversy. But, at least we should have a good working definition and find some approaches that shed light on it and lead us forward.

Since human and animal indeed have dynamic actions, so we trace literature of biology, sociology, and philosophy to see anything closely related. In fact, the concept very closely related to dynamic action is agency. According to wikiPedia[3]: "In social science, agency is defined as the capacity of individuals to act independently and to make their own free choices. By contrast, structure are those factors of influence (such as social class, religion, gender, ethnicity, ability, customs, etc.) that determine or limit an agent and their decisions." Also in wikiPedia[4]: "Agency is the capacity of an actor to act in a given environment." And, according to Open Education Sociology Dictionary[5]: "Definition of Agency is: The capacity of an individual to actively and independently choose and to affect change; free will or self-determination." These descriptions shed us light.

However, we would like to cite one sentence from a web article "Life with purpose" written beautifully by Philip Ball[6]: "One of biology's most enduring dilemmas is how it dances around the issue at the core of such a description: agency, the ability of living entities to alter their environment (and themselves) with purpose to suit an agenda." This sentence very well reflects our thoughts on dynamic action: it is a kind of ability of machine to see/sense/view/do/etc. by itself, which can make the machine's situation better.

So, we would like to use the above sentence as our working definition for dynamic action of a machine. We do not use the term agency (though we could). For machine, dynamic action is a better term, since it is more direct, more clear and sounds mechanic.

With such a working definition, we can consider further: Is it possible for a machine to have dynamic action? If so, how to realize it in a machine? This certainly is very controversial. Many people would argue: A machine can be reduced to a Turing machine. According to the computational theory, what a Turing machine can do is just a computable function. There is no way dynamic action could be generated by a clearly defined mathematical function. Such arguments are correct in some sense. But, we do not agree with such statements. We will discuss these issues by focus on machine's ability (by behaviorism approach). We will also touch a little on mechanism of dynamic action.

In the working definition, there are several critical parts. One is "by itself", another is "make better". Let's see them more closely.

How will a machine do *by itself*? Actually, this is a dilemma not only to machine, but also to many living things (as the sentence we cited above). But, if we concentrate on behavior, we can see things more clear. To illustrate, let's consider this simple and straightforward example. Consider a machine M and

---

[3] https://en.wikipedia.org/wiki/Agency_(sociology).

[4] https://en.wikipedia.org/wiki/Agency_(philosophy).

[5] https://sociologydictionary.org.

[6] https://aeon.co/essays/the-biological-research-putting-purpose-back-into-life.

there are some programs inside it. Consider two scenarios. Case A: M makes a judgment based on program C; Case B: same as case A, but in some situation, program D will be activated that modifies the parameters of C, thereby the behavior of M appears to have a special response to the situation by itself. So, for observers outside the machine, M in case B seems has higher subjectivity than in case A. This shows that even the behaviors of the machine are fully determined by some preset programs, the behaviors could show different levels of subjectivity. Therefore, from the perspective of behaviorism approach, a machine indeed can demonstrate ability to do "by itself". This is dynamic action to observer. Thus, machine can posses dynamic action, although it is done by preset programs.

For "*make better*", we can consider this example. Machine M, inside M there is a program C, which has 2 branches, A and B, and it is crucial to choose. There is a program D will be activated just before to choose branches, which will modify the parameters of C, thereby to help to choose branch. If D can get information about situation, and react according to the situation, then M appears to have ability to make situation better. So, to observers outside the machine, M demonstrates dynamic action to make situation better.

Above, we use examples to illustrate that a machine indeed can show subjectivity, "by itself", and "make better", if we stay outside machine to observe its behavior. But, we can go deeper into mechanism as well.

People often argue that although a machine could demonstrate dynamic action to outside observers, inside the machine it is actually a preset programs at working, so there is no dynamic action at all. They think that the computational theory denies dynamic action. It is not so simple. Let's see dynamic action from view of computational theory.

A persistent question lasting for decades in computational theory is whether Turing machines can perform non-computational tasks. Although it sounds very conflicting, that is, what a Turing machine can do is computable task, how can it complete non-computational tasks? But if we consider it carefully, the reason for objection is not sufficient, because we can use more than one Turing machine and use one Turing machine over and over again for the task. There are some previous works in this direction. In [8], Kugel's argument is: for a task, even it is not computable, it still is possible to use a Turning machine. He proposed a way to use Turing machines in order to treat tasks outside the computable range. He called such way to use Turing machines as Putnam-Gold machine. Such a way could be summarized as: to use Turning machine over and over again while increasing the critical capacity of Turing machine each time. Wang Pei [11] proposed another way. He propose to use several different procedures together, and these procedures are under resource competition, the procedures can be terminated not according to any preset program but according to the concurrent resource situation, thus forming a non-predetermined result. In [5], I proposed another way: Inside a learning machine, there is a container called conceiving space, and inside which, there are many X-forms to do information processing. Those X-forms actually are surging, i.e. to combine, cut, renew, etc.

Thus, such surging can produce completely non-predetermined effects, such as inspiration.

The methods mentioned above (Putnam-Gold machine, resource competition in NARS, the surging of X-form in conceiving space, and certainly more if more researches are done) have a common property: It has more than one information processing working together, and the collective result of these information processing will give out non-predetermined effects. Even though all pieces are preset, and the behavior of a single information processing is well understood, however, collectively, dynamic action is generated. This is the theoretical foundation of dynamic action. The research in this direction is still at the very early stages.

We can see dynamic action from another view, i.e. oracle machine proposed by Turing in 1939. According to Soare's discussion [10], o-machine M can be defined like following:

$$\delta : Q \times S_1 \times S_2 \to Q \times S_2 \times \{R, L\}, \quad where \quad \delta(q, a, b) = (p, c, X)$$

In the definition, S_1 is the so called oracle. Taking S_1 away, this machine M becomes a usual Turing machine M^*. S_1 is a oracle, it is not in the Turing machine M^*, but outside M^*. With S_1, the behavior of M^* is greatly or even fundamentally modified. In this way, o-machine M models Turing machine M^* with dynamic action (here is S_1).

In o-machine model, S_1 is just insert there and we do not know how to get it and what it is like, and needs to be specified. One way to do so is to use the above methods (such as surging in conceiving space) to get S_1.

Above, we used 2 simple examples to illustrate dynamic action. The 2 examples also can serve as very simple demonstration about the effects of several programs working together. We would like to point out: Once multiple programs working together, specially interact together, the outcome is very complicated. Also, such complexity of outcome will increase very quickly, when we add programs and/or increase layers of interaction. Such complicated situation is the source of dynamic action.

How to arrange multiple information processing working together to produce effective dynamic action? This is not easy. More researches in this direction are waiting for us. We would like to give our perspective: Such machine could hardly be achieved via programming purely by human. The best approach to get such machine is via learning. With a well designed framework, a machine can obtain its suitable dynamic action and other abilities via learning and experiences. This is a very different path than programming in the past. This indicates one very exciting research and development area. We started working in this direction [5–7].

Summary

The subjectivity of the machine is the collective property of all the programs inside the machine, which have important effects on the capability of machine.

Unlike the consciousness of a machine that is out of reach now, the subjectivity of a machine can be approached by solid technical methods. The subjectivity of a machine has 4 major aspects: a priori and cogitating, active perception to outside, awareness of self, and dynamic action. We can understand the aspects by behaviorism approach and mechanism approach. In principle, we can measure the degree of subjectivity. Among the 4 major aspects, dynamic action of machine is particularly important. We argue that the dynamic action of machine can be formed by multiple programs working together, if they are effectively arranged. How to form the dynamic action is an important topic for further studies. We are the creator of machine and we inject our subjectivity into machine as machine's when we build it. However, if a machine is a learning machine, it could eventually establish its subjectivity from experiences.

Acknowledgment. WeChat Group "Singularity0-Information" held a series of forums for AI and other topics in summer of 2020. The discussions in the forum promoted me to write down my thoughts on the subjectivity of a machine over long time in past few years. Thanks to participants in the forum discussion.

References

1. Xiong, C.: Discussion on Mechanical Learning and Learning Machine (2016). http://arxiv.org/pdf/1602.00198.pdf
2. Xiong, C.: Descriptions of Objectives and Processes of Mechanical Learning (2017). http://arxiv.org/pdf/1706.00066.pdf
3. Xiong, C.: Principle, Method, and Engineering Model for Computer Doing Universal Learning (in Chinese), researchage.net (2018). https://www.researchgate.net/profile/Chuyu_Xiong/research
4. Xiong, C.: Universal learning machine - principle, method, and engineering model. In: International Conference of Intelligence Science, Beijing (2018)
5. Xiong, C.: A Rudimentary model for Noetic Science (in Chinese), researchage.net (2019). https://doi.org/10.13140/RG.2.2.31596.72328
6. Xiong, C.: Sampling and Learning for Boolean Function (2020). http://arxiv.org/pdf/2001.07317.pdf
7. Xiong, C.: Chinese patent application # 201710298481.2
8. Kugel, P.: Thinking may be more than computing. 10.1.1.297.2677
9. Kugel, P.: You Don't Need a Hypercomputer to Evaluate an Uncomputable Function. https://www.researchgate.net/publication/
10. Soare, R.: Turing Oracle Machines Online Computing and Three Displacements in Computability Theory. http://www.people.cs.uchicago.edu/soare/History/turing.pdf
11. Wang, P.: Computer does not only do computing and Turing machine is not a machine (in Chinese). https://mp.weixin.qq.com/s/JQbYrKjqtmpBIoFQkHtThQ

Hexagon of Intelligence

Jean-Yves Beziau[(⊠)]

Brazilian Research Council, Brazilian Academy of Philosophy,
Federal University of Rio de Janeiro, Rio de Janeiro, RJ, Brazil
jyb@ufrj.br

Abstract. In this paper we discuss the nature of artificial intelligence (AI) and present a hexagon of opposition (generalization of the square of opposition) to characterize what intelligence is, its relation with computability, creativity, understanding and undecidability.

In a first part, we make some general comments about the history, development and objectives of AI. In a second part, we present two diametrically opposed ways of reasoning, one computational, one creational. In a third part, we talk about the relation between AI and logic, emphasizing that reasoning can be described or/and performed by different logical systems, mentioning the fact that non-monotonic logical systems have been promoted by AI researchers. In a fourth part, we present the theory of oppositions, with the three notions of opposition that are used to build squares and hexagons of opposition, and we then we present a hexagon of intelligence.

Keywords: Intelligence · Reasoning · Logic · Square of opposition · Computability · Chess · Creativity · John McCarthy · Aristotle

1 "Artificial Intelligence" and the Challenge of the Correlated Field

The expression "Artificial Intelligence" is attributed to John McCarthy (1927–2011) in the mid-1950s (cf. [13]) and it has become since then a major field of research. An expression does not necessarily lead to a field of research and a field of research may have no fixed and definite name, for example *Physics* was previously named "natural philosophy" (philosophia naturalis). But in the case of AI there is a narrow connection between the two.

McCarthy and Hayes (1969) [14] say that we can consider that the starting point of AI are two papers published shortly before the expression was coined: one by Turing [19] and one by Shannon [18], both in 1950. The expression "Artificial Intelligence" can be compared to "Cybernetics" and "Cognitive Science"; the three correlated fields being interrelated. The choice of "Artificial Intelligence" was made by McCarthy in some way to replace or improve "Cybernetics".

The expression "Artificial Intelligence" is compound of two words. The word "artificial" means created by humans and is opposed to "natural": a plane, a

© IFIP International Federation for Information Processing 2021
Published by Springer Nature Switzerland AG 2021
Z. Shi et al. (Eds.): ICIS 2020, IFIP AICT 623, pp. 25–34, 2021.
https://doi.org/10.1007/978-3-030-74826-5_3

building, a piano, a contraceptive, a computer are artificial; a tree, a cat, the sun are natural. The adjective "artificial" may have a negative connotation, when considering a failed or fake replication.

The challenge of artificial intelligence is to develop something which is similar to human intelligence or even better. Human beings have fully succeeded to create many artificial devices. A plane, inspired by natural birds, is going at a speed higher that any bird. And it makes sense to say that human beings can now fly.

Flying is something pretty clear, intelligence is more difficult to define. Some years ago a man able to quickly perform mentally a multiplication of two big numbers could have appear as very intelligent, but nowadays any calculator can do this better than a human being and a calculator is not generally considered as a symbol of intelligence.

The objective of AI is to perform more complicated tasks, typical examples since the beginning of AI (cf. [18]) are:

- Playing chess.
- Translating a language into another one.
- Orchestrating a melody.
- Proving a theorem.

Fig. 1. Translation: a hard AI task

Turing, Shannon, McCarthy and many other AI researchers have worked on developing programs that play chess and after some years a program was able to beat the best human chess player, Garry Kasparov. Although there are already lots of programs able to approximately translate a language into another one, it is not clear at all, up to now, if it will be possible one day that a program can perform translation is a satisfactory way (Fig. 1). This is an open problem related to the question whether a program can think or/and reason.

2   Two Different Kinds of Reasoning

Reasoning has many different aspects. Let us present here two diametrically opposed ways of reasoning, one computational, one creational.

Let us consider the following example: we have a board with 64 boxes; excluding the two boxes indicated in the diagram below, is it possible to place 31 dominos in the remaining boxes? (Fig. 2).

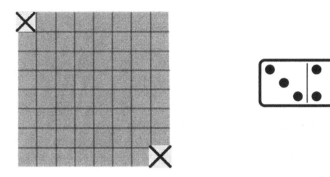

Fig. 2. 62 boxes and 21 dominos

It is not arithmetically impossible, since a domino occupies two boxes, and therefore 31 dominos occupy 62 boxes.

To check this possibility one may build a program (using for example LISP created by McCarthy) that will enumerate all the possibilities. This is in some sense what can be called a "step by step procedure". On the other hand there is a more ethereal reasoning, something that a program cannot necessarily perform.

Considering the black and white coloring of the board below (Fig. 3), we see that the two excluded boxes are white, so that at the end we have 30 white boxes and 32 black boxes. Since a domino necessarily occupies a white and a black boxes, we immediately see that it is not possible to place 31 dominos in the remaining boxes.

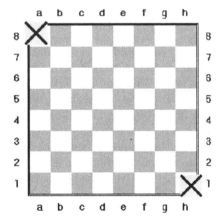

Fig. 3. Black and white coloring of the board

This proof depends on the idea of black and white coloring. How can a computer have such an idea? A computer may be able to better play chess than a human being, but it is not clear that he may have the intelligence of coloring a 64 box board into a chess board.

3   Logic(s) and Artificial Intelligence

Artificial Intelligence is deeply related to logic. Logic is one of the oldest fields of investigation but its name and its scope have been varying. Moreover there is a fundamental ambiguity surrounding logic: it can be considered as reasoning and/or as the theory of reasoning (cf. [2]). In ancient Greece, human beings were considered as "logical animals" ("rational animals" is the Latin transposition of this expression, see [6]). Human beings are reasoning. Logic, as a theory of reasoning, is a way to understand this capacity but also to correct or improve it. Logic since the beginning has a strong normative aspect.

Logic changed dramatically with the work of Boole in the mid XIXth century, in particular with his book entitled *The laws of thought* (1854) [10]. Boole's objective was not to reject the famous system of Aristotle, *Syllogistic*, but to improve it using mathematical tools. However it led to a new era of the science of reasoning called "mathematical logic" or "modern logic".

In modern logic there are many different systems. The most famous one is called "classical logic". But classical logic is not only one system of logic, it is a family of systems: classical propositional logic, first-order classical logic, second-order classical logic, etc. Simultaneously were developed lots of different systems commonly called "non-classical logics": many-valued logic, intuitionistic logic, paraconsistent logic, relevant logic, fuzzy logic, linear logic, etc. AI researchers have developed various systems of logic, most notably the so-called "non-monotonic logics" (see e.g. [15]).

When we have a system of logic SL, we can ask:

- Does SL properly describe reasoning?
- Is SL a good tool for developing/performing reasoning? (Fig. 4).

Due to the problematic double descriptive/normative aspect of logic, it is not clear how a system of logic should be assessed. Some people have rejected classical logic considering that it does not properly describe the way that we naturally reason. But this natural way of reasoning can be seen as limited, in the same was as a natural way of counting according to which there is *one, two, three, many* and that's all, can be seen as rather limited.

The objective of AI researchers is not to simulate these limitations, but to catch some features of human reasoning which are not necessarily those of mathematical reasoning. For example mathematical reasoning is monotonic in the sense that when something has been deduced from a set of hypotheses, it would remain valid if we add further hypotheses. The idea of non-monotonic logic is to reject this monotonicity considering for example that at a certain stage we can infer that all birds fly, but the day we meet penguins, we revise this conclusion.

Fig. 4. Non-monotonic logic, symbolized by the penguins

This is related to what has been called "belief revisions" (see [1] and subsequent works). The idea is to construct a system of logic that can explain how we can systematically do that and such a system of logic can lead to the development of programs that can also do that.

A system of logic can give a better understanding of what human intelligence is and moreover help to develop human intelligence. Such a system can be considered as artificial as any scientific theory, since it is a product of humans but it can also be considered as developing an artificial intelligence in the sense that it helps to develop an intelligence which is not naturally there right at the start, like in fact other mathematical theories.

A program that can also perform such kind of artificial intelligence is another step which is not necessarily straightforward. In particular we have to keep in mind that many systems of logic are not decidable even if they are recursive, the typical case being classical first-order logic.

4   The Theory of Oppositions

To have a better understanding of intelligence, it is useful to develop a theory of intelligence and this can be done using a simple logical tool like the theory of oppositions. According to this perspective, logic is used at a meta-level, not to directly perform intelligence but to model it. The theory of oppositions goes back to Aristotle. From his ideas was developed the square of opposition (Fig. 5) which is a structure based on implication (below in black) and three notions of opposition defined as follows.

Two propositions are said to be:

- Contradictories, when they cannot both be true and cannot both be false.
- **Contraries, when they can both be false but cannot both be true.**
- Subcontraries, when they can both be true but cannot both be false.

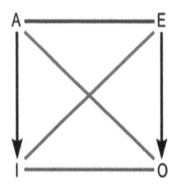

Fig. 5. The square of opposition

These oppositions were originally defined for propositions but they can naturally be applied to concepts. Below (Fig. 6) on the left the original square presented by Apuleius and on the right a square describing the relations between various classes of numbers [4]:

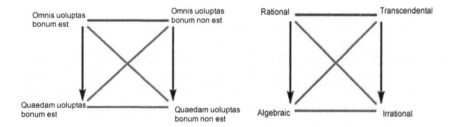

Fig. 6. Two examples of square of opposition

The square was generalized into a hexagon of oppositions by Robert Blanché [9], adding two additional "corners", defined as follows (Fig. 7):

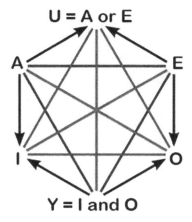

Fig. 7. The hexagon of oppositions

This hexagon can be used to understand many different concepts, ranging from quantification, to music, economy, painting, theory of colours, etc. (see [3,7,8,11,12]). In (Fig. 8) one of the most famous one, the deontic hexagon.

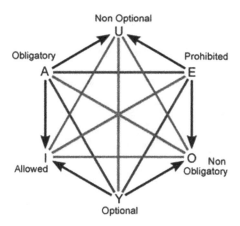

Fig. 8. The deontic hexagon of oppositions

It can even be applied to the theory of oppositions itself, as illustrated by the hexagon of oppositions below (Fig. 9).

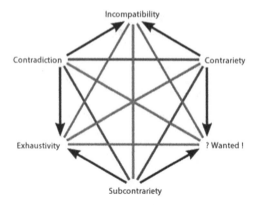

Fig. 9. The hexagon of oppositions of oppositions

As illustrated by this example (see [5]), it not necessarily obvious to find a positive determination for each of the corners of a hexagon. The O-corner in the above hexagon can be defined purely negatively as "non-contradiction". But what would be a good name for it that would help to develop a positive understanding of the related notion? That's not clear.

In any case, before presenting a hexagon of intelligence, let us emphasize that a hexagon of oppositions is based on a logical structure and that it shall not be confused with some artificial constructs, like the hexagon below (Fig. 10) designed by G.A Miller [16] to describe cognitive science. The arrows and edges of this hexagon do not correspond to logical relations.

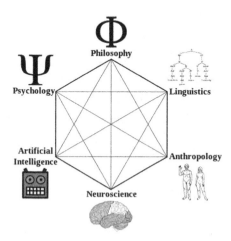

Fig. 10. The hexagon of cognitive science

One possible characterization of intelligence can be given through the following hexagon of oppositions (Fig. 11):

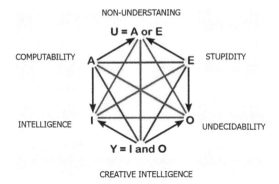

Fig. 11. The hexagon of oppositions of intelligence

We have two pairs of contradictory opposites, each having positive intuitive readings, for both sides: intelligence vs. stupidity and computability vs. undecidability. Undecidability may be seen rather negatively but recursion theory gives to it a precise definition.

The contradictory of creative intelligence, which according to the structure of the hexagon is the exclusive disjunction of computability and stupidity, may also appear as rather negative under the label of "Non-understanding". But we have tried to define it not literally as "Non-creative-intelligence". It is based in part on the claim by John Searle: "... we can see that the computer and its program do not provide sufficient conditions of understanding since the computer and the program are functioning, and there is no understanding" [17].

We hope this hexagon will provide inspiration for future developments of artificial intelligence aiming at catching creative intelligence.

Acknowledgements. I would like to thank the anonymous referees for their comments and helpful suggestions, as well as Mihir Chakraborty for inviting me to present this paper at *The Fourth International Conference on Intelligence Science* (ICIS2020) and Ivan Varzinczak for questions about formatation of the final version of this paper.

References

1. Alchourrón, C., Gärdenfors, P., Makinson, D.: On the logic of theory change: partial meet contraction and revision functions. J. Symb. Log. **50**, 510–530 (1985)
2. Beziau, J.-Y.: Logic is not logic. Abstracta **6**, 73–102 (2010)
3. Beziau, J.-Y.: The power of the hexagon. Log. Univers. **6**, 1–43 (2012)
4. Béziau, J.-Y.: There is no cube of opposition. In: Béziau, J.-Y., Basti, G. (eds.) The Square of Opposition: A Cornerstone of Thought. SUL, pp. 179–193. Birkhäuser, Basel (2017). https://doi.org/10.1007/978-3-319-45062-9_11

5. Beziau, J.-Y.: Disentangling contradiction form contrariety via incompatibility. Log. Univers. **10**, 157–170 (2016)
6. Beziau, J.-Y.: Being aware of rational animals. In: Dodig-Crnkovic, G., Giovagnoli, R. (eds.) Representation and Reality in Humans, Other Living Organisms and Intelligent Machines. SAPERE, vol. 28, pp. 319–331. Springer, Cham (2017). https://doi.org/10.1007/978-3-319-43784-2_16
7. Béziau, J.-Y., Basti, G. (eds.): The Square of Opposition: A Cornerstone of Thought. SUL, pp. 3–12. Birkhäuser, Basel (2017). https://doi.org/10.1007/978-3-319-45062-9
8. Beziau, J.-Y., Payette, G. (eds.): The Square of Opposition - A General Framework for Cognition, pp. 9–22. Peter Lang, Bern (2012)
9. Blanché, R.: Structures intellectuelles. Essai sur l'organisation systématique des concepts. Vrin, Paris (1966)
10. Boole, G.: An Investigation of the Laws of Thought on Which are Founded the Mathematical Theories of Logic and Probabilities. MacMillan, London (1854)
11. Chantilly, C., Beziau, J.-Y.: The hexagon of paintings. South Am. J. Log. **3**, 369–388 (2017)
12. Jaspers, D.: Logic and colour. Log. Univers. **6**, 227–248 (2012)
13. McCarthy, J., Minsky, M., Rochester, N., Shannon, C.E: A Proposal for the Dartmouth Summer Research Project on Artificial Intelligence, August 1955
14. McCarthy, J., Hayes, P.: Some philosophical problems from the standpoint of artificial intelligence. In: Meltzer, B., Michie, D. (eds.) Machine Intelligence 4, pp. 463–502. Edinburgh University Press, Edinburgh (1969)
15. McCarthy, J.: Circumscription: a form of non-monotonic reasoning. Artif. Intell. **13**, 23–79 (1980)
16. Miller, G.A.: The cognitive revolution: a historical perspective. Trends Cogn. Sci. **7**, 141–144 (2003)
17. Searle, J.R.: Minds, brains, and programs. Behav. Brain Sci. **3**, 417–457 (1980)
18. Shannon, C.: Programming a computer for playing chess. Phil. Mag. **41** (1950)
19. Turing, A.M.: Computing machinery and intelligence. Mind **59**, 433–460 (1950)

Uncertain Theory

Interactive Granular Computing Model for Intelligent Systems

Soma Dutta[1] and Andrzej Skowron[2,3]

[1] University of Warmia and Mazury in Olsztyn, Olsztyn, Poland
`soma.dutta@matman.uwm.edu.pl`
[2] Systems Research Institute, Polish Academy of Sciences, Warsaw, Poland
`skowron@mimuw.edu.pl`
[3] Digital Science and Technology Centre, UKSW, Warsaw, Poland

Abstract. The problem of understanding intelligence is treated, by some prominent researchers, as the greatest problem of this century. In this article we justify that a decision support systems to be intelligent there is a need for developing new reasoning tools which can take into account the significance of the processes of sensory measurement, experience and perception about the concerned situations; i.e., understanding the process of perceiving a situation is also required for making relevant decisions. We discuss how such reasoning, called adaptive judgment, can be performed over objects interacting in the physical world using Interactive Granular Computing Model (IGrC). The basic objects in IGrC are called the complex granules (c-granules, for short). A c-granule is designed to link the abstract and physical worlds and to realize the paths of judgments starting from sensory measurement, experience to perception. Some c-granules are extended by information layers, called informational c-granules (ic-granules, for short); they can create the basis for modeling a notion of *control* conducting the whole process of computation over the c-granules.

Keywords: Complex granule (c-granule) · Informational c-granule (ic-granule) · Control of c-granule · Perception of situation · Interactive granular computing (IGrC)

1 Introduction

In contrast to the world of pure mathematics, isolated from the real physical world, the present needs of Intelligent Systems (IS) are not met by using only static knowledge; it demands the ability of dynamically learning new information and updating reasoning strategies based on interactions with the dynamical real physical environment. IS often deal with complex phenomena of the physical world. The simplified models, designed for these complex phenomena, are obtained by ignoring the complexities and thus properties derived from such

Z. Shi et al. (Eds.): ICIS 2020, IFIP AICT 623, pp. 37–48, 2021.
https://doi.org/10.1007/978-3-030-74826-5_4

models often do not match with the data gathered by IS in interaction with the environment. This happens as the essence of the complexities is ignored [1]. This is one of the reasons for developing new computing model for IS. Other reasons follow from the emerging new application areas of IS related to, *e.g.*, Society 5.0 [17] or Wisdom Web of Things (W2T) [21]. It is assumed that in Society 5.0 various social challenges can be resolved by incorporating innovations of the fourth industrial revolution (*e.g.*, Internet of Things (IoT), big data, artificial intelligence (AI), robot and the sharing economy). Similar thought is reflected in [21], where authors described the need for a new area of research, called Wisdom Web of Things (W2T), emphasizing a practical way to realize the harmonious symbiosis of humans, computers, and things in the emerging hyper world, that uses data to connect humans, computers, and things.

So, the object of study no more remains a pure theoretical construct; rather it is a complex system, as stated in [5], connecting abstract information with physical objects: *Complex system: the elements are difficult to separate. This difficulty arises from the interactions between elements. Without interactions, elements can be separated. But when interactions are relevant, elements co-determine their future states. Thus, the future state of an element cannot be determined in isolation, as it co-depends on the states of other elements, precisely of those interacting with it.*

Thus, to understand and reason with complex system or phenomenon we need a new computing model. A model which can (i) continuously monitor (through interactions) some basic properties of the respective real physical configuration associated to the complex system, (ii) learn and predict (more compound or finer) properties/rules for the seen and/or unseen cases based on already stored knowledge, (iii) control the interaction process, as a part of a physical procedure, to reach a desired goal, and (iv) update new information in the knowledge base. The additional concern is that we can only partially perceive these elements and their dynamics; as a result we have only partial description of the states representing these elements and the transition relation representing their dynamics. So, for a new model of computation the main challenges are as follows.

In usual context, for a given family of sets $\{X_i\}_{i \in I}$ by a transition relation we mean a relation $tr_i \subseteq X_i \times X_i$. In the present context, we need to incorporate the components which can specify (i) how elements of X_i are perceived in the real physical environment, and (ii) how the transition relation tr_i is implemented in the real physical world. Existing approaches to soft computing, such as rough sets, fuzzy sets, and other tools used in machine learning lack in considering the above two components. There are two prevalent traditions of mathematical modeling. One is purely mathematical where it is considered that the sets are given. The second is called constructive, where it is assumed that objects are perceived by means of some features or attributes, and only a partial information about these objects in the form of vectors of attribute values is available. Both the tradition of modeling do not take into account how the process of perceiving attribute values is realised, where and how to access the concerned objects in the physical space, and why those attributes are selected. Hence, clearly the perception and action are out of the scope of such practices. However, this is crucial for many tasks dealing with complex phenomena in the real physical world. Thus, *e.g.*, characterization of the state of the complex physical phenomena by a priori given set of attributes becomes irrelevant. From a similar concern, the researchers in [11] proposed to extend Turing test by embedding into it the challenges related to action and perception.

So, how a function representing a particular vague concept is learned from the uses of the community, as well as which parameters to be considered crucial in defining a vague concept and how the values for these parameters are observed or measured, incorporating such information in the model is important for an intelligent agent; otherwise a non-human system cannot derive information about unseen cases. So, we need an extension of the existing approaches where apart from the information about a physical object, a specification of how the information label of a physical object is physically linked to the actual object also can be incorporated.

Keeping in mind the above needs we endorse an approach, called *Interactive Granular Computing* (IGrC). *Interactive* symbolizes *interaction between the abstract world and the real physical world*, and *Granular Computing* symbolizes *computation over imperfect, partial, granulated information abstracted about the real physical world*. In IGrC [2,3,6,7,18] computations are performed on complex granules (c-granules, for short) which are networks of more basic structures including c-granules with additional information layer (called ic-granules, for short), grounded in the physical reality. A brief description of c-granules is presented later.

This paper aims to present basic intuitions behind the new computing model; the target is rather to present the idea, without technicalities, through examples explaining different crucial components of the model. Section 2 presents some basic preliminaries about c-granules and ic-granules. The control of a c-granule is explained in Sect. 3. In the same section we illustrate how with the help of the control of a c-granule a computation process runs and reaches the goal. The paper ends with conclusions and possible further research directions.

2 Basis of Interactive Granular Computing

The rough sets, introduced by Zdzisław Pawlak [12], play a crucial role in the development of Granular Computing (GrC) [14,15,19]. The extension of GrC to IGrC (initiated by Skowron and co-workers [6,18]; see also publications about IGrC listed at https://dblp.uni-trier.de/pers/hd/s/Skowron:Andrzej), requires more generalization of the basic concepts of rough sets and GrC. For instance, it is needed to shift from *granules* to *complex granules* (including both physical and abstract parts), information (decision) systems to interactive information (decision) systems as well as methods of inducing hierarchical structures of information (decision) systems to methods of inducing hierarchical structures of interactive information (decision) systems. IGrC takes into account the granularity of information as used by humans in problem solving, as well as interactions with (and within) the real physical world. The computations are realized on the interactive complex granules and that evolve based on the consequences of the interactions occurring in the physical world. Hence, the computational models in IGrC cannot be constructed solely in an abstract mathematical space. In this context, the following quote of Immanuel Kant ([16], p. 4) is relevant to ponder over: [...] *cognition is the result of the interaction of two independent agents, the mind and the real object.*

The proposed model of computation based on complex granules seems to be of fundamental importance for developing intelligent systems dealing with complex phenomena, in particular in such areas as Data Science, Internet of Things, Wisdom Web of Things, Cyber Physical Systems, Complex Adaptive Systems, Natural Computing, Software Engineering, applications based on Blockchain Technology, etc. [2,6]. Our proposal is consistent with the thought envisaged in [9]: [...] *cognition is possible only when computation is realized physically, and the physical realization is not the same thing as its description. This is because we also need to account for how the computation is physically implemented.*

We assume that physical objects exist in the physical space as parts of it, and they are interacting with each other. Thus, some collections of physical objects create dynamical systems in the physical space. Properties of these objects and interactions among them can be perceived by so called complex granules (c-granules). To design the c-granules with the ability of perceiving physical objects and their interactions, which is required to achieve the goal of the computation, we use a notion control of a c-granule based on informational complex granules (ic-granules) and a special kind of reasoning, called judgment. Informational complex granules (ic-granules) are constructed over two basic sets of entities: abstract and physical; we may count these two sets of entities respectively as informational (I) and physical (P) objects.

Abstract entities of the ic-granules are families of formal specifications of spatio-temporal windows labelled by information in a formal/natural language specific for a given c-granule or a family of c-granules. The information may be of different kinds and may have different forms. One of the ic-granules also encodes in its information layer the local (discrete) time clock and enables the model to perceive features of physical objects at different moments of time and to reason about their changes. The information layer may contain formulas and their degrees of satisfiability at a given moment of time on some physical objects, as well as the formal specifications of the spatio-temporal windows indicating the location and time of those physical objects.

The physical layer of any ic-granule is called the c-granule and is divided into three parts: soft_suit, link_suit and hard_suit. Each of these parts is a collection of physical objects. The hard_suit consists of the physical objects that are to be perceived. The soft_suit is considered to have those objects which are directly accessible at a particular point of time. The objects in the link_suit create, in a sense, a physical pointer that links objects from the soft_suit to the hard_suit; this in turn helps to propagate interactions among physical objects of hard_suit and soft_suit. Directly accessible objects are those for which some features or their values can be directly measurable, or their changes in successive moments of local time of the c-granule can be directly measurable, or some features can be directly changed by the control; this is discussed in the next section.

Intuitively during the process of computation the behavior of an ic-granule g is modeled cyclically by the control (localised in the ic-granule controlling g). Each cycle starts from a current family, called configuration, of ic-granules containing a distinguished ic-granule with information representing the perception of the current situation. Each cycle consists of several steps such as modification, deletion, suspension of ic-granules or generation of some new ic-granules from the current configuration. In this process a special role is played by the so called implementational ic-granules. Once a new configuration of ic-granules is created the control measures features of some new physical objects in the scope of the newly developed ic-granules and/or matches or aggregates information with that of the previous ic-granules. After gathering perception about the current situation it takes relevant action based on the goal of the computation process.[1]

An ic-granule, in its information layer, contains a formal specification of its scope, i.e., specification of the spatio-temporal window referring to the physical space where the perception process of the ic-granule is localized. Information perceived by the ic-granule can be obtained either by (i) measuring features of the directly accessible objects of the ic-granule, or by (ii) applying reasoning on the already perceived information about the physical objects and their interactions in the scope of the ic-granule and the domain knowledge. The reasoning has to be robust with respect to the interactions from outside of the given scope. The robustness can be up to a degree depending on the formal specification of the scope and is specified in the information layer of the ic-granule.

The task of the above discussed ic-granule is to perceive properties of some part of the physical space lying in its scope. Such ic-granules, denoted as g_s, are perception oriented ic-granules. An ic-granule also can come into play in order to generate/modify new ic-granules. In such case the information layer of the concerned ic-granule contains the information related to the formal specification of the ic-granules to be generated/modified; such an ic-granule is called a planner ic-granule. The formal specifications of such a planner ic-granule can be (i) constraints on specifications of required spatio-temporal windows, satisfiability of which is necessary for activation of the granule to be generated/modified, (ii) specification of procedure for activating new granule, (iii) conditions concerning the expected behaviour of the granule to be generated expressed, (iv) acceptable variations of the expected properties of the ic-granule to be generated, etc.

Another special kind of ic-granule is responsible for storing relevant knowledge required for a process of computation. The relevant information about some contextual part of the computation can be encoded in an object in the soft_suit of this ic-granule; the soft_suit here can be considered as buffer or an internal memory. The link_suit is constructed out of the physical objects creating

---

[1] The case when some ic-granules from configuration have their own control will be considered elsewhere.

transmission channel to the hard_suit, which contains the hard disk where the information may have been be stored.

The ic-granules pertaining to perception, plan and knowledge still relate to the abstract part of an computation. However, in contrast to other approaches the ic-granules are themselves made of both abstract and physical entities. Now the above mentioned implementational ic-granules (g_i) come to play for implementation part of a computation process. Based on the perception of the environment available from g_s, general laws available from g_{kb}, a general plan is specified at the information layer of the planner ic-granule at some time point say t_0. Let us call this planner ic-granule evolved at t_0 as g_0. The abstract plan available at g_0 now gets translated by a relevant implementational ic-granule to a low level or implementation level language.[2] This low level language can be different based on context. For example, in a computer-run method the translation to a low level language can be translation of a program code, written in a language, to binary code. Thus, the implementational ic-granule carries the abstract description of a computation to a real physical realization.

Let us consider an example of an ic-granule g_s whose scope corresponds to a configuration of objects containing a blind person [10] or a robot with a stick and the objects lying in the surrounding environment. The person/robot and the top part of the stick are directly accessible and belong to the soft_suit. The part of the stick, which is distant from the direct touch of the person/robot, belongs to the link_suit; it links the objects beyond direct accessibility, such as holes, stones lying in the surrounding environment; i.e., in the hard_suit of g_s. The already perceived information about the objects in g_s is stored in an information layer attached to the soft_suit of $g_s{}^3$.

3   Control and Computation over ic-Granules

The control of a c-granule (or control of a computation process) aims to satisfy the current needs of the c-granule. For a given moment of time of the c-granule, the control has access to a family of ic-granules of the c-granule; thus, it has access to the information layers of those ic-granules, using which it directs a kind of complex game among these ic-granules as well as the environment to generate a new configuration of ic-granules from the existing one.

Formal specification of many complex tasks may be thought of as a complex game [6,18] consisting of a family of complex vague concepts, labeled by actions or plans that to be performed when the concepts are satisfied to a satisfactory degree. These complex vague concepts can be invariants that should be preserved to a satisfactory degree, conditions representing risk perceived in the environment, safety properties of the computation, quality of the current

---

[2] This decomposition process is related to information granulation and the Computing with Words paradigm introduced by Lotfi Zadeh [20] as well as to the challenge discussed by Judea Pearl in [13].

[3] In [6] this example is elaborated using c-granules without informational layers where encoding information from soft_suit is made by an external observer.

path toward achieving the goals, or risk indicating current needs are no longer achievable. These complex vague concepts (usually described in a fragment of a natural language) should be learned from data and domain knowledge with the use of physical laws. Moreover, the concepts as well as the labels in a complex game can evolve in time. Hence, control should have some adaptive strategies allowing relevant modification of the complex game.

3.1   Two Kinds of Transition Properties of ic-Granules Realised by Control

Properties of physical objects in the scope of a c-granule and their interactions are perceived by the control of the c-granule. At a given moment of local time t, the behaviour of the control depends on the existing ic-granules belonging to its scope, often called the configuration of ic-granules at t.

The control of a c-granule at time point t performs a reasoning using the current configuration of ic-granules which includes a distinguished ic-granule g_s incorporating information of the perceived situation. In the information layers of these ic-granules there are formal specifications of the precondition α that to be satisfied during the generation/modification of a new ic-granule and a postcondition β describing expected properties to be satisfied after generation/modification of the new ic-granule. The pairs of the form (α, β) create the family of expected transition properties, denoted as $\mathcal{R}_I$. After initiation of the ic-granule generation/modification process through some implementation ic-granule and embedding the implementation process through the real physical objects, a property of the new configuration, say γ is derived from data gathered about the behaviour of new configuration. Then control performs reasoning (using β, γ and domain knowledge) to estimate the degree of matching of the expected condition β with the observed property γ. The pairs of the form (α, γ) create the family of real (physical) transition properties, denoted as $\mathcal{R}_P$. The pair $\mathcal{R} = (\mathcal{R}_I, \mathcal{R}_P)$ describes the expected and actual transition properties among the ic-granules. The information associated to different ic-granules included in the control determines the dynamics of the transition among the ic-granules of a particular c-granule. Hence, $\mathcal{R}_I$ specifies a piece (or a set of pieces) of information that is expected to describe the next state of the computation.

Let us consider that at some time t the computation is taking place at the ic-granule g_{sub}. The information label of g_{sub}, say $inf_{g_{sub}}$, specifies the perceptual properties of the state of the environment. The information $inf_{g_{sub}}$ along with the relevant knowledge $inf_{g_{kb}}$ from g_{kb} points to the information $inf_{g'_{sub}}$, describing the next possible state. So, $(\{inf_{g_{sub}}, inf_{g_{kb}}\}, inf_{g'_{sub}})$ is an outcome of $\mathcal{R}_I$ indicating g'_{sub} is the next expected configuration of the ongoing computation. Contrary to $\mathcal{R}_I$, $\mathcal{R}_P$ specifies a relation among the actual objects lying in the scope of the current ic-granule g_{sub}. Let this actual real physical interrelation among the objects of g_{sub} at time t be encoded by $inf_{g_{sub}^{S-L-H}}$; this refers to the information concerning objects in the three suits of g_{sub} which becomes available after initiation of some interactions. Let after initiation of interactions the obtained configuration at time point t_n be described by the information

$inf_{g_{sub}}^{t_n}$. The properties of obtained ic-granule $g_{sub}^{t_n}$ can be different from the expected ic-granule g'_{sub}. Thus, unlike the transition relation in an automata the transition from one ic-granule to another is not completely defined a priori, and depends on both $\mathcal{R}_I$ (specification of the expected transition properties) and $\mathcal{R}_P$ (specification of the real transition properties).

We know that control of a c-granule is responsible for generating new configuration of ic-granules from the current ones. This dynamic process of changing from one configuration to other is carried out by the control in a cyclic order by using the relevant information localized in the currently existing ic-granules in the control and based on the aggregation of $\mathcal{R}_I$ and $\mathcal{R}_P$. But to create a new configuration of ic-granules it is needed to embed the above relevant information, gathered from the ic-granules at time t, to an implementational ic-granule g_i. Thus, from the formal specifications of ic-granules the physical realizations process is initiated by some implementational ic-granules. This process may involve updating, canceling or suspending some of the existing ic-granules from time t. Once a new configuration of ic-granules is created the process of perceiving the properties of the new state of the environment, verifying the degree of matching of the generated configuration with the expected one, and coupling them with relevant domain knowledge, starts. This marks the starting of a new cycle of the control at the next time point. The configurations of ic-granules at each cycle are represented by different layers of the process of computation. Through these layers the decomposition of formal specification is realised in a step-by-step manner to make it closer to the real physical environment of the computation.

3.2   Computation over the ic-Granules Directed by Control

This section presents how during a computation process different parts of different ic-granules participate and how based on both $R_\mathcal{I}$ and $R_\mathcal{P}$ the computation moves from one layer to another layer. We assume that at time t_0 some properties of a part $S \subseteq \mathcal{P}$ are available in $\mathcal{I}$. The information related to the perceived properties of objects corresponding to the window specifications describing the space-time hunks of S form the informational layer of an ic-granule, denoted as g_s; the physical objects surrounding S form the soft_suit, link_suit and the hard_suit of g_s. The objects which are directly accessible, or about which already some information is perceived, belong to the soft_suit of g_s. The objects in S, about which some information can be gathered only after performing some physical interactions with them, belong to the link_suit and hard_suit of g_s. Thus, objects between soft_suit and hard_suit are connected by a collection of objects forming a link from directly accessible to not directly accessible objects. The information part is attached to the soft_suit of g_s, and works as a label of g_s. Further steps of the computation are as follows.

Layer-0

(i) At t_0, the beginning of the control's cycle, g_s is labelled with the perceived information of the directly accessible objects of S. The target is to create a communication channel through the directly accessible objects so that objects lying in the hard_suit of g_s can be accessed to move forward the purpose of the computation. An example of g_s can be regarded as the ic-granule having in its scope a blind person or robot with a stick and the objects lying in the surrounding.

(ii) The description of the goal is attached as the information layer of g_0, a planner ic-granule at time t_0. In the context of our example, the goal can be moving forward avoiding the obstructions such as holes/stones in the hard_suit of g_s. So, g_0 consists of those particular cells of a human brain that have analytical functionalities. In case of a robot, g_0 is the part where the goal description of the robot is set.

(iii) To have perception about objects from the link_suit and hard_suit of g_s relevant knowledge about the environment from g_{kb} is sought for. The information layer of g_{kb} is labelled with the address to those relevant properties of the fragment of S. The soft_suit of g_{kb} has those objects which are like outer box of the memory location whose address is attached to the information layer; in order to access the detailed information about S some more inner boxes are to be opened. Keeping analogy to a computer memory, we can think of an outer folder containing some inner folders directing to the main folder. The name of the outer folder along with its path address is attached as the label of g_{kb}. In regard to the above example, g_{kb} can be considered as the part of the brain related to memory locations consisting of previous experiences of such environment.

(iv) Aggregating the perceptual information labelled at g_s, goal labelled at g_0, and information pertaining to experiences labelled at g_{kb}, the plan available at g_0 is decomposed in detail. This detailed plan is labelled at the ic-granule g_1 at the next time point t_1. For a visual representation the readers are referred to Fig. 1.

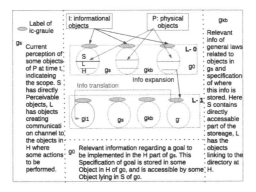

Fig. 1. Computation over ic-granules passing from layer-0 to layer-1

Layer-1

(i) The new plan attached to g_1 is a result of the informational relation $\mathcal{R}_I$ applied on the information associated to g_s, g_{kb} and g_0; that is plan in g_1 specifies the next state description of the plan. In the case of the example, at t_0 the information label of g_0 encodes the goal of the person that primarily registered in the brain, the soft_suit of g_0; the hard_suit of g_0, such as more deep analytical brain cells, remains still unaccessed. Combining information of g_s and g_{kb} at time t_1 the person digs into those analytical cells; thus interaction with the previously unaccessed part of g_0 happens. Gradually hard_suit of g_0 becomes accessible, and the hard_suit of g_0 becomes the soft_suit of g_1, labelled with further detailed plan.

(ii) Now in order to implement the abstract description of the plan available at g_1 through a real physical action, the plan needs to be transformed from the abstract level to an implementational level language. From the perspective of our example, this can be translation of the plan from the person's analytical brain cells to a language of actuators, like hands, legs, and the stick of the person. So, a new ic-granule is manifested at this layer. We call it as g_{i_1}, an implementational ic-granule. To be noted that g_{i_1} does not concern about the actual actuators; rather it is like another hard-drive in the brain of the person where the action plan can be stored in the language of actuators. The information layer of g_{i_1} also contains the specification of conditions for initiating the implementation plan through a real actuator.

Layer-2

(i) The specification of implementation plan of g_{i_1} is now realized through a physical object at time t_2. Let this object belong to the scope of the ic-granule g_2. In case of the example, it can be the stick of the blind person on which the abstract implementation plan is embedded, and g_2 represents the ic-granule containing the stick in its scope. Once the $inf_{g_{i_1}}$, the specification stored in g_{i_1}, is embedded on a real physical object, namely the stick, the role of $\mathcal{R}_P$ comes into play. The physical interaction of the stick with other objects in g_2 is encoded in the relational language of $\mathcal{R}_P$ in the information layer of g_2. If inf_{g_2} matches to a significant level to the condition for initiating implementation plan stored at $inf_{g_{i_1}}$ then an action compilation signal is passed to the next implementation granule g_{i_2}.

(ii) With the action compilation specification of g_{i_2} the objects lying in its link_suit and hard_suit propagates actions to realize a desired configuration in the hard suit of g_s. In the context of our example, g_{i_2} represents the ic-granule which specifies how to move the stick forward until it touches a stone on its way. This chain of objects between the stick and a stone creates a communication channel.

(iii) Through this communication channel the computation process enters into the hard_suit of g_s, which was unaccessible at time t_0. The initiation of the action compilation via g_{i_2} creates a link to the hard_suit of g_s. This new interaction gives access to the hard_suit of g_s which was previously unaccessible.

(iv) A new cycle starts by perceiving properties of the newly accessible part of g_s.

4   Reasoning in the Context of Complex Granules: Future Directions

In order to realize the above model, different kinds of reasoning strategies need to be incorporated. As the model couples the abstract information with its real physical semantics, the reasoning methods cannot only focus on deriving information from information; it needs to perform reasoning based on sensory measurements and perception too. Moreover, as ic-granules contain different heterogeneous forms of information, we need different forms of reasoning apart from abduction, deduction, induction [4,8].

For example, in the process of connecting a specified spatio-temporal window with its real physical semantics and transiting from one configuration of ic-granules to another the control needs to decide (i) which windows from the current configuration should remain active and which is to be suspended, (ii) when and how a new window need to be opened and implemented in the real physical space, (iii) how much variation between the real and expected information can be allowed, (iv) how to generate a window specification from measured values of attributes, (v) how to induce a relevant set of attributes classifying a window, etc. There can be many other aspects of reasoning related to hierarchically learning and improving each step of the computation process evolving from one layer of ic-granules to other. All these directions need a further exploration and expansion in order to develop an intelligent agent which is not restricted to behave just based on what it is taught once; rather can learn to adopt new strategies based on continuous interaction with the real physical environment.

Acknowledgement. Andrzej Skowron was partially supported by the ProME (Prognostic Modeling of the COVID-19 Epidemic) grant.

References

1. Brooks, F.P.: The Mythical Man-Month: Essays on Software Engineering. Addison-Wesley, Boston (1975). (extended Anniversary Edition in 1995)
2. Dutta, S., Jankowski, A., Rozenberg, G., Skowron, A.: Linking reaction systems with rough sets. Fundamenta Informaticae **165**, 283–302 (2019). https://doi.org/10.3233/FI-2019-1786
3. Dutta, S., Skowron, A., Chakraborty, M.K.: Information flow in logic for distributed systems: extending graded consequence. Inf. Sci. **491** 232–250 (2019). https://doi.org/10.1016/j.ins.2019.03.057
4. Gerrish, S.: How Smart Machines Think. MIT Press, Cambridge (2018)
5. Gershenson, C., Heylighen, F.: How can we think the complex? In: Richardson, K. (ed.) Managing Organizational Complexity: Philosophy, Information Age Publishing, Theory and Application, pp. 47–61 (2005)
6. Jankowski, A.: Interactive Granular Computations in Networks and Systems Engineering: A Practical Perspective. LNNS, vol. 17. Springer, Cham (2017). https://doi.org/10.1007/978-3-319-57627-5

7. Jankowski, A., Skowron, A.: A wistech paradigm for intelligent systems. In: Peters, J.F., Skowron, A., Düntsch, I., Grzymała-Busse, J., Orłowska, E., Polkowski, L. (eds.) Transactions on Rough Sets VI. LNCS, vol. 4374, pp. 94–132. Springer, Heidelberg (2007). https://doi.org/10.1007/978-3-540-71200-8_7

8. Martin, W.M. (ed.): Theories of Judgment. Psychology, Logic, Phenomenology. Cambridge University Press, Cambridge (2006)

9. Miłkowski, M.: Explaining the Computational Mind. The MIT Press, Cambridge (2013)

10. Nöe, A.: Action in Perception. MIT Press, Cambridge (2004)

11. Ortiz Jr., C.L.: Why we need a physically embodied Turing test and what it might look like. AI Mag. **37**, 55–62 (2016). https://doi.org/10.1609/aimag.v37i1.2645

12. Pawlak, Z.: Rough sets. Int. J. Comput. Inf. Sci. **11**, 341–356 (1982)

13. Pearl, J.: Causal inference in statistics: an overview. Stat. Surv. **3**, 96–146 (2009). https://doi.org/10.1214/09-SS057

14. Pedrycz, W., Skowron, S., Kreinovich, V. (eds.): Handbook of Granular Computing. John Wiley, Hoboken (2008)

15. Pedrycz, W.:Granular computing for data analytics: a manifesto of human - centric computing. IEEE/CAA J. Automatica Sinica **5**, 1025–1034 (2018). https://doi.org/10.1109/JAS.2018.7511213

16. Pietarinen, A.V.: Signs of Logic Peircean Themes on The Philosophy of Language, Games, and Communication. Springer, Heidelberg (2006). https://doi.org/10.1007/1-4020-3729-5

17. Salgues, B.: Society 5.0. Industry of the Future, Technologies, Methods and Tools. ISTE. Wiley, Hoboken (2018)

18. Skowron, A., Jankowski, A., Dutta, S.: Interactive granular computing. Granular Comput. **1**(2), 95–113 (2016). https://doi.org/10.1007/s41066-015-0002-1

19. Zadeh, L.: Toward a theory of fuzzy information granulation and its centrality in human reasoning and fuzzy logic. Fuzzy Sets Syst. **90**, 111–127 (1997). https://doi.org/10.1016/S0165-0114(97)00077-8

20. Zadeh, L.: Computing with Words: Principal Concepts and Ideas. Studies Fuzziness Soft Computing, vol. 277. Springer, Heidelberg (2012). https://doi.org/10.1007/978-3-642-27473-2

21. Zhong, N., et al.: Research Challenges and perspectives on wisdom web of things (W2T). J. Supercomput. **64**(3), 862–882 (2013). https://doi.org/10.1007/s11227-010-0518-8

A Framework for the Approximation of Relations

Piero Pagliani$^{(\boxtimes)}$

Rome, Italy

Abstract. The paper proposes a foundation to the approximation of relations by means of relations. We discuss necessary, possible and sufficient approximations and show their links with other topics, such as refinement and simulation. The operators introduced in the paper has been tested on computers.

1 Introduction

Given a subset X of a universe U, approximation techniques are required when *for some reason* the membership in X of an arbitrary element a of U is not sharply decidable. Generally speaking, the disorienting factor is that a is perceived within a *contour*, $\mathfrak{C}(a)$, of other elements of U. We use the terms "to perceive" to refer to any modality of acquiring and transforming uninterpreted *data* into interpreted pieces of *information*. To this end Rough Set Theory provides a pair of approximation operators: an element a belongs to the *lower approximation* of X, if $\mathfrak{C}(a) \subseteq X$, while a belongs to the *upper approximation* of X if $\mathfrak{C}(a) \cap X \neq \emptyset$. The relations $\subseteq$ and $\cap$ are decided by the classical $0, 1$-characteristic function (other possibilities are studied in the literature, such as fuzzy membership functions).

Rough Set Theory provides a 1-tier (Boolean) approximation mechanism. In a n-tier mechanism, any element a' of $\mathfrak{C}_n(a)$ comes with its own contour $\mathfrak{C}_{n-1}(a')$, so that one can put that a belongs to the n-lower approximation of X if all or a sufficient number of the elements of its contour belong to the $n-1$ lower approximation of X, and so on. If $n = 2$, for instance, one definition would prescribe that a belongs to the 2-*strict* lower approximation of X if $\forall a' \in \mathfrak{C}_2(a), \mathfrak{C}_1(a') \subseteq X$. A more general situation is given when the inclusion between a contour $\mathfrak{C}_2(a)$ and another contour $\mathfrak{C}_1(b)$ has to be decided for a and b belonging to different spaces *dynamically* linked in some way φ. Eventually, different criteria ψ can be applied to decide inclusion, besides the classical one. We call this pattern *pseudo-continuity* because if $\mathfrak{C}_1(b)$ and $\mathfrak{C}_2(a)$ are topological open sets, the inclusion is the identity map and ψ is the preimage f^{-1} of some function f, it represents the usual notion of "continuity":

© IFIP International Federation for Information Processing 2021
Published by Springer Nature Switzerland AG 2021
Z. Shi et al. (Eds.): ICIS 2020, IFIP AICT 623, pp. 49–65, 2021.
https://doi.org/10.1007/978-3-030-74826-5_5

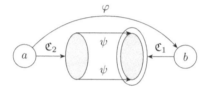

Pseudo-continuity: the contour $\mathfrak{C}_2(a)$ is ψ-included in the contour $\mathfrak{C}_1(\varphi(a))$

In addition, in the expression $\mathfrak{C}(a) \subseteq X$, a belongs to the set of inputs of the perception process, while $\mathfrak{C}(a)$ is a subset of the outputs, as well as X. This is a subtle distinction is evident when X is the output of another perception process. The proposed approach deals with the above considerations from a basic point of view.

2   Formalizing the Problem

We assume that contours are the results of the application of sequences of *constructors*. A constructor maps an element (formally a singleton) onto a set of elements connected to it by a binary relation R. If $R \subseteq A \times B$, A is called the *domain* of R, $dom(R)$, and B the *range*, $ran(R)$. If $A = B$ then R is called an *endorelation* or a *homogeneous relation*, otherwise it is called *heterogeneous*. If $X \subseteq A$, then $R(X) = \{b : \exists a \in X \wedge \langle a, b \rangle \in R\}$ is called the R-*neighbourhood* of X. If $X = \{x\}$ we write $R(x)$ instead of $R(\{x\})$. Thus, $\langle a, b \rangle \in R$ is also written $b \in R(a)$. R-neighbourhoods are the most elementary instances of 1-element sequence. In (generalized) Rough Set Theory, the contours are given by an endorelation R on a universe A. Given $X \subseteq A$, the 1-tier lower approximation of X via R, $(lR)(X)$, is the set of elements a of A such that $R(a)$ is classically included in X: $(lR)(X) = \{a : R(a) \subseteq X\}$ (or $(lR)(X) = \bigcup\{Y : R(Y) \subseteq X\}$ since neighbourhood formation is additive). In this definition both the result $(lR)(X)$ and X are subsets of A. But we have seen that there is a subtlety: $(lR)(X)$ is a subset of A *qua domain* of R, while X is a subset of A *qua range* of R. The difference is hidden because R is an endorelation and because, literally speaking, X is a *datum* (meaning: what is *given* to our perception). Such an elusive difference stands out if $R \subseteq A \times B$, for $A \neq B$. Now $(lR)(X)$ is a subset of A, while X is a subset of B. To overcome the problem one could change the definition as follows: $(lR)^*(X) = \bigcup\{R(a) : R(a) \subseteq X\}$. Incidentally, when R is an endorelation, $(lR)(X) = (lR)^*(X)$ if and only if R is a preorder. If $A \neq B$ we maintain that the *approximating* set should be a subset of the *domain* of R wile the *approximated* set is a subset of the *range*.

Moreover, we consider X not as a *datum* but as the result of some other perception mechanism. The most general situation is described by the relational schema

$$RS = (R \subseteq A \times B, Q \subseteq C \times D, W \subseteq A \times C, Z \subseteq B \times D)$$

which we call "Vitruvian" after Leonardo. We claim that this is sufficiently general a framework, because a number of additional situations can be accommodated in it.

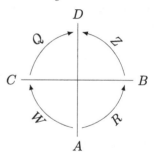

The present study deals with three main situations:

A) We have to find a subrelation $R^* \subseteq R$ which is coherent with the input-output processing of the *reference relation* Q, which runs parallel to R, according to a criterion with parameters in W and Z which provides a subset X of $dom(R)$. We set $R^* = R \restriction_X$ (R restricted to X), and call this task: *unilateral approximation of relations*.

B) The relation R is unknown. We have to reconstruct it according to a criterion with parameters in W, Z and Q. This task will be called *reconstruction of relations*.

C) The relation R is known and we have to approximate it with a relation R^* by means of a criterion which provides simultaneously the inputs and the outputs of R^*. We call this task *bilateral approximation of relations*.

We shall introduce a number of elementary operators described as follows:

Specifications: A formula made up of logical and set-theoretic operators. If any set-theoretical expression of the type $X \subseteq Y$ is expressed by the first order formula $\forall x(x \in X \implies x \in Y)$ the term *pre-specification* is used.

Pseudocodes: The translation of a specification into a formula made up of operations on relations. It is a sort of high level description of the computing procedure.

Procedures: The translation of a pseudocode into a formula whose ingredients are only operations which can be performed as manipulations of Boolean matrices.

2.1   Operations on Binary Relations

We assume that the reader is familiar with the usual properties of binary relations.

In order to manipulate relations it is important the cardinalities of their domains and ranges. We say that R is of type $A \times B$, written $R : A \times B$, when the domain has cardinality $|A|$ and the range $|B|$. It is assumed that when $|dom(R)| = |dom(Q)|$ then $dom(R)$ and $dom(Q)$ are linked by an implicit 1-1 function so that the two domains can be identified with each other. Similarly for the ranges. This way the performability of an operation depends just on the cardinalities of the dimensions and not on the names of the elements and there is no need to distinguish between $R \subseteq A \times B$ and $R : A \times B$. Therefore, $R \subseteq A \times B$ and $Z \subseteq C \times A$ will mean either that $dom(R) = ran(Z)$, or that $|dom(R)| = |ran(Z)|$ while on B and Z there are no special assumptions, and so on).

Definition 1. *Let A, B, C be three sets. In what follows, a^* is a dummy element of A and b^* a dummy element of B and so on (that is, they represent any element of the set they belong to). Let $R \subseteq A \times B$:*

1. *$-, \cap$ and $\cup$ are the usual set-theoretic operations. $R^{\smile}$ denotes the* converse of R: $R^{\smile} := \{\langle y, x \rangle : \langle x, y \rangle \in R\}$. *$\mathbf{1}_{A \times B}$ is the top element of the set of relations of that type. $\mathbf{0}_{A \times B} = -\mathbf{1}_{A \times B}$ is the bottom element.*

2. Let $X \subseteq A$. Then $R(X) = \{b : \exists x (x \in X \wedge \langle x, b \rangle \in R\}$ - the Peirce product of R and X, or R − neighbourhood of X, or R − granule of X

3. If $Q \subseteq B \times C$, then $R \otimes Q := \{\langle a, c \rangle : \exists b \in B(\langle a, b \rangle \in R \wedge \langle b, c \rangle \in Q)\}$ - the right composition of R with Q. $Q(R(X)) = (R \otimes Q)(X)$ and $(R \otimes Q)^{\smile} = Q^{\smile} \otimes R^{\smile}$.

4. $Id_A := \{\langle a, a \rangle : a \in A\}$ (same for Id_B). If $R \subseteq A \times A$, we eventually write Id_R instead of Id_A and call it the identity or diagonal of R.

5. If $X \subseteq A$ then $X_A := \{\langle a, a \rangle : a \in X\}$ is called a test of X. It is a way to represent sets as relations. Therefore, $Id_A \cap Q$ is the test X_A where $X = \{a : \langle a, a \rangle \in Q\}$.

6. Let $a*$ and $b*$ be dummy elements of A, resp. B. If $X \subseteq A$, then $X_R^{\rightarrow} := \{\langle a, b^* \rangle : a \in X \wedge b^* \in B\}$ is called the R-right cylinder of X. It is the relational embedding of X in R. If $Y \subseteq B$, then $Y_R^{\leftarrow} := \{\langle a^*, y \rangle : y \in Y \wedge a^* \in A\}$ is the R-left cylinder of Y. It is the relational embedding of Y in R.

   If A provides the dummy elements a^* of a cylinder, then it will be also denoted by A^*. Therefore, $X_R^{\rightarrow}(B^*) = X_R^{\rightarrow}(b^*) = X$. Symmetrically, $Y_R^{\leftarrow}(A^*) = Y_R^{\leftarrow}(a^*) = Y$.

7. Given $R \subseteq A \times B$ and $Z \subseteq A \times C$ the right residual of R and Z id defined as

$$R \longrightarrow Z = \{\langle b, c \rangle : \forall a (\langle a, b \rangle \in R \Longrightarrow \langle a, c \rangle \in Z)\} = \{\langle b, c \rangle : R^{\smile}(b) \subseteq Z^{\smile}(c)\} \quad (1)$$

It is the largest relation $K : B \times C$ such that $R \otimes K \subseteq Z$:

$$R \otimes K \subseteq Z \text{ iff } K \subseteq R \longrightarrow Z. \quad (2)$$

If $R \subseteq A \times B$ and $W \subseteq C \times B$ the left residual of R and W is

$$W \longleftarrow R = \{\langle c, a \rangle : \forall b (\langle a, b \rangle \in R \Longrightarrow \langle c, b \rangle \in W)\} = \{\langle c, a \rangle : R(a) \subseteq W(c)\} \quad (3)$$

It is the largest relation $K : C \times A$ such that $K \otimes R \subseteq W$:

$$K \otimes R \subseteq W \text{ iff } K \subseteq W \longleftarrow R. \quad (4)$$

The above operations are usually presented within some algebraic structure (see [1]). But the carriers of these algebras consists of endorelations, while in our study we are interested mainly in heterogeneous relations. The following results can be easily proved (see [8]):

Lemma 1. Given $R \subseteq A \times B$, $W \subseteq A \times C$, $Q \subseteq C \times D$ and $Z \subseteq B \times D$:

(a) $R \longrightarrow W = -(R^{\smile} \otimes -W)$ (b) $Q \longleftarrow Z = -(-Q \otimes Z^{\smile})$

(c) $R \longrightarrow W = -R^{\smile} \longleftarrow -W^{\smile}$ (d) $Q \longleftarrow Z = -Q^{\smile} \longrightarrow -Z^{\smile}$

Corollary 1. Let R, W, Q and Z be as above. Then:

(a) $(R \longrightarrow W)^{\smile} = W^{\smile} \longleftarrow R^{\smile}$ (b) $(Q \longleftarrow Z)^{\smile} = Z^{\smile} \longrightarrow Q^{\smile}$

(c) $Z \longrightarrow (R \longrightarrow W) = (R \otimes Z) \longrightarrow W$ (d) $(Q \longleftarrow Z) \longleftarrow R = Q \longleftarrow (R \otimes Z)$

Corollary 2. Given $R \subseteq A \times B$, $R \longrightarrow R$ and $R^{\smile} \longleftarrow R^{\smile}$ are preorders on B, $R^{\smile} \longrightarrow R^{\smile}$ and $R \longleftarrow R$ are preorders on A.

2.2   Perception Constructors

Now we introduce three pairs of operators which are defined by means of a binary relation.

Definition 2. *Let $R \subseteq A \times B$ be a binary relation, $X \subseteq A$, $Y \subseteq B$. The operators decorated with $\rightarrow$ transform subsets of $\wp(A)$ into subsets of $\wp(b)$, the operators decorated with $\leftarrow$ go the other way around. Then:*

1. $\langle\rightarrow\rangle(X) = \{b : \exists a(a \in X \wedge \langle a, b\rangle \in R)\} = R(X)$ - the right possibility *of X*.
2. $\langle\leftarrow\rangle(Y) = \{a : \exists b(b \in Y \wedge \langle a, b\rangle \in R)\} = R^{\smile}(Y)$ - the left possibility *of Y*.
3. $[\rightarrow](X) = \{b : \forall a(\langle a, b\rangle \in R \Longrightarrow a \in X)\} = R \longrightarrow X$ - the right necessity *of X*.
4. $[\leftarrow](Y) = \{a : \forall b(\langle a, b\rangle \in R \Longrightarrow b \in Y)\} = Y \longleftarrow R$
   - the left necessity *of Y*.
5. $[[\rightarrow]](X) = \{b : \forall a(a \in X \Longrightarrow \langle a, b\rangle \in R)\} = X \longrightarrow R$ - the right sufficiency *of X*.
6. $[[\leftarrow]](Y) = \{a : \forall b(b \in Y \Longrightarrow \langle a, b\rangle \in R)\} = R \longleftarrow Y$ - the left sufficiency *of Y*.

The above terminology is after Kripke models for modal logic and the basic logical readings of $A(x) \Longrightarrow B(x)$: "in order to be A it is necessary to be B" or "it is sufficient to be A in order to be B". If R is left total, $[\leftarrow]_R(Y) \subseteq \langle\leftarrow\rangle_R(Y)$. If R is right total, $[\rightarrow]_R(X) \subseteq \langle\rightarrow\rangle_R(X)$.

We call the operators $\langle\bullet\rangle$, $[\bullet]$ and $[[\bullet]]$ *constructors*, where $\bullet$ is either $\rightarrow$ or $\leftarrow$. Left (resp. right) operators will be collectively denoted with $op^{\leftarrow}$ (resp. $op^{\rightarrow}$), eventually with the index "R". If $X = \{x\}$ we shall write $op(x)$ instead of $op(\{x\})$. The $\leftarrow$ decorated constructors "*extensional*" will be called "*extensional*" and the $\rightarrow$ decorated ones "*intensional*".

For a historical account of the above constructors in Rough Set Theory see [9]. For their connections with pointless topology and Intuitionistic Formal Spaces (see [10]). However, they were long known in Category Theory as examples of Galois Adjunctions (see [3]). This important and useful fact will be explained in the next Section.

Depending on the context, if op_R is one of the above constructors, by extension we consider op_R as a relation $dom(R) \times ran(R)$ defined as $\{\langle a, b\rangle : a \in dom(R) \wedge b \in op_R(a)\}$.

2.3   Perception Constructors and Galois Adjunctions

Definition 3 (Galois adjunctions). *Let $\mathbf{O}$ and $\mathbf{O}'$ be two pre-ordered sets with order $\leq$, resp. $\leq'$ and $\sigma : \mathbf{O} \longmapsto \mathbf{O}'$ and $\iota : \mathbf{O}' \longmapsto \mathbf{O}$ be two maps such that for all $p \in \mathbf{O}$ and $p' \in \mathbf{O}'$*

$$\iota(p') \leq p \ \ iff \ \ p' \leq' \sigma(p) \tag{5}$$

then σ is called the upper adjoint *of ι and ι is called the* lower adjoint *of σ. This fact is denoted by $\mathbf{O}' \dashv^{\iota,\sigma} \mathbf{O}$ and we say that the pair $\langle\iota, \sigma\rangle$ forms a* Galois adjunction *or an* axiality.

From, (2) and (4) one immediately has that $\mathfrak{R}_{\mathrm{ran}}$ $\dashv^{\otimes_R, \leftarrow_R}$ $\mathfrak{R}_{\mathrm{ran}}$ and $\mathfrak{R}_{\mathfrak{dom}}$ $\dashv^{R \otimes, \rightarrow_R}$ $\mathfrak{R}_{\mathfrak{dom}}$, where $\mathfrak{R}_{\mathrm{ran}}$ ($\mathfrak{R}_{\mathfrak{dom}}$) is the set of relations with same range (domain), for $_R \otimes (-) = R \otimes (-)$, $\longrightarrow_R (-) = R \longrightarrow (-)$, $(-) \otimes _R = (-) \otimes R$ and $(-) \longleftarrow_R = (-) \longleftarrow R$.

The contravariant version, i.e. $\iota(p') \geq p$ *iff* $p' \leq' \sigma(p)$ is called a *Galois connection* and $\langle \iota, \sigma \rangle$ a *polarity*. Galois connections from binary relations were introduced in [6] and applied to data analysis in Formal Concept Analysis (FCA) (see [12]). Galois adjunctions have been introduced in classical Rough Set Theory in [4] with the name "dual Galois connections".

In what follows the decorations $\uparrow$ and $\downarrow$ inside the constructors denote opposite directions.

Lemma 2. *For any relation $R, R' \subseteq A \times B$, $R \subseteq R'$, $X \subseteq A$, $Y \subseteq B$, for $D, D' \in \wp(A)$ or $\wp(B)$ with $D \subseteq D'$ and $a \in A$, $b \in B$:*

$$\wp(B) \dashv^{\langle \uparrow \rangle, [\downarrow]} \wp(A); \quad \wp(B) \dashv^{[[\uparrow]][[\downarrow]]} \wp(A)^{op}; \quad If \dashv^{\triangleleft, \triangleright} \text{ then } \triangleleft \triangleright \triangleleft = \triangleleft \text{ and } \triangleleft \triangleright \triangleleft \triangleright = \triangleleft \triangleright \quad (6)$$

$$\langle \bullet \rangle(D) \subseteq \langle \bullet \rangle(D') \text{ and } [\bullet](D) \subseteq [\bullet](D'); \quad D \subseteq D' \text{ implies } [[\bullet]](D') \subseteq [[\bullet]](D). \quad (7)$$

$$\langle \bullet \rangle \text{ and } [[\bullet]] \text{ are monotone in } R : \langle \bullet \rangle_R(D) \subseteq \langle \bullet \rangle_{R'}(D) \text{ and } [[\bullet]]_R(D) \subseteq [[\bullet]]_{R'}(D). \quad (8)$$

$$[\bullet] \text{ is antitone in } R : [\bullet]_{R'}(D) \subseteq [\bullet]_R(D). \quad (9)$$

$$If D = \{d\} \text{ then } [[\bullet]](D) = \langle \bullet \rangle(D) \quad (10)$$

$$b \in \langle \rightarrow \rangle(X) \text{ iff } R^{\smile}(b) \cap X \neq \emptyset, \quad a \in \langle \leftarrow \rangle(Y) \text{ iff } R(a) \cap Y \neq \emptyset \quad (11)$$

$$b \in [\rightarrow](X) \text{ iff } R^{\smile}(b) \subseteq X, \quad a \in [\leftarrow](Y) \text{ iff } R(a) \subseteq Y \quad (12)$$

$$b \in [[\rightarrow]](X) \text{ iff } X \subseteq R^{\smile}(b), \quad a \in [[\leftarrow]](Y) \text{ iff } Y \subseteq R(a) \quad (13)$$

where $\wp(X)^{op}$ is $\wp(X)$ with reverse inclusion order.

Proof. Let us prove (6) (for the complete proof see [8]). Probably the most general proof comes from the fact that in Topos Theory, given a relation R, there are arrows which correspond (internally) to the constructors $\langle \bullet \rangle_R$ and $[\bullet]_R$ in the category of sets and total functions (see, for instance, [3], ch. 15). Together with the classical striking observation by William Lawvere in [5] that $\exists$ and $\forall$ are adjoint functors (via the intermediation of a substitution function), it is possible to prove the adjunction properties for the constructors. However, the set-theoretic proof is elementary: in view of (12) and additivity of Pierce products, $\langle \rightarrow \rangle(X) \subseteq Y$ iff $R(X) \subseteq Y$ iff $X \subseteq [\leftarrow](Y)$.

3   Approximating Heterogeneous Relations

The main topic of the paper is the *lower* approximation of relations organized in a Vitruvian schema by means of the pseudo-continuity pattern. Different modes of pseudo-continuity are definable by different concatenations of perception constructors, which we call *sequences*. A sequence $\mathbb{C}_R^X$ is an algebraic combination of relations starting with a constructor defined by the relation R, with input in $dom(R)$ and output in the set X.

Example 1. The following Vitruvian schema RS will be used throughout the paper:

					D				
1	1	1	1	l	0	0	1		
0	1	0	1	i	0	1	1		
1	0	0	1	h	1	0	1		
C	λ	κ	ι	θ	$\dfrac{Q\ \mid\ Z}{W\ \mid\ R}$	a	b	c	B
0	0	0	1	α	1	0	1		
1	0	1	0	β	0	0	1		
0	1	1	1	γ	1	1	0		
0	1	0	0	δ	0	1	1		
				A					

4   Unilateral Lower Approximation Operators

The overall idea is to identify a subset X of elements of $dom(R)$ whose outputs through R are coherent with those of their counterparts in $dom(Q)$, according to a given criterion ρ. The restriction of R to X, $R \upharpoonright_X := X_A \otimes R$, represents the approximation of R as it is reflected by Q according to ρ. If $dom(R) \neq dom(Q)$ a criterion which compares R and Q by connecting their domains using W will be called *nominal*, *generic* otherwise. Both of them can be existentially or universally quantified on C. These operators will output subsets of $dom(R) \times dom(R)$, say R^-, which will be transformed by means of the operation $Id_A \cap R^-$ into a test X_A where X is the set of elements $a \in A$ for which the criterion ρ holds. If $app(X) = \{\langle a, a \rangle ...\}$ we shall write $a \in app(X)$ instead of $\langle a, a \rangle \in app(X)$, for app an approximation operator. Now, for any formula $X \subseteq Y$ a question arises as to for how many elements of X or how many elements of Y does the inclusion hold. The possible answers are: 1) *at least one*, 2) *at least all*, 3) *at most all*. Therefore the operators will be classified on the basis of their modal characters: necessity, possibility, sufficiency.

Definition 4. *Suppose $R \subseteq A \times B$, $X \subseteq A$ and $Y \subseteq B$:*

Definition	Denotation	Definition	Denotation
$X \subseteq [\leftarrow]_R(Y)$	$X \subseteq_{[\leftarrow]_R} Y$ *necessary inclusion*	$[\rightarrow]_R(X) \subseteq Y$	$X \subseteq^{[\rightarrow]_R} Y$ *inverse necessary inclusion*
$X \subseteq \langle \leftarrow \rangle_R(Y)$	$X \subseteq_{\langle \leftarrow \rangle_R} Y$ *possible inclusion*	$\langle \rightarrow \rangle_R(X) \subseteq Y$	$X \subseteq^{\langle \rightarrow \rangle_R} Y$ *inverse possible inclusion*
$X \subseteq [[\leftarrow]]_R(Y)$	$X \subseteq_{[[\leftarrow]]_R} Y$ *sufficient inclusion*	$[[\rightarrow]]_R(X) \subseteq Y$	$X \subseteq^{[[\rightarrow]]_R} Y$ *inverse sufficient inclusion*

The following equivalences hold because of adjunction or connection relations:

$$(1)\ X \subseteq_{[\leftarrow]_R} Y \equiv X \subseteq^{\langle\rightarrow\rangle_R} Y. \qquad (2)\ X \subseteq_{[[\leftarrow]]_R} Y \equiv Y \subseteq^{[[\rightarrow]]_R} X.$$

For $S \subseteq B \times A$ symmetric modalized inclusions can be defined, such as $X \subseteq [\rightarrow]_S(Y)$, that is, $X \subseteq_{[\rightarrow]_S} Y$ (*co-necessary inclusion*). From now on we shall deal just with inclusions decorated by R. Thus, by default the subscript of the constructors will be R.

Lemma 3.

(a) $X \subseteq_{[\leftarrow]} Y$ iff $R(X) \subseteq Y$	(b) $X \subseteq^{[\rightarrow]} Y$ iff $\forall b(R^\smile(b) \subseteq X \Longrightarrow b \in Y)$
(c) $X \subseteq_{\langle\leftarrow\rangle} Y$ iff $X \subseteq R^\smile(Y)$	(d) $X \subseteq^{\langle\rightarrow\rangle} Y$ iff $R(X) \subseteq Y$
(e) $X \subseteq_{[[\leftarrow]]} Y$ iff $Y \subseteq \bigcap\{R(a)\}_{a\in X}$	(f) $X \subseteq^{[[\rightarrow]]} Y$ iff $\forall b(X \subseteq R^\smile(b)) \Longrightarrow b \in Y)$

Proof. (a): By adjunction $X \subseteq [\leftarrow](Y)$ iff $\langle\rightarrow\rangle(X) \subseteq Y$ which in turn is trivially equivalent by definition to $R(X) \subseteq Y$. (b): Because $b \in [\rightarrow](Y)$ iff $R^\smile(b) \subseteq X$. (c): By definition. (d): By definition or because by adjunction it is equivalent to (a). (e): $X \subseteq [[\leftarrow]](Y)$ iff $\forall a \in X, Y \subseteq R(a)$ iff $Y \subseteq \bigcap\{R(a)\}_{a\in X}$. (f): By reversing the inclusion of (b).

As to the comparative strength of the above modalized inclusions, see [8], where also the isotone and antitone behaviours of the modalized inclusion operators are discussed.

Coding the modalized inclusions. In what follows we shall deal only with 1-element sequences so that contours are simply relational neighbourhoods. The following lemma exhibits the skeleton of the pseudo-codes of the approximation operators we will discuss:

Lemma 4. *Let RS be as above. Then:*

$$R(a) \subseteq_{[\leftarrow]_Z} Q(c) \qquad iff \qquad \langle c, a \rangle \in (Q \longleftarrow Z) \longleftarrow R) \tag{14}$$

$$R(a) \subseteq_{\langle\leftarrow\rangle_Z} Q(c) \qquad iff \qquad \langle c, a \rangle \in (Q \otimes Z^\smile) \longleftarrow R \tag{15}$$

$$R(a) \subseteq_{[[\leftarrow]]_Z} Q(c) \qquad iff \qquad \langle c, a \rangle \in (Q^\smile \longrightarrow Z^\smile) \longleftarrow R \tag{16}$$

$$R(a) \subseteq^{[\rightarrow]_Z} Q(c) \qquad iff \qquad \langle c, a \rangle \in Q \longleftarrow (R \longleftarrow Z^\smile) \tag{17}$$

$$R(a) \subseteq^{\langle\rightarrow\rangle_Z} Q(c) \qquad iff \qquad \langle c, a \rangle \in (Q \longleftarrow Z) \longleftarrow R \tag{18}$$

$$R(a) \subseteq^{[[\rightarrow]]_Z} Q(c) \qquad iff \qquad \langle c, a \rangle \in Q \longleftarrow (R^\smile \longrightarrow Z) \tag{19}$$

Proof. We prove just (14): $R(a) \subseteq_{[\leftarrow]_Z} Q(c)$ iff $\forall b (b \in R(a) \implies (Z(b) \subseteq Q(c)))$. From (3) this holds iff $R(a) \subseteq (Q \longleftarrow Z)(c)$ iff $\langle c, a \rangle \in (Q \longleftarrow Z) \longleftarrow R$.

4.1 Nominal Unilateral Lower Approximation Operators

The philosophy of nominalization suggests applying the quantifiers only to C, through $W(a)$, because its meaning is "for all possible C-avatar of a ...". We obtain 12 approximation operators, which produce subsets of the domain of R.

Specifications

Definition 5. *Let RS be as above.*

$$(lnunQ)_{W,Z}(R) \quad := \quad \{a : \forall c (c \in W(a) \implies (R(a) \subseteq_{[\leftarrow]_Z} Q(c))\} \tag{20}$$

$$(lnupQ)_{W,Z}(R) \quad := \quad \{a : \forall c (c \in W(a) \implies R(a) \subseteq_{\langle\leftarrow\rangle_Z} Q(c)\} \tag{21}$$

$$(lnusQ)_{W,Z}(R) \quad := \quad \{a : \forall c (c \in W(a) \implies (R(a) \subseteq_{[[\leftarrow]]_Z} Q(c))\} \tag{22}$$

$$(linunQ)_{W,Z}(R) \quad := \quad \{a : \forall c (c \in W(a) \implies (R(a) \subseteq^{[\rightarrow]_Z} Q(c))\} \tag{23}$$

$$(lnenQ)_{W,Z}(R) \quad := \quad \{a : \exists c (c \in W(a) \wedge (R(a) \subseteq_{[\leftarrow]_Z} Q(c))\} \tag{24}$$

$$(lnepQ)_{W,Z}(R) \quad := \quad \{a : \exists c (c \in W(a) \wedge (R(a) \subseteq_{\langle\leftarrow\rangle_Z} Q(c))\} \tag{25}$$

$$(lnesQ)_{W,Z}(R) \quad := \quad \{a : \exists c (c \in W(a) \wedge (R(a) \subseteq_{[[\leftarrow]]_Z} Q(c))\} \tag{26}$$

$$(linepQ)_{W,Z}(R) \quad := \quad \{a : \exists c (c \in W(a) \wedge R(a) \subseteq^{\langle\rightarrow\rangle_Z} Q(c))\} \tag{27}$$

The naming rule is the following: "l" stands for "lower"; before Q we have: "n" for "necessary", p for "possible" and "s" for "sufficient". The "n" after the "l" stands for "nominal", "i" for "inverse", "u" for "universal" and "e" for "existential". The term "direct" is not used either in the names or in the acronyms. Here are two examples:

$(lnunQ)_{W,Z}(R)$	*(direct) nominal universal necessary lower approximation* of R via Q
$(linepQ)_{W,Z}(R)$	*inverse nominal existential possible lower approximation* of R via Q.

Notice, for example, that since $X \subseteq Y \wedge X \subseteq Y'$ if and only if $X \subseteq Y \cap Y'$ one has that $a \in (lnunQ)_{W,Z}(R)$ if and only if $Z(R(a)) \subseteq \bigcap\{Q(c)\}_{c \in W(a)}$. The existential version of this operator fulfils the same relation but relaxed to single results of Q. This situation is illustrated by the picture on the right where the existential pattern is indicated by $\exists$ and the universal one by $\forall$.

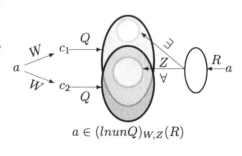

$$a \in (lnunQ)_{W,Z}(R)$$

The restriction of R to $X \subseteq A$ is defined as $R \upharpoonright_X = \{a \in X \wedge \langle a, b \rangle \in R\}$. In view of Definition 2 and Lemma 2, the following sample reading is straightforward:

If $X = (lnunQ)_{W,Z}(R)$ then $\langle a, b \rangle \in R \upharpoonright_X$ iff $Z(R(a)) \subseteq Q(c)$, any $c \in W(a)$

From Lemma 4 the pseudo-codes can be easily derived:

Theorem 1.

$$(lnunQ)_{W,Z}(R) = (Id_A \cap (W^{\smile} \longrightarrow (Q \longleftarrow Z) \longleftarrow R))(A) \qquad (28)$$

$$(lnupQ)_{W,Z}(R) = (Id_A \cap (W^{\smile} \longrightarrow (Q \otimes Z^{\smile}) \longleftarrow R))(A) \qquad (29)$$

$$(lnusQ)_{W,Z}(R) = (Id_A \cap (W^{\smile} \longrightarrow (Q^{\smile} \longrightarrow Z^{\smile} \longleftarrow R))(A) \qquad (30)$$

$$(linunQ)_{W,Z}(R) = (Id_A \cap (W^{\smile} \longrightarrow Q \longleftarrow (R \longleftarrow Z^{\smile})))(A) \qquad (31)$$

$$(lnenQ)_{W,Z}(R) = (Id_A \cap (W \otimes ((Q \longleftarrow Z) \longleftarrow R)))(A) \qquad (32)$$

$$(lnepQ)_{W,Z}(R) = (Id_A \cap (W \otimes ((Q \otimes Z^{\smile}) \longleftarrow R)))(A) \qquad (33)$$

$$(lnesQ)_{W,Z}(R) = (Id_A \cap (W \otimes (Q^{\smile} \longrightarrow Z^{\smile} \longleftarrow R)))(A) \qquad (34)$$

$$(linepQ)_{W,Z}(R) = (Id_A \cap (W \otimes ((Q \longleftarrow Z) \longleftarrow R)))(A) \qquad (35)$$

Proof. We give just one example (straightforward consequence of Lemma 4):

$$(lnunQ)_{W,Z}(R) = \{\langle a, a \rangle : \forall c (c \in W(a) \implies (R(a) \subseteq_{[\leftarrow]_Z} Q(c))\}$$
$$= \{\langle a, a \rangle; \forall c (\langle a, c \rangle \in W \implies \langle c, a \rangle \in (Q \longleftarrow Z) \longleftarrow R)\}$$
$$= \{\langle a, a \rangle : \langle a, a \rangle \in W^{\smile} \longrightarrow ((Q \longleftarrow Z) \longleftarrow R)\}$$
$$= Id_A \cap (W^{\smile} \longrightarrow ((Q \longleftarrow Z) \longleftarrow R))$$

The $\exists \forall$ definition of nominal existential lower approximation operators is a strong requirement: a single element must fit different paths. The weaker requirement $\forall \exists$ is provided by inverse inclusion operators applied to the composition of W and Q, which we call Π-operators (after the arithmetical hierarchy) or *infix existential operators*:

$$R(a) \subseteq_W^{op\overrightarrow{Z}} Q(c) \text{ iff } \forall d (d \in op\overrightarrow{Z}(R(a)) \implies \exists c (c \in W(a) \wedge d \in Q(c)))$$
$$\text{iff } R(a) \subseteq^{op\overrightarrow{Z}} Q(W(a)) \qquad (36)$$

An interesting example is: $(ilinepQ)_{W,Z}(R) := \{a : R(a) \subseteq_W^{\langle\rightarrow\rangle z} Q(c)\}$ whose pseudo-code is $\{a : \langle a, a \rangle \in (W \otimes Q) \longleftarrow (R \otimes Z)\}$. By adjointness it coincides with $(ilnenQ)_{W,Z}(R) = \{a : R(a) \subseteq_W^{[\leftarrow]z} Q(c)\} := \{a : R(a) \subseteq_{[\leftarrow]z} Q(W(a))\} = \{a : \forall b(b \in R(a) \Longrightarrow Z(b) \subseteq Q(W(a)))\}$.

In turn, $(ilnesQ)_{W,Z}(R) := \{a : R(a) \subseteq_W^{[[\leftarrow]]z} Q(c)\} = \{a : R(a) \subseteq_{[[\leftarrow]]z} Q(W(a))\}$ which is equal to $(lnusQ)_{W,Z}(R)$. Finally, $(ilnepQ)_{W,Z}(R) = \{a : R(a) \subseteq (W \otimes Q \otimes Z^{\smallsmile})(a)\} = \{a : \langle a, a \rangle \in (W \otimes Q \otimes Z^{\smallsmile}) \longleftarrow R\}$. It is a simple operator which states that $R(a)$ is a subset of the complete tour of a along the Vitruvian schema. This kind of existential operators exhibit interesting geometric patterns which are part of the link between the present theory of approximation of relations and Process Algebra.

4.2   Generic Unilateral Lower Approximation Operators

Let RS be as above. The *generic* versions of the above approximation operators are obtained by substituting "$\forall c(...)$" and "$\exists c(...)$" for $\forall c(c \in W(a) \Longrightarrow ...)$" and, respectively, $\exists c(c \in W(a) \wedge ...)$". Their names are the same as their corresponding nominal companions, with n (nominal) replaced by g (generic). For instance $(lgunQ)_Z(R) := \{a : \forall c(R(a) \subseteq_{[\leftarrow]z} Q(c))\}$. Hence the relation W does not play any role and can be replaced with $\mathbf{1}_{A \times C}$ and $\mathbf{1}_{C \times A}$. Therefore, one obtains $(lgunQ)_Z(R) = (Id_A \cap (\mathbf{1}_{C \times A} \longrightarrow ((Q \longleftarrow Z) \longleftarrow R)))(A)$.

4.3   Computing Procedure

The outputs of the above operators are tests X_A for $X \subseteq A$, that is, they are subsets of the diagonal of $A \times A$, while the actual inclusions are provided by the basic equations (14), (15) and (16), which form what we call the inner part of the *core formulas* of the operators, which, in turn, are the expressions on the right of "$Id_A \cap$". Note that the pseudo-codes and specifications of the main core formulas are (for nominal operators):

$$W^{\smallsmile} \longrightarrow (Q \longleftarrow Z) \longleftarrow R = \{\langle a, a' \rangle : \forall c(c \in W(a) \Longrightarrow (R(a') \subseteq_{[\leftarrow]z} Q(c)))\}$$
$$W^{\smallsmile} \longrightarrow (Q \otimes Z^{\smallsmile}) \longleftarrow R = \{\langle a', a \rangle : \forall c(c \in W(a) \Longrightarrow (R(a') \subseteq_{\langle\leftarrow\rangle z} Q(c)))\}$$
$$(R^{\smallsmile} \longrightarrow Z \longleftarrow Q) \longleftarrow W = \{\langle a', a \rangle : \forall c(c \in W(a) \Longrightarrow (R(a') \subseteq_{[[\leftarrow]]z} Q(c)))\}$$
$$((R \otimes Z) \longleftarrow Q) \longleftarrow W = \{\langle a', a \rangle : \forall c(c \in W(a) \Longrightarrow (R(a') \subseteq^{[[\leftarrow]]R} Q(c)))\}$$

This explains why one has to select the elements of the form $\langle x, x \rangle$ out of the results of the core formulas, by applying $Id_A \cap$.

Now we show step by step the computing procedure of a lower approximation operator, illustrated using the relational schema of Example 1. Consider $(lnunQ)_{W,Z}(R)$.

1) Pseudo-code: $W^{\smallsmile} \longrightarrow (Q \longleftarrow Z) \longleftarrow R$.
2) Procedure: $-(W \otimes -Q \otimes Z^{\smallsmile} \otimes R^{\smallsmile})$.

3) Checking the type of the relational expression:

$$
\begin{array}{ccccccc}
W & \otimes & Q & \otimes & Z^{\smile} & \otimes & R^{\smile} \\
\hline
A \times C & \otimes & C \times D & \otimes & D \times B & \otimes & B \times A \\
\hline
& A \times D & & \otimes & & D \times A & \\
\hline
& & & A \times A & & &
\end{array}
$$

The steps of computation run as follows:

$W \otimes -Q$	h i l
α	0 0 0
β	1 1 0
γ	1 1 0
δ	1 0 0

$W \otimes -Q \otimes Z^{\smile}$	a b c
α	0 0 0
β	1 1 1
γ	1 1 1
δ	1 0 1

$W \otimes -Q \otimes Z^{\smile} \otimes R^{\smile}$	α β γ δ
α	0 0 0 0
β	1 1 1 1
γ	1 1 1 1
δ	1 1 1 1

$-(W \otimes -Q \otimes Z^{\smile} \otimes R^{\smile})$	α β γ δ
α	1 1 1 1
β	0 0 0 0
γ	0 0 0 0
δ	0 0 0 0

$Id_A \cap -(W \otimes -Q \otimes Z^{\smile} \otimes R^{\smile})$	α β γ δ
α	1 0 0 0
β	0 0 0 0
γ	0 0 0 0
δ	0 0 0 0

It follows that $(lnunQ)_{W,Z}(R) = \{a\}$.

4.4   Unilateral Upper Approximation Operators

Unilateral upper approximation operators are given by considering "$R(a) \subseteq$" as a function F with "$op(Q(c))$" its argument X and tacking its dual $-(F(-X))$. Let us see some results:

Definition 6.

1. $(unupQ)_{W,Z}(R) := \{a : \forall c(c \in W(a) \implies R(a) \cap_{\langle \leftarrow \rangle z} Q(c) \neq \emptyset)\}$
2. $(unusQ)_{W,Z}(R) := \{a : \forall c(c \in W(a) \implies R(a) \cap_{[[\leftarrow]]z} Q(c) \neq \emptyset)\}$
3. $(unepQ)_{W,Z}(R) := \{a : \exists c(c \in W(a) \wedge R(a) \cap_{\langle \leftarrow \rangle z} Q(c) \neq \emptyset)\}$
4. $(uinupQ)_{W,Z}(R) := \{a : \forall c(c \in W(a) \implies R(a) \cap^{\langle \rightarrow \rangle z} Q(c) \neq \emptyset)\}$
5. $(uinusQ)_{W,Z}(R) := \{a : \forall c(c \in W(a) \implies R(a) \cap^{[[\rightarrow]]z} Q(c) \neq \emptyset)\}$
6. $(uinepQ)_{W,Z}(R) := \{a : \exists c(c \in W(a) \wedge R(a) \cap^{\langle \rightarrow \rangle z} Q(c) \neq \emptyset)\}$
7. $(uinesQ)_{W,Z}(R) := \{a : \exists c(c \in W(a) \wedge R(a) \cap^{[[\rightarrow]]z} Q(c) \neq \emptyset)\}$

We present the pseudo-codes of the upper approximation operators in a slightly different way: instead of setting $Id_A \cap$ *core formula* and project the result on the domain A, we state "$\{a : \langle a, a \rangle \in$ *core formula*$\}$".

Lemma 5.

(a) $(unupQ)_{W,Z}(R) = \{a : \langle a, a \rangle \in (R \otimes Z \otimes Q^{\smile}) \longleftarrow W\}$
(b) $(unusQ)_{W,Z}(R) = \{a : \langle a, a \rangle \in ((R \otimes (Z \longleftarrow Q))) \longleftarrow W\}$
(c) $(unepQ)_{W,Z}(R) = \{a : \langle a, a \rangle \in R \otimes Z \otimes Q^{\smile} \otimes W^{\smile}\}$

(d) $(uinupQ)_{W,Z}(R) = (unupQ)_{W,Z}(R)$
(e) $(uinusQ)_{W,Z}(R) = \{a : \langle a, a \rangle \in ((R^{\smile} \longrightarrow Z) \otimes Q^{\smile})) \longleftarrow W\}$
(f) $(uinepQ)_{W,Z}(R) = (unepQ)_{W,Z}(R)$
(g) $(uinesQ)_{W,Z}(R) = \{a : \langle a, a \rangle \in ((R^{\smile} \longrightarrow Z) \otimes Q^{\smile} \otimes W^{\smile}\}$

4.5   The Injective Cases

By "injective case" we intend when the connections between A and C or B and D are given by possibly partial injective functions $\varphi : A \longmapsto C$ and, respectively, $\psi : B \longmapsto D$. We consider as domains or ranges of R and Q the images and pre-images of these functions. We can get rid of the functions, therefore, by renaming any element $c \in C$ with the unique member of its pre-image $\varphi^{-1}(c)$ or $\psi^{-1}(d)$. Thus φ or ψ can be substituted by the identity relations Id_A, respectively Id_B and we write $A =_{Id} C$ or $B =_{Id} D$.

The resulting approximation operators depend on a *cancellation rule* which is detailed in [8] and which can be described as follows: one or both the parameters W and Z can be deleted according to the fact that for any R_1 and R_2 of the appropriate type $R_1 \otimes Id = R_1$ and $Id \otimes R_2 = R_2$ (cf. Definition 1.(4.)) and $Id^{\smile} = Id$. On the contrary, $R_1 \otimes -Id \neq R_1$ and $-Id \otimes R_2 \neq R_2$. There are three possible situations:

$B =_{Id} D$ **and** $Z = Id_B$: **deleting** Z
One result of the cancellation rule is, for instance:

$$(lnunQ)_W(R) = (lnupQ)_W(R) = (linunQ)_W(R) \doteq (linupQ)_W(R) =$$
$$= (Id_A \cap (W^{\smile} \longrightarrow Q \longleftarrow R))(A)$$
$$= \{a : \forall c (c \in W(a) \Longrightarrow (R(a) \subseteq Q(c))\} = \{a : W(a) \subseteq_{[[\leftarrow]]_Q} R(a)\}$$
$$\tag{37}$$

$A =_{Id} C$ **and** $W = Id_A$: **deleting** W
In nominal operators the parameter W can be freely cancelled because it appears always as the premise of a residuation or as an argument of a composition. For instance:

$$(lnunQ)_Z(R) = (lnenQ)_Z(Q) = (linupQ)_Z(R) = (lnenQ)_Z(R) = (linepQ)_Z(R)$$
$$= (Id_A \cap ((Q \longleftarrow Z) \longleftarrow R))(A) = \{a : R(a) \subseteq_{[\leftarrow]_Z} Q(a)\}$$
$$\tag{38}$$

Generic approximation operators do not use the parameter W and remain unchanged.

2-tier approximations. Consider now two relations $R \subseteq A \times B$ and $Z \subseteq B \times D$. An element a of A belongs to the 2-strict lower approximation of a set $X \subseteq D$ if for all elements $b \in R(a)$, $Z(b) \subseteq X$. Since neighbourhoods are additive, this leads to the set $\{a : Z(R(a)) \subseteq X\}$. It follows from (38) and Lemma 3 that if $X = Q(x)$ for some relation Q with range D, the 2-strict lower approximation is given by $(lnunQ)_Z(R)$ and its equivalent fellows.

Example 2. In our sample relational frame, $Z(R(\alpha)) = Z(R(\beta)) = Z(R(\gamma)) = Z(R(\delta)) = \{h, i, l\}$. Hence, an element of A belongs to the 2-strict lower approximation of $Q(\theta)$ only.

$A =_{Id} C$ **and** $B =_{Id} D$: **deleting** W **and** Z When $RS = (R \subseteq A \times B, W = Id_A, Q \subseteq A \times B, Z = Id_B)$ we simply have:

$$(lnunQ)(R) = (lnupQ)(R) = (lnenQ)(R) = (lnepQ)(R) = (lgunQ)(R) = (genQ)(R)$$
$$= \{a : R(a) \subseteq Q(a)\} = (Id_A \cap (Q \longleftarrow R))(A)$$

The "generic" operators are computed as in the case of the deletion of Z but without the use of the normalizing parameters $1_{A \times C}$ and $1_{C \times A}$.

5   Reconstruction of a Relation

If we are not given the relation R, we can reconstruct it by postulating that the elements of A and B are related by R if their W and, respectively, Z counterparts are related by Q:

$$\{\langle a, b \rangle : \forall c, d(c \in W(a) \wedge d \in Z(b) \Longrightarrow d \in Q(c)))\} \tag{39}$$

From (39) we obtain:

$$\{\langle a, b \rangle : \forall c(c \in W(a) \Longrightarrow (Z(b) \subseteq Q(c)))\} = W^{\smile} \longrightarrow Q \longleftarrow Z \tag{40}$$

The pseudo-code of (40) is the converse of (16) with R replaced by W, Z by Q and Q by Z. This story teaches three things. First, it brings to the notion of a *lower sufficient reconstruction* of R via W and Z as lateral relations and Q as "reference" relation:

$$(lsrWZ)_Q(R) := \{\langle a, b \rangle : W(a) \subseteq_{[[\leftarrow]]_Q} Z(b)\} \tag{41}$$

Second, let us set the following *binary hyper-relation*:

$$H_{WZ} = \{\langle \langle a, b \rangle, \langle c, d \rangle \rangle : \langle a, c \rangle \in W \wedge \langle b, d \rangle \in Z\} \tag{42}$$

then (39) is equivalent to $\{\langle a, b \rangle : H_{WZ}(\langle a, b \rangle) \subseteq Q\}$. So $(lsrWZ)_Q(R)$ is a sort of lower approximation. Actually, it is a generalisation/specialisation of the classical approach to rough relations proposed in [11] because it takes into account *arbitrary* but *binary* relations.

Third, it is evident that $(lsrWZ)_Q(R)$ is the part of the core formula of $(lnunQ)_{W,Z}(R)$ which is on the left of $\longleftarrow R$. This observation suggests that when R is detachable from the core formula of a nominal approximation operator, a kind of reconstruction is obtained. We say that a relation R is detachable from a formula $F(R, W, Z, Q)$ if the remaining part is still computable (that is, the types of the remaining part are coherent). Let us examine the seven independent results, including $(lsrWZ)_Q(R)$.

From $(lnupQ)_{W,Z}(R)$ one obtains $W^\smile \longrightarrow (Q \otimes Z^\smile) = \{\langle a, b \rangle : \forall c (c \in W(a) \Longrightarrow \exists d (d \in Z(b) \wedge d \in Q(c)))\}$. Its converse pseudo-code, $(Z \otimes Q^\smile) \longleftarrow W$, is an instance of (15), that is, a possibility inclusion. Hence we set the following definition:

Lower possible reconstruction of R: $(lprWZ)_Q(R) := \{\langle a, b \rangle : W(a) \subseteq_{\langle \leftarrow \rangle_Q} Z(b)\}$

From $(lnusQ)_{W,Z}(R)$ one obtains $W^\smile \longrightarrow (Q^\smile \longrightarrow Z^\smile) = \{\langle a, b \rangle : \forall d \exists c (c \in W(a) \wedge d \in Q(c) \Longrightarrow d \in Z(b)))\}$. Its converse formulation $(Z \longleftarrow Q) \longleftarrow W$ is equivalent to $Z \longleftarrow (W \otimes Q)$. From (14) it turns out to be a necessity inclusion, so that we define:

Lower necessary reconstruction of R: $(lnrWZ)_Q(R) := \{\langle a, b \rangle : W(a) \subseteq_{[\leftarrow]_Q} Z(b)\}$

From $(lnenQ)_{W,Z}(R)$ let us detach $W \otimes (Q \longleftarrow Z) = \{\langle a, b \rangle : \exists c (c \in W(a) \wedge \forall d (d \in Z(b) \Longrightarrow d \in Q(c)))\}$. It is an instance of a sufficient intersection so that we define:

Upper sufficiency reconstruction of R: $(usrWZ)_Q(R) := \{\langle a, b \rangle : W(a) \cap_{[[\leftarrow]]_Q} Z(b) \neq \emptyset\}$

From $(lnesQ)_{W,Z}(R)$ one obtains $W \otimes (Q^\smile \longrightarrow Z^\smile) = \{\langle a, b \rangle : \exists c (c \in W(a) \wedge \forall d (c \in Q^\smile(d) \Longrightarrow b \in Z^\smile(d)))\}$. It is an instance of a necessary intersection, therefore we define:

Upper necessary reconstruction of R: $(unrWZ)_Q(R) := \{\langle a, b \rangle : W(a) \cap_{[\leftarrow]_Q} Z(b) \neq \emptyset\}$

In turn, $(ilnenQ)_{W,Z}(R)$ gives $(W \otimes Q) \longleftarrow Z = \{\langle a, b \rangle : \forall d (d \in Z(b) \Longrightarrow \exists c (c \in W(a) \wedge d \in Q(c)))\}$. Thus a sort of co-possible inclusion is obtained and we set:

Lower co-possible reconstruction of R: $(lcprZW)_Q(R) := \{\langle a, b \rangle : Z(b) \subseteq_{\langle \rightarrow \rangle_Q} W(a)\}$

Notice the use of the co-inclusion operator $\subseteq_{\langle \rightarrow \rangle_Q}$.

Other reconstructions are definable by detaching $R \otimes$ from upper approximations, with some surprises. For instance, if $R \otimes$ is detached from the core formula of $(unepQ)_{W,Z}(R)$ or $(uinepQ)_{W,Z}(R)$, or $R \longrightarrow$ is detached from the core formula of $(uinesQ)_{W,Z}(R)$ then one obtains $Z \otimes Q^\smile \otimes W^\smile$ which is of type $B \times C$. We transpose it and define:

$$(dirWZ)_Q(R) = W \otimes Q \otimes Z^\smile = \{\langle a, b \rangle : W(a) \cap^{\langle \rightarrow \rangle_Q} Z(b) \neq \emptyset\} \qquad (43)$$

Therefore, $(dirWZ)_Q(R)$ can be called *inverse possible reconstruction of R*. $(dirWZ)_Q(R)$ is also obtained by detaching $\longleftarrow R$ from $(ilnepQ)_{W,Z}(R)$ and $(lprWZ)_Q(R)$ can be obtained by detachment of $R \otimes$ from $(unupQ)_{W,Z}(R)$ (or $(uinupQ)_{W,Z}(R)$) and so on.

We shall see that the above reconstruction operators are linked to Process Algebra.

Approximation of Cylinders. The above machinery can be used to approximate sets represented by cylinders in a homogeneous relational framework. Thus, assume we are given the relational schema $(W \subseteq A \times A, Z \subseteq A \times A, Q \subseteq A \times A)$. By replacing $Q(a)$ by a set in guise of a cylinder one obtains $\{a : \forall a' (a' \in$

$W(a) \implies (Z(a') \subseteq X))\} = ((X_Q^{\leftarrow} \longleftarrow Z) \longleftarrow W)(a^*)$ If $Q = W = Z := R$ for R a preorder, the above formula turns into:

$$((X_R^{\leftarrow} \longleftarrow R) \longleftarrow R)(a^*) = (X_R^{\leftarrow} \longleftarrow R)(a^*) = (lR)(X) \qquad (44)$$

Thus, the usual approximations of sets are recovered (for other ways see [8] and cfr. [7]).

5.1 Bilateral Approximation of a Relation with Reference Relation

Let us just mention the topic. Given the complete relational schema RS the purpose is to approximate the relation R by procedures which output ordered pairs $\langle a, b \rangle \in A \times B$, with the aid of the reference relation Q. It is natural to combine in some way the expressions "$\langle a, b \rangle \in R$" and "$\langle c, d \rangle \in Q$" through formulas in W and Z. An immediate solution is the combination of "$\langle a, b \rangle \in R$" with some kind of reconstruction of R itself. For instance:

$$\{\langle a, b \rangle : \langle a, b \rangle \in R \wedge \langle a, b \rangle \in (lsrWZ)_Q(R)\} = R \cap W^{\smile} \longrightarrow Q \longleftarrow Z.$$

5.2 Refinement and Simulation

Simulation is a relation $R \subseteq F(W, Z, Q)$, where $F(W, Z, Q)$ is an algebraic combination of the arguments. Let us consider the following types of simulation (see for instance [2]):

A) *Strong simulation.* It is defined as $W^{\smile} \otimes R \otimes Z \subseteq Q$. Applying (4) after (2) we obtain $R \otimes Z \subseteq W^{\smile} \longrightarrow Q$ and finally $R \subseteq W^{\smile} \longrightarrow Q \longleftarrow Z$. Therefore, from (41) strong simulation states that $R \subseteq (lsrWZ)_Q(R)$, that is, R must be a refinement of its own sufficiency lower reconstruction. Since $Id_A \otimes R = R$ the condition states that $Id_A \otimes R \subseteq W^{\smile} \longrightarrow Q \longleftarrow Z$ which is equivalent to $Id_A \subseteq (W^{\smile} \longrightarrow Q \longleftarrow Z) \longleftarrow R$. Otherwise stated: R strongly simulates Q iff $\forall a \in A, a \in (lnunQ)_{W,Z}(R)$.

B) *Down simulation*: $W^{\smile} \otimes R \subseteq Q \otimes Z^{\smile}$. It transforms into $R \subseteq (lprWZ)_Q(R)$. It follows that R down simulates Q iff $\forall a \in A, a \in (lnupQ)_{W,Z}(R)$.

C) *Up simulation*: $R \otimes Z \subseteq W \otimes Q$. Hence, $R \subseteq (lcprZW)_Q(R)$. As a consequence: R up simulates Q iff $\forall a \in A, a \in (ilnenQ)_{W,Z}(R)$.

D) *Weak simulation*: $R \subseteq W \otimes Q \otimes Z^{\smile}$. It is immediate that $R \subseteq (dirWZ)_Q(R)$. Thus: R weakly simulates Q iff $\forall a \in A, a \in (lnepQ)_{W,Z}(R)$.

Only the cited works are reported. For a more comprehensive bibliography see [8].

References

1. Brink, C., Britz, K., Schmidt, R.A.: Peirce algebras. Formal Aspects Comput. **6**(3), 339–358 (1994)
2. Fokkinga M.: A Summary of Refinement Theory., : Unregistered Technical Report. University of Twente, Enschede, Netherlands (1994)

3. Goldblatt R.: Topoi: The Categorial Analysis of Logic. North- Holland (1984)
4. Järvinen, J.: Knowledge Representation and Rough Sets. Ph.D. Dissertation, University of Turku (1999)
5. Lawvere, F.W.: Adjointness in foundations. Dialectica **23**(3/4), 281–296 (1969)
6. Ore, O.: Galois Connexions. Trans. AMS **55**, 493–531 (1944)
7. Pagliani P.: Modalizing Relations by means of Relations: a general framework for two basic approaches to Knowledge Discovery in Database. In: Proceedings of the IPMU 1998, LaSorbonne, France, Paris France, Paris, 6–10 July, pp. 1175–1182 (1998)
8. Pagliani, P.: A foreword to a general theory of approximation of relations. A coronavirus paper (2020). Published at www.academia.edu and www.researchgate.net
9. Pagliani, P., Chakraborty, M.K.: A Geometry of Approximation, Trends in Logic, vol. 27. Springer, Cham (2008)
10. Sambin G.: Intuitionistic formal spaces - a first communication. In: Skordev, D. (ed.) Mathematical Logic and Its Applications, pp. 187–204. Plenum Press (1987)
11. Skowron, A., Stepaniuk, J.: Approximation of relations. In: Ziarko, W.P. (ed.) Rough Sets, Fuzzy Sets and Knowledge Discovery, pp. 161–166. Springer (1994)
12. Wille, R.: Restructuring lattice theory. In: Rival, I. (ed.) Ordered Sets. NATO ASI Series, vol. 83, pp. 445–470. Reidel (1982)

Logical Treatment of Incomplete/Uncertain Information Relying on Different Systems of Rough Sets

Tamás Mihálydeák$^{(\boxtimes)}$

Department of Computer Science, Faculty of Informatics,
University of Debrecen, Holló László utca 6, Debrecen 4034, Hungary
mihalydeak@unideb.hu

Abstract. The different systems of rough set theory treat uncertainty in special ways. There is a very important common property: all systems rely on given background knowledge and we cannot say more about an arbitrary set or about its members than its lower and upper approximations make possible. New membership relations have to be introduced. The semantics of classical logic is based on classical set theory. An important question: Is it possible to build logical systems with semantics relying on different versions of the theory of rough sets by using new membership relations? The paper gives an overview of first-order logical systems with partial semantics in order to show the influence of incomplete/uncertain information in logically valid inferences.

Keywords: Rough set theory · Partial first-order logic · Multivalued logic

1 Introduction

Pawlak's original theory of rough sets (see in e.g. [8,9,11]), covering systems relying on tolerance relations [12], general covering systems [10,15], decision-theoretic rough set theory [14], general partial approximation spaces [1], similarity based approximation spaces [7] are different systems of rough set theory. There is a very important common property: all systems rely on given background knowledge and one cannot say more about an arbitrary set (representing a 'new' property) or about its members than the lower and upper approximations of the set make possible. The system of base sets represents background knowledge.

The different versions of rough set theory have been focusing on objects: What can one say (what can one know)—relying on the knowledge embedded in an information system—about an object: whether it belongs to an arbitrary given set? But background knowledge embedded in an information system contains

© IFIP International Federation for Information Processing 2021
Published by Springer Nature Switzerland AG 2021
Z. Shi et al. (Eds.): ICIS 2020, IFIP AICT 623, pp. 66–78, 2021.
https://doi.org/10.1007/978-3-030-74826-5_6

much more information: base sets represent properties, and therefore they circumscribe a conceptual structure. An arbitrary set of objects may be considered as the representation of a property. A new important question appears: What can one say—relying on the knowledge embedded in an information system—about the property represented by a set of objects? What is more: What can one say about a relation between objects on the basis of background knowledge?

In the semantics of classical first-order logic classical set theory plays a crucial role. Classical first-order logic is not a logic of set theory, it is a logic *on* sets, i.e. a logic with semantics relying *on* sets. Classical first-order logic reveals the logical laws of general (not singular) propositions, therefore it gives possibilities to get general knowledge (it shows how to achieve new general knowledge) by using quantified propositions. The most important tool in the semantics of classical first-order logic is the membership relation of classical set theory. The truth values of atomic formulae containing predicate parameter(s) can be defined with the help of the membership relation of classical set theory. In various versions of rough set theory different characteristic functions, and so different rough membership relations can be introduced. These functions rely on background knowledge represented by the systems of base sets, and if they are used in the semantics of first-order logic, then the logical system is able to reveal the logical laws of general propositions relying on embedded background knowledge, the nature and the consequences of conceptual structure (conceptual knowledge) embedded in the system of base sets. The main goal of the paper is to give a general and flexible first-order logical system based on generalized characteristic functions of theories of rough sets.

After giving a general picture of approximation spaces the role of base sets is surveyed. The influences of embedded knowledge on membership relations result in generalized characteristic functions. Finally, a partial first-order logical system relying on different generalized characteristic functions is presented precisely.[1]

2   Background

2.1   General Approximation Space

Definition 1. *The ordered 5-tuple* $\langle U, \mathfrak{B}, \mathfrak{D}_{\mathfrak{B}}, \mathsf{l}, \mathsf{u} \rangle$ *is a general approximation space if*

1. U *is a nonempty set;*
2. $\mathfrak{B} \subseteq 2^U \setminus \emptyset$, $\mathfrak{B} \neq \emptyset$ *($\mathfrak{B}$ is the set of base sets);*
3. $\mathfrak{D}_{\mathfrak{B}}$ *is the set of definable sets and it is given by the following inductive definition:*
   (a) $\emptyset \in \mathfrak{D}_{\mathfrak{B}}$;

---

[1] The logical properties of different systems of rough sets are investigated extensively (e.g. [2] and [13]). Typically the investigations are on propositional level and focus on the general properties of rough sets. Arguments, inferences, consequences of embedded knowledge are not in the center of considerations.

(b) $\mathfrak{B} \subseteq \mathfrak{D}_{\mathfrak{B}}$;
(c) if $D_1, D_2 \in \mathfrak{D}_{\mathfrak{B}}$, then $D_1 \cup D_2 \in \mathfrak{D}_{\mathfrak{B}}$

4. $\langle \mathsf{l}, \mathsf{u} \rangle$ *is a Pawlakian approximation pair i.e.*
   (a) $\mathsf{l}(S) = \cup \mathcal{C}^{\mathsf{l}}(S)$, *where* $\mathcal{C}^{\mathsf{l}}(S) = \{B \mid B \in \mathfrak{B} \text{ and } B \subseteq S\}$;
   (b) $\mathsf{u}(S) = \cup \mathcal{C}^{\mathsf{u}}(S)$ *where* $\mathcal{C}^{\mathsf{u}}(S) = \{B \mid B \in \mathfrak{B} \text{ and } B \cap S \neq \emptyset\}$.

Definition 2. *Different types of general approximation space* $\langle U, \mathfrak{B}, \mathfrak{D}_{\mathfrak{B}}, \mathsf{l}, \mathsf{u} \rangle$ *are the followings:*

1. *A general approximation space is Pawlakian if* $\mathfrak{B}$ *is a partition of* U.
2. *A general approximation space is a covering approximation space generated by a tolerance relation* $\mathcal{R}$ *if* $\mathfrak{B} = \{[u]_{\mathcal{R}} \mid u \in U\}$, *where* $[u]_{\mathcal{R}} = \{u' \mid u \mathcal{R} u'\}$.
3. *A general approximation space is a covering approximation space if* $\bigcup \mathfrak{B} = U$.
4. *A general approximation space is a partial approximation space if* $\bigcup \mathfrak{B} \neq U$.
5. *A general approximation space is a similarity based (partial) approximation space generated by a tolerance relation* $\mathcal{R}$ *if*

$$\mathfrak{B} = \{B \mid B \in \mathfrak{Cl}(\mathcal{R}) \text{ and } B \text{ is not a singleton}\},$$

where $\mathfrak{Cl}(\mathcal{R})$ *is the set of clusters determined by a correlation clustering process based on the tolerance relation* $\mathcal{R}$.

2.2 Role of Base Sets

From the philosophical point of view the most important question of the different versions of rough set theory is the following: What can one say (what can one know)—relying on the knowledge embedded in an information system—about an object: whether it belongs to an arbitrary given set? Knowledge embedded in an information system is the background knowledge behind all propositions, which can be made by using the information system. Therefore background knowledge plays a crucial role in rough set theory: background knowledge specifies 'the world of certainty'. Background knowledge is represented by a system of base sets. It means that the membership relation of classical set theory can only be used with base sets, and so background knowledge restricts possibilities: the members of a base set have to be treated in the same way at least in a special sense.)

The next step is to make clear the 'nature', the usage, and the influences of background (and embedded) knowledge. The 'meanings' (which determine the usage) of base sets are the followings:

1. In Pawlak's original system: if an object belongs to a given base set, then it is indiscernible from the members of the base set. In this case, the limit of represented knowledge appears explicitly: base sets consist of indiscernible objects, there is no way to distinguish them from each other.
2. In a covering approximation space generated by a tolerance relation: if an object belongs to a given base set, then it is in the tolerance relation with the generator member of the base set (similar to the generator member of the base set), therefore it may (or has to) be treated in the same way as the other objects which are similar to the generator member of the base set.

3. In a covering approximation space: if an object belongs to a given base set, then it has a common property represented by the base set with the members of the base set. Objects with the same property (members of a base set) may (or have to) be handled in the same way.[2]

4. General partial spaces are similar to general covering ones, but it is not supposed that all objects have at least one property represented by a base set. In practical cases information systems are not total, there is no relevant information about an object: it may be in our database but some information is missing, and so it does not have any property represented by a base set.

5. In a similarity based (partial) approximation space: if an object belongs to a given base set, then it is similar (at least according to a correlation clustering process) to the members of the base set. In general case similarity based approximation spaces are partial and the base sets are pairwise disjoint ones. Therefore there is no way to distinguish the members of a base set from each other.

The definition of definable sets expands 'the word of certainty' by showing the possible usage of base sets and so the possible usage of background knowledge. An arbitrary set of objects belonging to an information system can be approximated by definable sets. One of the most important points of rough set theory is its approximative aspect.

3   Influences of Embedded Knowledge on Membership Relations

From the theoretical point of view the system of base sets represents a sort of limit of our knowledge embedded in an information system. In some situation, it makes our judgment of the membership relation uncertain—making a set vague—because a decision about a given object affects the decision about all other objects which are in the same base set containing the given object.

The main source of uncertainty is in our background knowledge, but according to represented background knowledge there are some special sets: uncertainty does not appear in their cases:

Definition 3. *Let $\langle U, \mathfrak{B}, \mathfrak{D}_\mathfrak{B}, \mathsf{l}, \mathsf{u} \rangle$ be a general approximation space, and $S \subseteq U$, $\overline{S} = U \setminus S$. Then*

1. *the set S is lower script if $S = \mathsf{l}(S)$, and upper script if $S = \mathsf{u}(S)$;*
2. *the set S is strictly lower/upper script if both S and $\overline{S}$ (i.e. the complement of set S) are lower/upper script respectively;*
3. *the set S is partially lower script if $S \cap \cup \mathfrak{B} = \mathsf{l}(S)$ and partially upper script if $S \cap \cup \mathfrak{B} = \mathsf{u}(S)$;*

---

[2] In point 2 and 3 'may or have to' are used: if an object is a member of more than one base sets, then a base set has to be chosen whose members are handled in the same way. It depends on the given situation.

4. the set S is strictly partially lower/upper script if both S and $\overline{S}$ are partially lower/upper script respectively.

In general approximation spaces all definable sets (and therefore $\mathsf{l}(S)$ and $\mathsf{u}(S)$ for all $S \subseteq U$) are lower script (partially lower script in partial spaces), but they are not strictly lower script (strictly partially lower script in partial spaces) necessarily, and if a set is (strictly) upper script, then it is (strictly) lower script. If the approximation space is Pawlakian, then if a set S is lower script, then it is strictly lower script.

Theorem 1. *A general approximation space is Pawlakian if and only if all lower script sets are strictly lower script.*

A general approximation space is one-layered (i.e. its base sets are pairwise disjoint ones) if and only if all partially lower script sets are strictly partially lower script.

Let S be a subset of U, and $x, y \in U$. What is the consequence of embedded and limited background knowledge? What can be said about y with respect to x?

1. In an original Pawlakian space relying on an equivalence relation $\mathcal{R}$:
   (a) If $x \in \mathsf{l}(S)$, then $y \in S$ for all $y, x\mathcal{R}y$ (i.e. x and y belong to the same base set).
   (b) If $x \in \mathsf{l}(\overline{S})$, where $\overline{S} = U \setminus S$, then $y \notin S$ for all $y, x\mathcal{R}y$.
   (c) If $x \notin \mathsf{l}(S)$ and $x \notin \mathsf{l}(\overline{S})$, then $x \in \mathsf{u}(S) \setminus \mathsf{l}(S)$, and there are y_1, y_2 such that $x\mathcal{R}y_1, y_1 \in S$, and $x\mathcal{R}y_2, y_2 \notin S$.
   (d) $x \in \mathsf{u}(S) \setminus \mathsf{l}(S)$, if and only if $x \notin \mathsf{l}(S)$ and $x \notin \mathsf{l}(\overline{S})$, therefore the point 1. (c) and this case are the same.
2. In a covering space generated by a tolerance relation $\mathcal{R}$:
   (a) If $x \in \mathsf{l}(S)$, then there is an $x' \in U$ such that $x \in [x']_{\mathcal{R}}$ and $[x'] \in \mathcal{C}^{\mathsf{l}}(S)$. Therefore $y \in S$ for all y if $y \in \cup\{[x']_{\mathcal{R}} \mid x \in [x']_{\mathcal{R}} \text{ and } [x']_{\mathcal{R}} \in \mathcal{C}^{\mathsf{l}}(S)\}$.
   (b) If $x \in \mathsf{l}(\overline{S})$, then there is an $x' \in U$ such that $x \in [x']_{\mathcal{R}}$ and $[x'] \in \mathcal{C}^{\mathsf{l}}(\overline{S})$. Therefore $y \notin S$ for all y, if $y \in \cup\{[x']_{\mathcal{R}} \mid x \in [x']_{\mathcal{R}} \text{ and } [x']_{\mathcal{R}} \in \mathcal{C}^{\mathsf{l}}(\overline{S})\}$.
   (c) If $x \notin \mathsf{l}(S) \cup \mathsf{l}(\overline{S})$, then
      i. there is no base set $[x']_{\mathcal{R}}$ such that $x \in [x']_{\mathcal{R}}$, and $[x']_{\mathcal{R}} \subseteq S$,
      ii. there is no base set $[x'']_{\mathcal{R}}$ such that $x \in [x'']_{\mathcal{R}}$, and $[x'']_{\mathcal{R}} \subseteq \overline{S}$,
      iii. there is a base set $[x''']_{\mathcal{R}}$ such that $x \in [x''']_{\mathcal{R}}$, $[x''']_{\mathcal{R}} \not\subseteq S$, $[x''']_{\mathcal{R}} \not\subseteq \overline{S}$
      but in this case one cannot say anything about any other object with respect to x connected with S or $\overline{S}$.
   (d) If $x \in \mathsf{u}(S) \setminus \mathsf{l}(S)$, then there is a base set $[x']_{\mathcal{R}}$ such that $x \in [x']_{\mathcal{R}}$, $[x']_{\mathcal{R}} \cap S \neq \emptyset$ and $[x']_{\mathcal{R}} \not\subseteq S$, but in this case one cannot say anything about any other object with respect to x connected with S or $\overline{S}$.
3. In a general covering space:
   (a) If $x \in \mathsf{l}(S)$, then there is a base set B, such that $x \in B$ and $B \in \mathcal{C}^{\mathsf{l}}(S)$. Therefore $y \in S$ for all $y \in \cup\{B \mid x \in B \text{ and } B \in \mathcal{C}^{\mathsf{l}}(S)\}$.
   (b) If $x \in \mathsf{l}(\overline{S})$, then there is a base set B such that $x \in B$ and $B \in \mathcal{C}^{\mathsf{l}}(\overline{S})$. Therefore $y \notin S$ for all $y \in \cup\{B \mid x \in B \text{ and } B \in \mathcal{C}^{\mathsf{l}}(\overline{S})\}$.

(c) If $U \setminus (\mathsf{I}(S) \cup \mathsf{I}(\overline{S})) \neq \emptyset$ (i.e. S is not stricly script) and $x \notin \mathsf{I}(S) \cup \mathsf{I}(\overline{S})$, then there is a base set B such that $x \in B$ and $B \cap S \neq \emptyset, B \not\subseteq S$, $B \cap \overline{S} \neq \emptyset, B \not\subseteq \overline{S}$. In this case one cannot say anything about any other object belonging to B with respect to x connected with S or $\overline{S}$.

4. In a general partial space:
   (a) If $x \in \mathsf{I}(S)$, then there is a base set B, such that $x \in B$ and $B \in \mathcal{C}^{\mathsf{l}}(S)$. Therefore $y \in S$ for all $y \in \cup\{B \mid x \in B$ and $B \in \mathcal{C}^{\mathsf{l}}(S)\}$.
   (b) If $x \in \mathsf{I}(\overline{S})$, then there is a base set B such that $x \in B$ and $B \in \mathcal{C}^{\mathsf{l}}(\overline{S})$. Therefore $y \notin S$ for all $y \in \cup\{B \mid x \in B$ and $B \in \mathcal{C}^{\mathsf{l}}(\overline{S})\}$.
   (c) If $x \in \cup\mathfrak{B}, x \notin \mathsf{I}(S)$ and $x \notin \mathsf{I}(\overline{S})$, then there is a base set B such that $x \in B, B \cap S \neq \emptyset$ and $B \not\subseteq S$, but in this case one cannot say anything about any other object with respect to x connected with S or $\overline{S}$.
   (d) Otherwise one does not know anything about x (i.e. there is no base set B such that $x \in B$), therefore one cannot say anything about any other object with respect to x.

3.1   Rough Membership Relations

Classical membership relations (with the answer yes/no) can only be used without any restriction in the case of definable sets (more precisely in the case of base sets, but definable sets are introduced relying on base sets), therefore a new membership relation, the rough membership relation ($\in^{rough}$) has to be introduced. The rough membership relation is not a yes/no relation, in the general case it is a relation with four possible answers:

Let $\langle U, \mathfrak{B}, \mathfrak{D}_{\mathfrak{B}}, \mathsf{I}, \mathsf{u} \rangle$ be a general approximation space, S be an arbitrary subset of the universe U. The answers for the question whether $x \in^{rough} S$ are the followings:

- 'yes' if $x \in \mathsf{I}(S)$ (i.e. x belongs to S necessarily);
- 'no' if $x \in \mathsf{I}(\overline{S})$ (i.e. x does not belong to S necessarily);
- 'maybe or maybe not' if $x \in \cup\mathfrak{B}, x \notin \mathsf{I}(S) \cup \mathsf{I}(\overline{S})$ (i.e. the embedded background knowledge says something about the object x, but it is not enough to decide whether x belongs to or does not belong to S necessarily);
- 'there is no information (embedded in base sets)' if $x \notin \cup\mathfrak{B}$ (it is possible only in partial cases).

Note that the set $\{x \mid x \in^{rough} S$ is 'maybe or maybe not'$\}$ is not a definable set necessarily, but the answer 'maybe or maybe not' can be determined by using only base sets (and the corresponding classical yes/no membership relation.) If $1, 0, 1/2, 2$ are used to represent 'yes', 'no', 'maybe or maybe not' and 'there is no information (embedded in base sets)' respectively, then a generalized characteristic function of set S (called rough characteristic function) can be introduced:

$$\mu^r(S, x) = \begin{cases} 1 & \text{if } x \in \mathsf{I}(S) \\ 0 & \text{if } x \in \mathsf{I}(\overline{S}) \\ 1/2 & \text{if } x \in \cup\mathfrak{B}, x \notin \mathsf{I}(S) \cup \mathsf{I}(\overline{S}) \\ 2 & \text{otherwise (i.e. } x \notin \cup\mathfrak{B}) \end{cases}$$

Connection between the rough membership relation $x \in^{rough} S$ and the rough characteristic function $\mu^r(S, x)$ is the following:

$$x \in^{rough} S : \begin{cases} \text{yes} & \text{if } \mu^r(S, x) = 1 \\ \text{no} & \text{if } \mu^r(S, x) = 0 \\ \text{maybe or maybe not} & \text{if } \mu^r(S, x) = 1/2 \\ \text{there is no information} & \text{if } \mu^r(S, x) = 2 \end{cases}$$

3.2 Rough Membership Measurement on Finite Universe

In practical applications, universes are finite non-empty sets. Therefore, as in DTRS, some special functions can be introduced. These functions show the measurements of 'belonging to a set', and say something about the level of 'maybe or maybe not' from the side of 'maybe'.

Relying on a given general approximation space $\langle U, \mathfrak{B}, \mathfrak{D}_\mathfrak{B}, \mathsf{l}, \mathsf{u} \rangle$ three different partial characteristic functions (μ^o for optimistic, μ^a for average and μ^p for pessimistic) can be introduced. For the sake of simplicity we use a null entity (the number 2) to show that a function is undefined for an object x, i.e. to represent partiality of characteristic functions.[3]

Let $\langle U, \mathfrak{B}, \mathfrak{D}_\mathfrak{B}, \mathsf{l}, \mathsf{u} \rangle$ be a general approximation space, U be a finite nonempty set, $S \subseteq U$, $x \in U$ and $\mathcal{C}(x) = \{B \mid B \in \mathfrak{B} \text{ and } x \in B\}$. Then

1. If $x \in \cup \mathfrak{B}$, then $\mathcal{V}(x, S) = \left\{ \frac{|B \cap S|}{|B|} \mid B \in \mathcal{C}(x) \right\}$.
2. Optimistic rough membership measurement:

$$\mu^o(S, x) = \begin{cases} 1 & \text{if } x \in \mathsf{l}(S) \\ 0 & \text{if } x \in \mathsf{l}(\overline{S}) \\ \max(\mathcal{V}(x, S)) & \text{if } x \in \cup \mathfrak{B}, x \notin \mathsf{l}(S) \cup \mathsf{l}(\overline{S}) \\ 2 & \text{otherwise (i.e. } x \notin \cup \mathfrak{B}) \end{cases}$$

3. Average rough membership measurement:

$$\mu^a(S, x) = \begin{cases} 1 & \text{if } x \in \mathsf{l}(S) \\ 0 & \text{if } x \in \mathsf{l}(\overline{S}) \\ \mathsf{avg}(\mathcal{V}(x, S)) & \text{if } x \in \cup \mathfrak{B}, x \notin \mathsf{l}(S) \cup \mathsf{l}(\overline{S}) \\ 2 & \text{otherwise (i.e. } x \notin \cup \mathfrak{B}) \end{cases}$$

4. Pessimistic rough membership measurement:

$$\mu^p(S, x) = \begin{cases} 1 & \text{if } x \in \mathsf{l}(S) \\ 0 & \text{if } x \in \mathsf{l}(\overline{S}) \\ \min(\mathcal{V}(x, S)) & \text{if } x \in \cup \mathfrak{B}, x \notin \mathsf{l}(S) \cup \mathsf{l}(\overline{S}) \\ 2 & \text{otherwise (i.e. } x \notin \cup \mathfrak{B}) \end{cases}$$

5. For the sake of simplicity it is useful to introduce the partial characteristic function:

$$\mu^c(S, x) = \begin{cases} 1 & \text{if } x \in S \cap \cup \mathfrak{B} \\ 0 & \text{if } x \in \cup \mathfrak{B} \setminus S \\ 2 & \text{otherwise} \end{cases}$$

---

[3] In the definition three different functions (on finite sets of numbers) are used: function min for the minimum, function max for the maximum and function avg for the average value of a finite set of numbers.

Optimistic, average, and pessimistic rough membership measurements are partial functions in the general case, and these are total ones when the approximation space is total. These functions are generalized characteristic functions of a set S relying on a given general approximation space.

Rough membership relation and rough membership measurements (as generalized characteristic functions) can be considered as generalizations of the membership relation of classical set theory.

4 Partial First-Order Logic (PFoL) Relying on Different Generalized Characteristic Functions

In the previous subsections, different characteristic functions, and so different rough membership relations have been introduced. These functions rely on background knowledge represented by the systems of base sets. If new characteristic functions are used in the semantics of first-order logic, then the logical system can reveal the logical laws of general propositions relying on embedded background knowledge, the nature and the consequences of conceptual structure (conceptual knowledge) embedded in the system of base sets.

4.1 Language of PFoL

At first, the language of PFoL has to be defined. The language of classical first-order logic must be modified to treat background knowledge represented by base sets. A finite non-empty set of one-argument predicates stands for the system of base sets. The members of this finite non-empty set $(\mathcal{T})$ are called tools. Two new logical constants are introduced: $^\uparrow, ^\downarrow$. Their intended meanings are to determine whether upper or lower approximations of a predicate is used.

Definition 4. *L is a language of PFoL with the set $\mathcal{T}$ of tools, if*

1. $L = \langle LC \cup \{^\uparrow, ^\downarrow\}, Var, Con, Term, \mathcal{T}, Form \rangle$
2. $L^{(1)} = \langle LC, Var, Con, Term, Form \rangle$ *is a language of classical first-order logic with the following extension of the set $Form$*
   - *If P is an n-argument $(n \geq 1)$ predicate parameter and $t_1, \ldots, t_n \in Term$, then $P^\uparrow(t_1, \ldots, t_n), P^\downarrow(t_1, \ldots, t_n) \in Form$*
3. $\mathcal{T}$ *is a finite non-empty set of one-argument predicate parameters.*

4.2 Semantics of PFoL

The semantic values of tools (i.e. the members of set $\mathcal{T}$) play a crucial role in giving different types of generalized characteristic functions because their semantic values (as sets) generate the base sets of a logically relevant general partial approximation space.

Definition 5. *Let L be a language of PFoL with the set $\mathcal{T}$ of tools. The ordered pair $\langle U, \varrho \rangle$ is a tool-based interpretation of L, if*

1. U is a finite nonempty set;
2. ϱ is a function such that $Dom(\varrho) = Con$ and
   (a) if $a \in \mathcal{N}$ ($\mathcal{N}$ is the set of name parameters), then $\varrho(a) \in U$;
   (b) if $p \in \mathcal{P}(0)$ ($\mathcal{P}(0)$ is the set of proposition parameters), then $\varrho(p) \in \{0,1\}$;
   (c) if $P \in \mathcal{P}(n)$ ($n = 1, 2, \ldots$) ($\mathcal{P}(n)$ is the set of n-argument predicate parameters), then $\varrho(P) \subseteq U^{(n)}$, where $U^{(1)} = U$, $U^{(n+1)} = U^{(n)} \times U$;
   (d) if $T \in \mathcal{T}$, then $\varrho(T) \neq \emptyset$.

In order to give semantic rules we only need the notions of assignment and modified assignment:

Definition 6.

1. Function v is an assignment relying on the interpretation $\langle U, \varrho \rangle$ if $v : Var \to U$.
2. Let v be an assignment relying on the interpretation $\langle U, \varrho \rangle$, $x \in Var$ and $u \in U$. Then $v[x : u]$ is a modified assignment of v, if $v[x : u]$ is an assignment, $v[x : u](y) = v(y)$ if $x \neq y$, and $v[x : u](x) = u$.

The semantic values of tools (the members of set $\mathcal{T}$) determine a general (maybe partial) approximation space for the given interpretation. The generated approximation space is logically relevant in the sense, that it gives the lower and upper approximations (what is more, the different generalized characteristic functions) of any predicate P to be taken into consideration in the definition of semantic rules.

Definition 7. Let L be a language of PFoL with the set $\mathcal{T}$ of tools and $\langle U, \varrho \rangle$ be a tool-based interpretation of L.
   The ordered 5-tuple

$$\mathsf{GAS}(\mathcal{T}) = \langle \mathcal{PR}(U), \mathfrak{B}(\mathcal{T}), \mathfrak{D}_{\mathfrak{B}(\mathcal{T})}, \mathsf{l}, \mathsf{u} \rangle$$

is a logically relevant general partial approximation space generated by set $\mathcal{T}$ of tools with respect to the interpretation $\langle U, \varrho \rangle$ if

1. $\mathcal{PR}(U) = \bigcup_{n=1}^{\infty} U^{(n)}$;
2. $\mathfrak{B}(\mathcal{T}) = \bigcup_{n=1}^{\infty} \mathfrak{B}_n(\mathcal{T})$ where $\mathfrak{B}_n(\mathcal{T}) = \{\varrho(T_1) \times \cdots \times \varrho(T_n) \mid T_i \in \mathcal{T}\}$;

The semantic values of tools (given by the interpretation) generate the set $\mathfrak{B}(\mathcal{T})$. It contains those sets by which the semantic value of any predicate parameter is approximated.

In the semantics of PFol, the semantic value of an expression depends on a given interpretation, and a given logically relevant general partial approximation space generated by set of tools with respect to the interpretation. For the sake of simplicity, we use a null entity to represent the partiality of semantic rules. We use number 0 for falsity, number 1 for truth, numbers greater than 0 and less than 1 for true degree and number 2 for the null entity. In many cases, five possibly different semantic values can be given: rough, optimistic, average, pessimistic, and crisp ones. The forms of semantic rules are similar in different cases and so the superscript $\star$ can be used to denote one of them ($\star \in \{^r, ^o, ^a, ^p, ^c\}$). The semantic value of an expression A is denoted by $[\![A]\!]_v^\star$.

The most important semantic rules are the following:

1. If $P \in P(n)$ ($n \neq 0$), i.e. P is an n-argument predicate parameter and $t_1, t_2, \ldots, t_n \in Term$, then
$$[\![P(t_1, \ldots, t_n)]\!]_v^\star = \mu^\star(\varrho(P), \langle [\![t_1]\!]_v^\star, \ldots, [\![t_n]\!]_v^\star \rangle);$$
$$[\![P^\uparrow(t_1, \ldots, t_n)]\!]_v^\star = \mu^\star(u(\varrho(P)), \langle [\![t_1]\!]_v^\star, \ldots, [\![t_n]\!]_v^\star \rangle);$$
$$[\![P^\downarrow(t_1, \ldots, t_n)]\!]_v^\star = \mu^\star(l(\varrho(P)), \langle [\![t_1]\!]_v^\star, \ldots, [\![t_n]\!]_v^\star \rangle).$$

2. If $A \in Form$, then
$$[\![\neg A]\!]_v^\star = \begin{cases} 2 & \text{if } [\![A]\!]_v^\star = 2 \\ 1 - [\![A]\!]_v^\star & \text{otherwise} \end{cases}$$

3. If $A, B \in Form$, then
$$[\![(A \wedge B)]\!]_v^\star = \begin{cases} 2 & \text{if } [\![A]\!]_v^\star = 2, \text{ or } [\![B]\!]_v^\star = 2; \\ \min\{[\![A]\!]_v^\star, [\![B]\!]_v^\star\} & \text{otherwise} \end{cases}$$
$$[\![(A \vee B)]\!]_v^\star = \begin{cases} 2 & \text{if } [\![A]\!]_v^\star = 2, \text{ or } [\![B]\!]_v^\star = 2; \\ \max\{[\![A]\!]_v^\star, [\![B]\!]_v^\star\} & \text{otherwise} \end{cases}$$
$$[\![(A \supset B)]\!]_v^\star = \begin{cases} 2 & \text{if } [\![A]\!]_v^\star = 2, \text{ or } [\![B]\!]_v^\star = 2; \\ \max\{[\![\neg A]\!]_v^\star, [\![B]\!]_v^\star\} & \text{otherwise} \end{cases}$$

4. If $A \in Form$, $x \in Var$ and $\mathcal{V}(A) = \left\{ [\![A]\!]_{v[x:u]}^\star \mid u \in U, [\![A]\!]_{v[x:u]}^\star \neq 2 \right\}$, then
$$[\![\forall x A]\!]_v^\star = \begin{cases} 2 & \text{if } \mathcal{V}(A) = \emptyset, \\ \min\{\mathcal{V}(A)\} & \text{otherwise} \end{cases}$$
$$[\![\exists x A]\!]_v^\star = \begin{cases} 2 & \text{if } \mathcal{V}(A) = \emptyset, \\ \max\{\mathcal{V}(A)\} & \text{otherwise} \end{cases}$$

From the logical point of view, flexibility is the main advantage of defined general logical framework: according to the first semantic rule different characteristic functions can be taken into consideration to determine the semantic value of an atomic formula. One of its consequences is that the semantic value of any formula can be determined by the help of different general characteristic functions and so (as it can be seen in the definition of models and consequence relations) the user can decide which general characteristic function is applied for a formula.

4.3   Central Semantic Notions

The notion of models plays a fundamental role in the semantic definition of the consequence relation. The standard notion of models has to be generalized:

1. In order to take into consideration different general characteristic functions for formulae, models are defined for ordered n-tuples of formulae (not for the set of formulae). Decision types determine which general characteristic function is used for a formula.
2. Apart from the null entity the values of general characteristic functions are between 0 and 1. Therefore only a parametrized notion of models is possible.

Definition 8. *Let L be a language of PFoL with the set T of tools, and $\Gamma = \langle A_1, A_2, \ldots, A_n \rangle$ be an ordered n-tuple of formulae ($A_1, A_2, \ldots, A_n \in Form$).*

1. *The ordered n-tuple $\Delta = \langle \delta_1, \ldots, \delta_n \rangle$ is a decision type of Γ if $\delta_1, \ldots, \delta_n \in \{r, o, a, p, c\}$.*
2. *Let $\Delta = \langle \delta_1, \ldots, \delta_n \rangle$ be a decision type Γ. Then $\langle U, \varrho, v \rangle$ is a Δ-type model of Γ with parameter α ($0 < \alpha \leq 1$), if*
   (a) $\langle U, \varrho \rangle$ is an interpretation of L; v is an assignment relying on $\langle U, \varrho \rangle$;
   (b) $[\![A_i]\!]_v^{\delta_i} \neq 2$ for all i ($i = 1, 2, \ldots, n$)
   (c) $[\![A_i]\!]_v^{\delta_i} \geq \alpha$ for all i ($i = 1, 2, \ldots, n$).

Definition 9. *Let L be a language of PFoL with the set T of tools, $\Gamma = \langle A_1, \ldots, A_n \rangle$ be an ordered n-tuple of formulae ($A_1, \ldots, A_n \in Form$) and $B \in Form$ be a formula.*

1. *$\Delta \to \delta$ is a decision driven consequence type from Γ to B if Δ is a decision type of Γ and δ is a decision type of $\{B\}$.*
2. *Let $\Delta \to \delta$ is a decision driven consequence type from Γ to B. B is a parametrized consequence of Γ driven by $\Delta \to \delta$ with the parameter pair $\langle \alpha, \beta \rangle$ if all Δ-type models of Γ with the parameter α are δ-type models of B with the parameter β ($\Gamma \models_{\Delta \to \delta}^{\langle \alpha, \beta \rangle} B$)*

The introduced notion of decision-driven consequence relations may be useful in many practical cases. It can be determined

1. how premises and conclusions are evaluated: all generalized characteristic functions can be used in the evaluation process;
2. the expected level of truth of premises and conclusions can be determined.

In this paper there is no enough space to give all proved theorems. In [3–6] the most important properties are investigated.

For example in [4], Aristotle's syllogisms of the first figure are investigated. These consequence relations represent the most typical usages of quantified propositions containing one-argument predicates. A general observation: in any decision driven consequence of Aristotle's syllogisms of the first figure the parameters which give the level of truth of premises have to be greater than $1/2$, therefore there is no valid consequence relation with parameters less than or equal to $1/2$. It means that in order to say something about the conclusion the premises have to be closer to truth than to falsity. Proved theorems show that the validity of the consequence relation (in the case of optimistic, average and pessimistic rough membership measurements) requires that:

– Barbara and Celarent syllogisms:
   1. evaluate premises and conclusion by using the same (optimistic, average or pessimistic) membership function and so there is no real freedom for making different decisions concerning two premises and the conclusion;
   2. suppose that the level of truth of premises is greater than $1/2$, i.e. the premises have to be closer to truth then falsity.

– Darii syllogism:
  1. the decision for the first premiss can not be stronger than the decision for the second one;
  2. the decisions for the first premiss and the conclusion have to be the same;
  3. the first premiss plays a more important role than the second one.
– Ferio syllogism:
  1. the decision for the first premiss can not be stronger than the decision for the second one;
  2. the decisions for the second premiss and the conclusion have to be the same;
  3. the second premiss plays a more important role than the first one.

5 Conclusion and Future Work

The main result of the paper is to give a (very general) partial first–order logic using generalized characteristic functions received from different systems of rough sets. An important advantage of the logical system is that in one integrated system any user can determine

– the evaluation processes of premisses and conclusions;
– the expected level of the truth of premisses and conclusions.

The next step is to use the introduced logical system in practice to solve some problems in data mining connected with incomplete/uncertain information.

References

1. Csajbók, Z., Mihálydeák, T.: A general set theoretic approximation framework. In: Greco, S., Bouchon-Meunier, B., Coletti, G., Fedrizzi, M., Matarazzo, B., Yager, R.R. (eds.) IPMU 2012. CCIS, vol. 297, pp. 604–612. Springer, Heidelberg (2012). https://doi.org/10.1007/978-3-642-31709-5_61
2. Golińska-Pilarek, J., Orłowska, E.: Logics of similarity and their dual tableaux a survey. In: Della Riccia, G., Dubois, D., Kruse, R., Lenz, H.J. (eds.) Preferences and Similarities, pp. 129–159. Springer, Vienna (2008). https://doi.org/10.1007/978-3-211-85432-7_5
3. Mihálydeák, T.: Partial first-order logic relying on optimistic, pessimistic and average partial membership functions. In: Proceedings of the 8th Conference of the European Society for Fuzzy Logic and Technology (EUSFLAT 2013), pp. 374–379. Atlantis Press, August 2013. https://doi.org/10.2991/eusflat.2013.53
4. Mihálydeák, T.: Aristotle's syllogisms in logical semantics relying on optimistic, average and pessimistic membership functions. In: Cornelis, C., Kryszkiewicz, M., Ślęzak, D., Ruiz, E.M., Bello, R., Shang, L. (eds.) RSCTC 2014. LNCS (LNAI), vol. 8536, pp. 59–70. Springer, Cham (2014). https://doi.org/10.1007/978-3-319-08644-6_6
5. Mihálydeák, T.: Logic on similarity based rough sets. In: Nguyen, H.S., Ha, Q.-T., Li, T., Przybyła-Kasperek, M. (eds.) IJCRS 2018. LNCS (LNAI), vol. 11103, pp. 270–283. Springer, Cham (2018). https://doi.org/10.1007/978-3-319-99368-3_21

6. Mihálydeák, T.: First-order logic based on set approximation: a partial three-valued approach. In: 2014 IEEE 44th International Symposium on Multiple-Valued Logic, pp. 132–137, May 2014. https://doi.org/10.1109/ISMVL.2014.31
7. Nagy, D., Mihálydeák, T., Aszalós, L.: Similarity based rough sets. In: Polkowski, L., et al. (eds.) IJCRS 2017. LNCS (LNAI), vol. 10314, pp. 94–107. Springer, Cham (2017). https://doi.org/10.1007/978-3-319-60840-2_7
8. Pawlak, Z.: Rough sets. Int. J. Parallel Program. **11**(5), 341–356 (1982)
9. Pawlak, Z., Skowron, A.: Rough sets and Boolean reasoning. Inf. Sci. **177**(1), 41–73 (2007)
10. Pawlak, Z., Skowron, A.: Rudiments of rough sets. Inf. Sci. **177**(1), 3–27 (2007)
11. Pawlak, Z., et al.: Rough sets: theoretical aspects of reasoning about data. In: System Theory, Knowledge Engineering and Problem Solving, vol. 9. Kluwer Academic Publishers, Dordrecht (1991)
12. Skowron, A., Stepaniuk, J.: Tolerance approximation spaces. Fundamenta Informaticae **27**(2), 245–253 (1996)
13. Vakarelov, D.: A modal characterization of indiscernibility and similarity relations in Pawlak's information systems. In: Ślęzak, D., Wang, G., Szczuka, M., Düntsch, I., Yao, Y. (eds.) RSFDGrC 2005. LNCS (LNAI), vol. 3641, pp. 12–22. Springer, Heidelberg (2005). https://doi.org/10.1007/11548669_2
14. Yao, J., Yao, Y., Ziarko, W.: Probabilistic rough sets: approximations, decision-makings, and applications. Int. J. Approximate Reasoning **49**(2), 253–254 (2008)
15. Yao, Y., Yao, B.: Covering based rough set approximations. Inf. Sci. **200**, 91–107 (2012). https://doi.org/10.1016/j.ins.2012.02.065. http://www.sciencedirect.com/science/article/pii/S0020025512001934

Possibility Theory and Possibilistic Logic: Tools for Reasoning Under and About Incomplete Information

Didier Dubois$^{(\boxtimes)}$ and Henri Prade

IRIT – CNRS, 118, route de Narbonne, 31062 Toulouse Cedex 09, France
{dubois,prade}@irit.fr

Abstract. This brief overview provides a quick survey of qualitative possibility theory and possibilistic logic along with their applications to various forms of epistemic reasoning under and about incomplete information. It is highlighted that this formalism has the potential of relating various independently introduced logics for epistemic reasoning.

1 Introduction

Possibility theory is a formal setting for uncertainty management that has been independently proposed by different authors since the late 1940's (e.g. [27,54, 63]). It has been later on fully developed in the last 40 years [34,40,66]. It offers a framework dedicated to the representation of *incomplete* information, that can be naturally extended to graded beliefs, and the ensuing forms of uncertain reasoning. Because of its simplicity and natural appeal, some basic parts of this setting are implicit in many theories of epistemic reasoning, so that possibility theory has a significant unifying power for these alternative formalisms.

This extended abstract presents a short survey of possibility theory, of possibilistic logic that is based on it, and of their applications (providing detailed references in each case).

2 Possibility Theory

Possibility theory has a remarkable situation among the settings devoted to the representation of imprecise and uncertain information.

First, it may be numerical or qualitative [36]. In the first case, possibility measures and the dual necessity measures can be regarded respectively as upper bounds and lower bounds of ill-known probabilities; they are also particular cases of Shafer [64] plausibility and belief functions respectively. In fact, possibility measures and necessity measures constitute the simplest, non trivial, imprecise probability system [65]. Second, when qualitative, possibility theory provides a natural approach to the grading of possibility and necessity modalities on finite

© IFIP International Federation for Information Processing 2021
Published by Springer Nature Switzerland AG 2021
Z. Shi et al. (Eds.): ICIS 2020, IFIP AICT 623, pp. 79–89, 2021.
https://doi.org/10.1007/978-3-030-74826-5_7

scales $\mathcal{S} = \{1 = \lambda_1 > \ldots \lambda_n > \lambda_{n+1} = 0\}$ where grades have an ordinal flavor. In the following, we focus on qualitative possibility theory.

In possibility theory, pieces of information, viewed as epistemic states, are represented in terms of sets or fuzzy sets viewed as possibility distributions. A possibility distribution π is a mapping from a set of states, or universe of discourse, denoted by U, to a totally ordered scale $\mathcal{S}$ ($\pi(u) = 0$ means that state u is rejected as impossible; $\pi(u) = 1$ means that state u is totally possible, i.e., plausible). When $\mathcal{S} = \{0, 1\}$, π is the characteristic function of a set E of possible mutually exclusive values or states of affairs, obtained by ruling out the impossible. Such a set represents the epistemic state of an agent and is called an epistemic set. In the general, graded, case, E is a fuzzy set.

Two increasing set functions, similar to probability functions, are induced from a possibility distribution, namely a possibility measure Π ($\Pi(A) = \sup_{u \in A} \pi(u)$) and a dual necessity measure N ($N(A) = 1 - \Pi(A^c)$, where $A^c = U \setminus A$ is the complement of A). $\Pi(A)$ evaluates to what extent A is consistent with the epistemic state E, and $N(A)$ to what extent A is certainly implied by E. Note that $N(A) > 0$ implies $\Pi(A) = 1$.

In the Boolean case, when the epistemic state E is a crisp subset of U, possibility and necessity functions are such that:

- $\Pi(A) = 1$ if $A \cap E \neq \emptyset$, and 0 otherwise; so $\Pi(A) = 0$ when proposition A is incompatible with the epistemic state.
- $N(A) = 1$ if $E \subseteq A$, and 0 otherwise; so $N(A) = 1$ when proposition A is true in all states of the world compatible with the epistemic state.
- When $\Pi(A) = 1$ and $N(A) = 0$ it corresponds to the case when the epistemic state E does not allow for deciding whether A is true or false (ignorance).

Possibility measures satisfy a characteristic "maxitivity" property in the form of the identity $\Pi(A \cup B) = \max(\Pi(A), \Pi(B))$, and necessity measures a "minitivity" property $N(A \cap B) = \min(N(A), N(B))$.

In contrast with the minitivity of necessity functions and the maxitivity of possibility functions, it is generally not true that $\Pi(A \cap B) = \min(\Pi(A), \Pi(B))$, nor that $N(A \cup B) = \max(N(A), N(B))$; for instance let $B = A^c$, and $\Pi(A) = \Pi(A^c) = 1$ (ignorance case), then $\Pi(A \cap B) = 0 \neq \min(\Pi(A), \Pi(B)) = 1$. The property $\Pi(A \cap B) = \min(\Pi(A), \Pi(B))$ may hold for some special pair of events for instance when the possibility distribution is multidimensional, i.e., $U = U_1 \times \cdots \times U_k$ and $\pi = \min_{i=1}^{k} \pi_i$, where π_i is a marginal possibility distribution on U_i. π is said to be decomposable and the variables are said to be non-interactive. Non-interactivity is a graded generalisation of logical independence.

Two other decreasing set functions can be associated with π:

i) a measure of *guaranteed possibility* or *strong* possibility [36]: $\Delta(A) = \inf_{u \in A} \pi(u)$ which estimates to what extent *all* states in A are possible according to evidence. $\Delta(A)$ can be used as a degree of evidential support for A;

ii) its dual conjugate $\nabla(A) = 1 - \Delta(A^c)$: it evaluates a degree of potential or *weak* necessity of A, as it is 1 only if some state u out of A is impossible.

In the Boolean case, it reduces to $\Delta(A) = 1$ if $A \subseteq E$ and 0 otherwise, while $\nabla(A) = 1$ if $A \cup E \neq U$. Interestingly enough, the four set function are necessary for describing the relative positions of two subsets such as A and E, see, e.g., [38]. The four set functions are weakly related by the constraint for all A, $\max(N(A), \Delta(A)) \leq \min(\Pi(A), \nabla(A))$ (provided that π and $1 - \pi$ are both normalized, i.e. they reach 1 for some $u \in U$).

3   Connections Between Possibility Theory and Other Representations of Incomplete Knowledge

Necessity and possibility functions reminds of modal logics, where these modalities, respectively denoted by $\square$ and $\Diamond$, are used to prefix logical sentences. In KD modal logics, which are also known as epistemic logics [49], asserting $\square\phi$ stands for declaring that an agent believes or knows that proposition ϕ is true. If ϕ is a propositional formula that is true for those and only those states of affairs in the set A, asserting $\square\phi$ is faithfully encoded by the identity $N(A) = 1$. Counterparts of properties of possibility functions are valid in KD modal logics, in particular minitivity of necessity functions ($\square(\phi \wedge \psi) \equiv \square\phi \wedge \square\psi$), duality ($\Diamond\phi$ stands for $\neg\square\neg\phi$ and encodes the identity $\Pi(A) = 1$), and maxitivity of possibility functions ($\Diamond(\phi \vee \psi) \equiv \Diamond\phi \vee \Diamond\psi$). These possibilistic semantics are the basis of a very simple modal logic (MEL [3]). However, the language of modal logics is more complex than the one of possibility theory because propositions prefixed by modalities may contain modalities. The semantics of modal logics is usually in terms of relations rather than epistemic states [48].

A semantics of Kleene three-valued logic can be devised in terms of multidimensional possibility distributions over a universe that is a Cartesian product $U = U_1 \times \cdots \times U_k$, where each subspace $U_i = \{a_i, \neg a_i\}$. Kleene logic uses 3 truth-values $\{T > I > F\}$. Assigning T (resp. F) to an atom a_i, i.e., $t(a_i) = T$ (resp. $t(a_i) = F$) expresses that a_i is surely true, i.e., $N^T(a_i) = 1$ (resp. surely false : $N^F(\neg a_i) = 1$) which corresponds to Boolean possibility distributions $\pi_i^T(a_i) = 1, \pi_i^T(\neg a_i) = 0$ (resp. $\pi_i^F(a_i) = 0, \pi_i^F(\neg a_i) = 1$); finally assigning I to a_i expresses a lack of knowledge about whether a_i is true or false ($\pi_i^I(a_i) = 1, \pi_i^I(\neg a_i) = 1$). In Kleene logic, knowledge is thus expressed on atoms only, and it handles epistemic states in the form of possibility distributions $\min_{i=1}^{k} \pi_i$ where $\pi_i \in \{\pi_i^T, \pi_i^F, \pi_i^I\}$. A truth assignment t is a partial model of the form $(\wedge_{i:t(a_i)=T} a_i) \wedge (\wedge_{j:t(a_j)=F} \neg a_j)$ encoded by the possibility distribution $\pi_t = \min(\min_{i:t(a_i)=T} \pi_i^T, \min_{i:t(a_i)=F} \pi_i^F)$ (π_i^I disappears, being equal to 1 everywhere). The truth-functionality of Kleene logic can be justified in terms of the non-interactivity of the Boolean variables. The possibilistic framework highlights the limited expressiveness of Kleene 3-valued logic, which cannot account for epistemic states modelled by subsets of U that are not partial models [26].

4   Possibilistic Logic and Its Applications

Possibilistic logic (PL) [31,39,41,42] amounts to a classical logic handling of certainty-qualified statements. Certainty is estimated in the setting of possibility theory as a lower bound of a necessity set-function. An elementary possibilistic formula (a, α) is made of a classical logic formula a associated with a certainty level $\alpha \in \mathcal{S} \setminus \{0\}$. Basic PL handles only conjunctions of such formulas, and PL bases can be viewed as classical logic bases layered in terms of certainty. Semantics is in terms of epistemic states represented by fuzzy sets of interpretations. A PL base Γ is associated with an inconsistency level above which formulas are safe from inconsistency (this level is defined by $inc(\Gamma) = \max\{\alpha | \Gamma \vdash (\bot, \alpha)\}$, which semantically corresponds to the lack of normalization of the possibility distribution associated with Γ).

Applications of possibilistic logic (and possibility theory) include

- *Bayesian-like possibilistic networks* [12,15], where possibilistic conditioning is defined using a Bayesian-like equation $\Pi(B \cap A) = \Pi(B \mid A) \star \Pi(A)$ where $\Pi(A) > 0$ and $\star$ is the minimum in the qualitative case or the product in the quantitative case; moreover $N(B \mid A) = 1 - \Pi(B^c \mid A)$. Several notions of independence make sense in the possibilistic setting [11,29]. Like joint probability distributions, joint possibility distributions can be decomposed into a conjunction of conditional possibility distributions (using $\star = $ minimum, or product), once an ordering of the variables is chosen, in a way similar to Bayes nets. Moreover, possibilistic nets can be directly translated into PL bases and vice-versa.
- *Reasoning with default rules* [17,18]: Possibility theory can be used for describing the normal course of things. A default rule "if a then generally b" is understood formally as the constraint $\Pi(a \wedge b) > \Pi(a \wedge \neg b)$ on a possibility measure Π describing the semantics of the available knowledge. It expresses that in the context where a is true it is more possible that b is true than the opposite. A default rule "if a_i then generally b_i" in a conditional knowledge base can be turned into a PL clause $(\neg a_i \vee b_i, N(\neg a_i \vee b_i))$, where the necessity N is computed from the set of constraints corresponding to the default rules in the base. We thus obtain a PL base encoding this set of rules. Then using PL inference on the PL base, augmented with propositional formulas describing a factual situation, enables us to perform non monotonic reasoning in agreement with a postulate-based approach (namely, rational closure in the sense of Lehmann and Magidor [53]).
- *Belief revision.* Since non monotonic reasoning and belief revision can be closely related, PL finds application also in belief revision. In fact, comparative necessity relations (a relational counterpart of qualitative necessity measures) [28] are nothing but the epistemic entrenchment relations [35] that underlie well-behaved belief revision processes [47]. This enables the PL setting to provide syntactic revision operators that apply to possibilistic knowledge bases, including the case of uncertain inputs [21,59].
- *Information fusion.* The combination of possibility distributions, by means of fuzzy set connectives such as min, can be equivalently performed syntactically in terms of PL bases [10,20]. Besides, this approach can be also applied to the syntactic encoding of the merging of *classical* logic bases based on Hamming distance (where distances are computed between each interpretation and the

different classical logic bases, thus giving birth to counterparts of possibility distributions) [16].

– *Preference modeling.* In this case, certainty is turned into priority: Each PL formula (a, α) represents a goal a to be reached with some priority level α. Beyond PL, interpretations (corresponding to different alternatives) can be compared in terms of vectors acknowledging the satisfaction or the violation of the formulas associated with the different goals, using suitable order relations. Thus, partial orderings of interpretations can be obtained [13].

– *Modeling desires*: In contrast with static beliefs, (positive) desires are such that endorsing $a \vee b$ as a desire means to desire a *and* to desire b. However, desiring both a and $\neg a$ does not sound rational. The modeling of desires can be achieved using a "desirability" distribution $\delta : U \rightarrow [0, 1]$ such that $\delta(u) = 0$ for some $u \in U$. The logic of desires is thus a inversed mirror image of classical logic. Just as belief revision relies on an epistemic entrenchment relation (and thus on a necessity measure), well-behaved desire revision relies on a guaranteed possibility function Δ [33].

– *Qualitative decision.* Possibility theory provides a valuable setting for qualitative decision under uncertainty where pessimistic and optimistic decision criteria have been axiomatized [44] and cast in possibilistic logic by means of two bases, one for expressing knowledge, the other for expressing goals [32].

Let us also briefly mention different extensions of basic PL, where

– *Lattice-valued possibilistic logic.* Examples are i) a timed PL with logical formulas associated with *fuzzy* sets of time instants where the formula is known to be certain to some extent; ii) a *logic of supporters* [52], where formulas a are associated with sets of arguments in their favor. Closely related to this latter logic is the idea of associating each formula with a set of distinct explicit sources that support its truth more or less strongly. This has led to the proposal of a "social" logic where formulas are of the form (a, A), where A denotes a subset of agents and the formula means that *at least all* the agents in A believe that a is true [6]. It can be extended to pieces of information of the form "at least all agents in A believe a at least at level α".

– *Symbolic PL.* Instead of using weights from a totally ordered scale, one may use pairs (p, x) where x is a symbolic entity that stands for an unknown weight. Then we can model the situation where only a partial ordering between ill-known weights is specified by means of inequality constraints [9,22].

– An extension of possibilistic inference has been proposed for handling paraconsistent (conflicting) information [19].

In a computational perspective, possibilistic logic has also impacted logic programming [1,5,56,57]. Besides, the possibilistic handling of uncertainty in description logic has been suggested [60,67]. Computational advantages of description logic can then be preserved for its PL extensions, in particular in the case of the *possibilistic DL-Lite* family [7,8]. Another application is the encoding of control access policies [14].

5   Generalized Possibilistic Logic

In the so-called generalized possibilistic logic (GPL) [46], negation and disjunction can be used to combine possibilistic formulas, on top of conjunction as in PL. GPL use graded necessity and possibility modalities, i.e., a PL formula (a, α), is encoded as $\square_\alpha a$ in GPL.

GPL can be viewed as both a generalization of PL and a generalization of Meta-Epistemic Logic (MEL) [3], the simplest logic of belief and partial ignorance - a fragment of modal logic KD where all formulas are modal, and modalities cannot be nested. See Fig. 1 [41], which points out how GPL extends propositional logic through MEL and PL. GPL is in fact just a two-tiered standard propositional logic, in which propositional formulas are encapsulated by weighted modal operators, forming higher order propositional formulas, interpreted in terms of uncertainty measures from possibility theory. GPL can be still extended to a logic involving both objective and non-nested multimodal formulas [4].

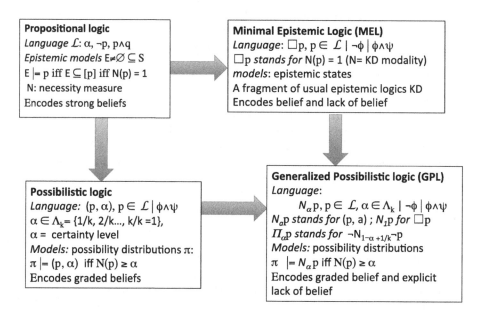

Fig. 1. From propositional logic to generalized possibilistic logic

GPL has applications such as:

– Reasoning about ignorance [46]: Some GPL formulas encode statements such as "All that is known is a". It means that a is known but nothing more (like, e.g., some b entailing a). It semantically corresponds to the use of the guaranteed possibility set function.

- Representing answer set programs [45,46]: GPL can encode answer set programs, adopting a three-valued scale $\mathcal{S} = \{0, 1/2, 1\}$. In this case, we can discriminate between propositions we are fully certain of and propositions we consider only more plausible than not. This is sufficient to encode non-monotonic ASP rules (with negation as failure) within GPL, which lays bare their epistemic semantics.
- Handling comparative certainty logic [23]. Rather than using weighted formulas, one might wish to represent at a syntactic level statements of the form "a is more certain than b". This kind of statements can be to some extent captured by GPL [46], which is thus more expressive than PL with symbolic weights.
- Encoding many-valued logics of uncertainty, such as Kleene and Lukasiewicz 3-valued logics [24], as well as 3-valued paraconsistent logics such as Priest logic of paradox [25]. MEL is enough for this task. GPL with 2 necessity modalities (one weak and one strong) is needed to encode the 5-valued equilibrium logic [45]. Many-valued logics with more truth-values could be encoded in GPL with a suitable number of certainty levels.

As a perspective, a logic of arguments, similar to GPL, has been outlined in [2]. The basic formulas are pairs (x, y) and stand for "y is a reason for believing x". Another perspective is reasoning about other agents' beliefs as in the muddy children problem [42].

6   Concluding Remarks

Generally speaking, the interest and the strength of PL and GPL relies on a sound alliance between classical logic and possibility theory which offers a rich representation setting allowing an accurate modeling of partial ignorance.

An interesting feature of possibility theory (involving the four set functions) lies in its capability of providing a *bipolar* representation of positive and negative information using pairs of distributions (δ, π). This view applies both to knowledge and preferences. Then, the distribution π describes the complement of the fuzzy set of values that are ruled out (being impossible or rejected), while the distribution δ describes a fuzzy subset of values that are actually possible to some extent or desired (positive information) [37].

Counterparts of the four possibility theory set functions also make sense in formal concept analysis, where formal concepts in the usual sense are defined from the counterpart of the Δ function and the scale $\mathcal{S}$ is replaced by a power set [38].

Possibility theory still has several other noticeable applications such as the design and the handling of uncertain databases [55,58], or the numerical computation with fuzzy intervals [30].

Lastly, possibility theory may have some potential in machine learning both on the quantitative side due to its link with imprecise probabilities and the existence of probability/possibility transformations, and on the qualitative side for its relation to logic and its use in the modeling of different types of if-then rules, which may be also of interest for explanation purposes; see [43] for a discussion and references. This is still to be further explored. Let us more particularly mention some works on the qualitative side in relation with possibilistic logic. PL can be applied to inductive logic programming (ILP). Indeed having a stratified set of first-order logic rules as an hypothesis in ILP is of interest for learning both rules covering normal cases and more specific rules for exceptional cases [62]. A different approach to the induction of possibilistic logic theories is proposed in [51]. It relies on the fact that any set of formulas in Markov logic [61] can be exactly translated into possibilistic logic formulas [46,50].

References

1. Alsinet, T., Godo, L., Sandri, S.: Two formalisms of extended possibilistic logic programming with context-dependent fuzzy unification: a comparative description. Electron. Notes Theor. Comput. Sci. **66**(5), 1–21 (2002)
2. Amgoud, L., Prade, H.: Towards a logic of argumentation. In: Hüllermeier, E., Link, S., Fober, T., Seeger, B. (eds.) SUM 2012. LNCS (LNAI), vol. 7520, pp. 558–565. Springer, Heidelberg (2012). https://doi.org/10.1007/978-3-642-33362-0_43
3. Banerjee, M., Dubois, D.: A simple logic for reasoning about incomplete knowledge. Int. J. Approx. Reason. **55**, 639–653 (2014)
4. Banerjee, M., Dubois, D., Godo, L., Prade, H.: On the relation between possibilistic logic and modal logics of belief and knowledge. J. Appl. Non-Class. Logics **27**(3–4), 206–224 (2017)
5. Bauters, K., Schockaert, S., DeCock, M., Vermeir, D.: Possible and necessary answer sets of possibilistic answer set programs. In: Proceedings of 24th IEEE International Conference on ICTAI, Athens, pp. 836–843 (2012)
6. Belhadi, A., Dubois, D., Khellaf-Haned, F., Prade, H.: Multiple agent possibilistic logic. J. Appl. Non-Class. Logics **23**, 299–320 (2013)
7. Benferhat, S., Bouraoui, Z.: Possibilistic DL-Lite. In: Liu, W., Subrahmanian, V.S., Wijsen, J. (eds.) SUM 2013. LNCS (LNAI), vol. 8078, pp. 346–359. Springer, Heidelberg (2013). https://doi.org/10.1007/978-3-642-40381-1_27
8. Benferhat, S., Bouraoui, Z., Loukil, Z.: Min-based fusion of possibilistic DL-Lite knowledge bases. In: Proceedings of IEEE/WIC/ACM International Joint Conferences on Web Intelligence (WI 2013), Atlanta, pp. 23–28 (2013)
9. Benferhat, S., Prade, H.: Encoding formulas with partially constrained weights in a possibilistic-like many-sorted propositional logic. In: Proceedings of 9th IJCAI, Edinburgh, pp. 1281–1286 (2005)
10. Boldrin, L., Sossai, C.: Local possibilistic logic. J. Appl. Non-Class. Logics **7**, 309–333 (1997)
11. Ben Amor, N., Benferhat, S., Dubois, D., Mellouli, K., Prade, H.: A theoretical framework for possibilistic independence in a weakly ordered setting. Int. J. Uncertainty Fuzziness Knowl. Based Syst. **10**, 117–155 (2002)
12. Ben Amor, N., Benferhat, S., Mellouli, K.: Anytime propagation algorithm for min-based possibilistic graphs. Soft Comput. **8**(2), 150–161 (2003)
13. Ben Amor, N., Dubois, D., Gouider, H., Prade, H.: Possibilistic preference networks. Inf. Sci. **460–461**, 401–415 (2018)
14. Benferhat, S., El Baida, R., Cuppens, F.: A possibilistic logic encoding of access control. In: Proceedings of 16th International Florida AI Research Society Conference (FLAIRS 2003), pp. 481–485. AAAI Press (2003)

15. Benferhat, S., Dubois, D., Garcia, L., Prade, H.: On the transformation between possibilistic logic bases and possibilistic causal networks. Int. J. Approx. Reason. **29**(2), 135–173 (2002)
16. Benferhat, S., Dubois, D., Kaci, S., Prade, H.: Possibilistic merging and distance-based fusion of propositional information. Ann. Math. Artif. Intell. **34**, 217–252 (2002). https://doi.org/10.1023/A:1014446411602
17. Benferhat, S., Dubois, D., Prade, H.: Nonmonotonic reasoning, conditional objects and possibility theory. Artif. Intell. **92**(1–2), 259–276 (1997)
18. Benferhat, S., Dubois, D., Prade, H.: Practical handling of exception-tainted rules and independence information in possibilistic logic. Appl. Intell. **9**(2), 101–127 (1998). https://doi.org/10.1023/A:1008259801924
19. Benferhat, S., Dubois, D., Prade, H.: An overview of inconsistency-tolerant inferences in prioritized knowledge bases. In: Fuzzy Sets, Logic and Reasoning About Knowledge, pp. 395–417. Kluwer (1999). https://doi.org/10.1007/978-94-017-1652-9_25
20. Benferhat, S., Dubois, D., Prade, H.: A computational model for belief change and fusing ordered belief bases. In: Williams, M.A., Rott, H., (eds.) Frontiers in Belief Revision, pp. 109–134. Kluwer (2001). https://doi.org/10.1007/978-94-015-9817-0_5
21. Benferhat, S., Dubois, D., Prade, H., Williams, M.A.: A framework for iterated belief revision using possibilistic counterparts to Jeffrey's rule. Fundam. Inform. **99**, 147–168 (2010)
22. Cayrol, C., Dubois, D., Touazi, F.: Symbolic possibilistic logic: completeness and inference methods. J. Logic Comput. **28**(1), 219–244 (2018)
23. Cayrol, C., Dubois, D., Touazi, F.: Possibilistic reasoning from partially ordered belief bases with the sure thing principle. IfCoLog J. Logics Appl. **5**(1), 5–40 (2018)
24. Ciucci, D., Dubois, D.: A modal theorem-preserving translation of a class of three-valued logics of incomplete information. J. Appl. Non-Class. Logics **23**, 321–352 (2013)
25. Ciucci, D., Dubois, D.: From possibility theory to paraconsistency. In: Beziau, J.-Y., Chakraborty, M., Dutta, S. (eds.) New Directions in Paraconsistent Logic. SPMS, vol. 152, pp. 229–247. Springer, New Delhi (2015). https://doi.org/10.1007/978-81-322-2719-9_10
26. Ciucci, D., Dubois, D., Lawry, J.: Borderline vs. unknown: comparing three-valued representations of imperfect information. Int. J. Approx. Reason. **55**, 1866–1889 (2014)
27. Cohen, L.J.: The Probable and the Provable. Clarendon, Oxford (1977)
28. Dubois, D.: Belief structures, possibility theory and decomposable measures on finite sets. Comput. AI **5**, 403–416 (1986)
29. Dubois, D., Fariñas del Cerro, L., Herzig, A., Prade, H.: A roadmap of qualitative independence. In: Fuzzy Sets, Logics and Reasoning about Knowledge, pp. 325–350. Kluwer (1999). https://doi.org/10.1007/978-94-017-1652-9_22
30. Dubois, D., Kerre, E., Mesiar, R., Prade, H.: Fuzzy interval analysis. In: Dubois, D., Prade, H. (eds.) Fundamentals of Fuzzy Sets, pp. 483–581. Kluwer, Boston (2000). https://doi.org/10.1007/978-1-4615-4429-6_11
31. Dubois, D., Lang, J., Prade, H.: Possibilistic logic. In: Gabbay, D.M., et al. (eds.) Handbook of Logic in Artificial Intelligence and Logic Programming, vol. 3, pp. 439–513. Oxford University Press, Oxford (1994)
32. Dubois, D., Le Berre, D., Prade, H., Sabbadin, R.: Using possibilistic logic for modeling qualitative decision: ATMS-based algorithms. Fundam. Inform. **37**, 1–30 (1999)

33. Dubois, D., Lorini, E., Prade, H.: The strength of desires: a logical approach. Mind. Mach. **27**(1), 199–231 (2017)
34. Dubois, D., Prade, H.: Possibility Theory. An Approach to Computerized Processing of Uncertainty. With the collaboration of H. Farreny, R. Martin-Clouaire, C. Testemale. Plenum, New York (1988)
35. Dubois, D., Prade, H.: Epistemic entrenchment and possibilistic logic. Artif. Intell. **50**, 223–239 (1991)
36. Dubois, D., Prade, H.: Possibility theory: qualitative and quantitative aspects. In: Smets, P. (ed.) Quantified Representation of Uncertainty and Imprecision. HDRUMS, vol. 1, pp. 169–226. Springer, Dordrecht (1998). https://doi.org/10.1007/978-94-017-1735-9_6
37. Dubois, D., Prade, H.: An overview of the asymmetric bipolar representation of positive and negative information in possibility theory. Fuzzy Sets Syst. **160**(10), 1355–1366 (2009)
38. Dubois, D., Prade, H.: Possibility theory and formal concept analysis: characterizing independent sub-contexts. Fuzzy Sets Syst. **196**, 4–16 (2012)
39. Dubois, D., Prade, H.: Possibilistic logic. An overview. In: Siekmann, J. (ed.) Handbook of The History of Logic. Computational Logic, North-Holland, vol. 9, pp. 283–342 (2014)
40. Dubois, D., Prade, H.: Possibility theory and its applications: where do we stand? In: Kacprzyk, J., Pedrycz, W. (eds.) Springer Handbook of Computational Intelligence, pp. 31–60. Springer, Heidelberg (2015). https://doi.org/10.1007/978-3-662-43505-2_3
41. Dubois, D., Prade, H.: A crash course on generalized possibilistic logic. In: Ciucci, D., Pasi, G., Vantaggi, B. (eds.) SUM 2018. LNCS (LNAI), vol. 11142, pp. 3–17. Springer, Cham (2018). https://doi.org/10.1007/978-3-030-00461-3_1
42. Dubois, D., Prade, H.: Possibilistic logic: from certainty-qualified statements to two-tiered logics – a prospective survey. In: Calimeri, F., Leone, N., Manna, M. (eds.) JELIA 2019. LNCS (LNAI), vol. 11468, pp. 3–20. Springer, Cham (2019). https://doi.org/10.1007/978-3-030-19570-0_1
43. Dubois, D., Prade, H.: From possibilistic rule-based systems to machine learning - a discussion paper. In: Davis, J., Tabia, K. (eds.) SUM 2020. LNCS (LNAI), vol. 12322, pp. 35–51. Springer, Cham (2020). https://doi.org/10.1007/978-3-030-58449-8_3
44. Dubois, D., Prade, H., Sabbadin, R.: Decision-theoretic foundations of qualitative possibility theory. Eur. J. Oper. Res. **128**(3), 459–478 (2001)
45. Dubois, D., Prade, H., Schockaert, S.: Stable models in generalized possibilistic logic. In: Brewka, G., Eiter, T., McIlraith, S.A. (eds.) Proceedings of 13th International Conference on Principles of Knowledge Representation and Reasoning (KR 2012), pp. 520–529. AAAI Press (2012)
46. Dubois, D., Prade, H., Schockaert, S.: Generalized possibilistic logic: foundations and applications to qualitative reasoning about uncertainty. Artif. Intell. **252**, 139–174 (2017)
47. Gärdenfors, P.: Knowledge in Flux, 2nd edn. College Publications (2008). MIT Press (1988)
48. Gasquet, O., Herzig, A., Said, B., Schwarzentruber, F.: Kripke's Worlds - An Introduction to Modal Logics via Tableaux. Studies in Universal Logic, Birkhäuser (2014)
49. Hintikka, J.: Knowledge and Belief: An Introduction to the Logic of the Two Notions. Cornell University Press (1962)

50. Kuzelka, O., Davis, J., Schockaert, S.: Encoding Markov logic networks in possibilistic logic. In: Meila, M., Heskes, T. (eds.) Proceedings of 31st Conference on Uncertainty in Artificial Intelligence (UAI 2015), Amsterdam, 12–16 July 2015, pp. 454–463. AUAI Press (2015)

51. Kuzelka, O., Davis, J., Schockaert, S.: Induction of interpretable possibilistic logic theories from relational data. In: Proceedings of IJCAI 2017, Melbourne, pp. 1153–1159 (2017)

52. Lafage, C., Lang, J., Sabbadin., R.: A logic of supporters. In: Bouchon-Meunier, B., Yager, R.R., Zadeh, L.A. (eds.) Information, Uncertainty and Fusion, pp. 381–392. Kluwer (1999). https://doi.org/10.1007/978-1-4615-5209-3_30

53. Lehmann, D., Magidor, M.: What does a conditional knowledge base entail? Artif. Intell. **55**(1), 1–60 (1992)

54. Lewis, D.L.: Counterfactuals. Basil Blackwell, Oxford (1973)

55. Link, S., Prade, H.: Relational database schema design for uncertain data. Inf. Syst. **84**, 88–110 (2019)

56. Nicolas, P., Garcia, L., Stephan, I., Lefèvre, C.: Possibilistic uncertainty handling for answer set programming. Ann. Math. Artif. Intell. **47**(1–2), 139–181 (2006). https://doi.org/10.1007/s10472-006-9029-y

57. Nieves, J.C., Osorio, M., Cortés, U.: Semantics for possibilistic disjunctive programs. In: Baral, C., Brewka, G., Schlipf, J. (eds.) LPNMR 2007. LNCS (LNAI), vol. 4483, pp. 315–320. Springer, Heidelberg (2007). https://doi.org/10.1007/978-3-540-72200-7_32

58. Pivert, O., Prade, H.: A certainty-based model for uncertain databases. IEEE Trans. Fuzzy Syst. **23**(4), 1181–1196 (2015)

59. Qi, G., Wang, K.: Conflict-based belief revision operators in possibilistic logic. In: Proceedings of the 26th AAAI Conference on Artificial Intelligence, Toronto, pp. 800–806 (2012)

60. Qi, G., Ji, Q., Pan, J.Z., Du, J.: Extending description logics with uncertainty reasoning in possibilistic logic. Int. J. Intell. Syst. **26**(4), 353–381 (2011)

61. Richardson, M., Domingos, P.M.: Markov logic networks. Mach. Learn. **62**(1–2), 107–136 (2006). https://doi.org/10.1007/s10994-006-5833-1

62. Serrurier, M., Prade, H.: Introducing possibilistic logic in ILP for dealing with exceptions. Artif. Intell. **171**(16–17), 939–950 (2007)

63. Shackle, G.L.S.: Decision, Order and Time in Human Affairs, 2nd edn. Cambridge University Press, Cambridge (1961)

64. Shafer, G.: A Mathematical Theory of Evidence. Princeton University Press, Princeton (1976)

65. Walley, P.: Measures of uncertainty in expert systems. Artif. Intell. **83**, 1–58 (1996)

66. Zadeh, L.A.: Fuzzy sets as a basis for a theory of possibility. Fuzzy Sets Syst. **1**, 3–28 (1978)

67. Zhu, J., Qi, G., Suntisrivaraporn, B.: Tableaux algorithms for expressive possibilistic description logics. In: Proceedings of IEEE/ACM International Conferences on Web Intelligence (WI 2013), Atlanta, pp. 227–232(2013)

Machine Learning

Similarity-Based Rough Sets with Annotation Using Deep Learning

Dávid Nagy$^{(\boxtimes)}$, Tamás Mihálydeák, and Tamás Kádek

Department of Computer Science, Faculty of Informatics,
University of Debrecen, Egyetem tér 1, Debrecen 4010, Hungary
{nagy.david,kadek.tamas}@inf.unideb.hu, mihalydeak@unideb.hu

Abstract. In the authors' previous research the possible usage of correlation clustering in rough set theory was investigated. Correlation clustering is based on a tolerance relation that represents the similarity among objects. Its result is a partition which can be treated as the system of base sets. However, singleton clusters represent very little information about the similarity. If the singleton clusters are discarded, then the approximation space received from the partition is partial. In this way, the approximation space focuses on the similarity (represented by a tolerance relation) itself and it is different from the covering type approximation space relying on the tolerance relation. In this paper, the authors examine how the partiality can be decreased by inserting the members of some singletons into base sets and how this annotation affects the approximations. This process can be performed by the user of system. However, in the case of a huge number of objects, the annotation can take a tremendous amount of time. This paper shows an alternative solution to the issue using neural networks.

Keywords: Rough set theory · Correlation clustering · Set approximation

1 Introduction

In our previous work, we examined whether the clusters, generated by correlation clustering, can be understood as a system of base sets. Correlation clustering is a clustering method in data mining that is based on a tolerance relation. Its result is a partition. The groups, defined by this partition, contain similar objects. In [12], we showed that it is worth to generate the system of base sets from the partition. In this way, the base sets contain objects that are typically similar to each other and they are also pairwise disjoint. The proposed approximation space is different from the tolerance-based covering approximation spaces. There can be some clusters that have only one member. These singletons represent very little information regarding the similarity. This is the reason why they are not treated as base sets. Without them, the approximation space becomes partial.

Z. Shi et al. (Eds.): ICIS 2020, IFIP AICT 623, pp. 93–102, 2021.
https://doi.org/10.1007/978-3-030-74826-5_8

In practice, partiality can cause issues in logical systems. That is why its degree should be minimized. In [13] we showed a possible way to decrease it by allowing the user to insert a member of a singleton into a base set. We called this process annotation. Its main problem is that it needs to be performed manually which takes a lot of time if there are a huge number of data points. In this paper, we propose an improved version of the annotation which performs the process using a neural network. Thus the annotation can be done automatically. So, the main problem of manual annotation is mitigated. The structure of the paper is the following: A theoretical background about the classical rough set theory comes first. In Sect. 4 we present our previous work and in Sect. 3 we define correlation clustering mathematically. Then, we show why decreasing partiality is important in logical systems. In Sect. 6 the annotation process is described. Finally, we conclude our results.

2   Theoretical Background

In general, a set is a collection of objects which is uniquely identified by its members. It means that if one would like to decide, whether an object belongs to a certain set, then a precise answer can be given (yes/no). A good example is the set of numbers that are divisible by 3 because it can be decided if an arbitrary number is divisible by 3 or not. Of course, it is required that one knows how to use the modulo operation. This fact can be considered as a background knowledge and it allows us to decide if a number belongs to the given set. Naturally, it is not necessary to know how to use the modulo operation for each number. Some second graders may not be able to divide numbers greater than 100. They would not be able to decide if 142 is divisible by 3 because they lack the required background knowledge. For them, 142 is neither divisible nor indivisible by 3. So there is uncertainty (vagueness) based on their knowledge. Rough set theory was proposed by Zdisław Pawlak in 1982 [14]. The theory offers a possible way to treat vagueness caused by some background knowledge. In data sciences, each object can be characterized by a set of attribute values. If two objects have the same known attribute values, then these objects cannot really be distinguished. The indiscernibility generated this way, gives the mathematical basis of rough set theory.

Definition 1. *The ordered 5–tuple $\langle U, \mathfrak{B}, \mathfrak{D}_{\mathfrak{B}}, \mathsf{l}, \mathsf{u} \rangle$ is a general approximation space if*

1. *U is a nonempty set;*
2. *$\mathfrak{B} \subseteq 2^U \setminus \emptyset$, $\mathfrak{B} \neq \emptyset$ ($\mathfrak{B}$ is the set of base sets);*
3. *$\mathfrak{D}_{\mathfrak{B}}$ is the set of definable sets and it is given by the following inductive definition:*
   (a) $\emptyset \in \mathfrak{D}_{\mathfrak{B}}$;
   (b) $\mathfrak{B} \subseteq \mathfrak{D}_{\mathfrak{B}}$;
   (c) if $D_1, D_2 \in \mathfrak{D}_{\mathfrak{B}}$, then $D_1 \cup D_2 \in \mathfrak{D}_{\mathfrak{B}}$
4. *$\langle \mathsf{l}, \mathsf{u} \rangle$ is a Pawlakian approximation pair i.e.*

(a) $Dom(\mathsf{l}) = Dom(\mathsf{u}) = 2^U$
(b) $\mathsf{l}(S) = \bigcup\{B \mid B \in \mathfrak{B} \text{ and } B \subseteq S\}$;
(c) $\mathsf{u}(S) = \bigcup\{B \mid B \in \mathfrak{B} \text{ and } B \cap S \neq \emptyset\}$.

The system of base sets represents the background knowledge or its limit. The functions l and u give the lower and upper approximation of a set. The lower approximation contains objects that surely belong to the set, and the upper approximation contains objects that possibly belong to the set.

Definition 2. *A general approximation space is Pawlakian [15,16] if $\mathfrak{B}$ is a partition of U.*

The indiscernibility modeled by an equivalence relation represents the limit of our knowledge embedded in an information system (or background knowledge). It has also an effect on the membership relation. In certain situations, it makes our judgment of the membership relation uncertain – thus making the set vague – as a decision about a given object affects the decision about all the other objects that are indiscernible from the given object. In practice, indiscernibility can be too strict as the attribute values of the objects must be completely the same. In these situations, the similarity among the objects can be enough to consider. Over the years, many new approximation spaces have been developed as the generalization of the original Pawlakian space [11]. The main difference between these kinds of approximation spaces (with a Pawlakian approximation pair) lies in the definition of the base sets (members of $\mathfrak{B}$).

Definition 3. *Different types of general approximation spaces $\langle U, \mathfrak{B}, \mathfrak{D}_{\mathfrak{B}}, \mathsf{l}, \mathsf{u} \rangle$ are as follows:*

1. *A general approximation space is a covering approximation space [17] generated by a tolerance relation $\mathcal{R}$ if $\mathfrak{B} = \{[u]_{\mathcal{R}} \mid u \in U\}$, where $[u]_{\mathcal{R}} = \{u' \mid u\mathcal{R}u'\}$.*
2. *A general approximation space is a covering approximation space if $\bigcup \mathfrak{B} = U$.*
3. *A general approximation space is a partial approximation space if $\bigcup \mathfrak{B} \neq U$.*

3 Correlation Clustering

Cluster analysis is an unsupervised learning method in data mining. The goal is to group the objects so that the objects in the same group are more similar to each other than to those in other groups. In many cases, the similarity is based on the attribute values of the objects. Although there are some cases when these values are not numbers, we can still say something about their similarity or dissimilarity. From the mathematical point of view, similarity can be modeled by a tolerance relation. Correlation clustering is a clustering technique based on a tolerance relation [5,6,18]. Bansal et al. defined correlation clustering for complete weighted graphs. Here $G = (V, E)$ is a graph and function $w : E \rightarrow \{+1, -1\}$ is the weight of edges. Weight $+1$ and -1 denotes the similarity/dissimilarity of the nodes of the edges. We always treat a node similar to itself. This graph defines a relation: xRy iff $w\big((x,y)\big) = +1$ or $x = y$. It is obvious, that this relation R is tolerance relation: it is reflexive and symmetric. Let p denote a clustering (partition) on this graph and let $p(x)$ be the set of vertices that are in the same cluster as x. In a partition p we call an edge (x,y) a conflict if $w\big((x,y)\big) = +1$ and $x \notin p(y)$ or $w\big((x,y)\big) = -1$ and $x \in p(y)$. The cost function is the number of these disagreements. Solving the correlation clustering is minimizing its cost function.

It is easy to check that we cannot necessarily find a perfect partition for a graph. Consider the simplest case, given three objects x, y and z, and x is similar to both y and z, but y and z are dissimilar. The number of partitions can be given by the Bell number [1], which grows exponentially. So the optimal partition cannot be determined in a reasonable time. In a practical case, a quasi-optimal partition can be sufficient, so a search algorithm can be used. The main advantage of the correlation clustering is that the number of clusters does not need to be specified in advance like in many clustering algorithms, and this number is optimal based on the similarity. However, since the number of partitions grows exponentially, it is an NP-hard problem.

In the original definition, a weight of an edge could be only $+1$ or -1. Naturally the function w can be the following as well: $w : E \rightarrow [-1, 1]$. If $w\big((x,y)\big) > 0$, then x and y are similar. If $w\big((x,y)\big) < 0$, then x and y are dissimilar and if the weight is 0, then they are neutral. In a partition p we call an edge (x,y) a conflict if $w\big((x,y)\big) > 0$ and $x \notin p(y)$ or $w\big((x,y)\big) < 0$ and $x \in p(y)$. A natural cost function is one in which an edge of weight w incurs a cost of $|w|$ when it is clustered improperly and a cost of 0 when it is correct.

4   Similarity-Based Rough Sets

When we would like to define the base sets, we use the background knowledge embedded in an information system. The base sets represent background knowledge (or its limit). In a Pawlakian system, we can say that two objects are indiscernible if all of their known attribute values are identical. The indiscernibility relation defines an equivalence relation. In some cases, it is enough to treat the similar objects in the same way. From the mathematical point of view, similarity can be described by a tolerance relation. Some covering systems are based on a tolerance relation. In these covering spaces, a base set contains objects that are similar to a distinguished member. This means that the similarity to a given element is considered and it generates the system of base sets. Using correlation clustering, we obtain a (quasi- optimal) partition of the universe (see in [2–4]). The clusters contain such elements which are typically similar to each other and not just to a distinguished member. In our previous research, we investigated whether the partition can be understood as a system of base sets (see in [12]). By our experiments, it is worth to generate a partition with correlation clustering. The base sets, generated from the partition, have several good properties:

- the similarity of objects relying on their properties (and not the similarity to a distinguished object) plays a crucial role in the definition of base sets;

- the system of base sets consists of disjoint sets, so the lower and upper approximation are closed in the following sense: Let S be a set and $x \in U$. If $x \in l(S)$, then we can say, that every $y \in U$ object which is in the same cluster as $x \in l(S)$. If $x \in u(S)$, then we can say, that every $y \in U$ object which is in the same cluster as $x \in u(S)$.
- only the necessary number of base sets appears (in applications we have to use an acceptable number of base sets);
- the size of base sets is not too small, or too big.

In the case of singleton clusters, their members cannot be considered as similar to any other objects without increasing the value of the cost function (see in Sect. 3). Therefore, they represent very little information about the similarity. This is the reason why these objects can be treated as outliers. In machine learning, outliers can impair the decisions and result in more inaccurate results. Singleton clusters, therefore, are not considered as base sets. Thus, the approximation space becomes partial (the union of the base sets does not cover the universe).

5 Partiality in Logical Systems

Classical first-order logic gives the necessary tools to prove the soundness of the inference chains. But what inferences could be derived from the background knowledge when it appears in a vague (rough) structure, and what kind of logical systems need to be used to verify the correctness of the information gained from a rough-set-based framework? In this section, we briefly introduce the rough-set-based semantics of first-order logic (or at least we will show one approach), then we will emphasize the threats hiding in partiality.

The semantics of classical first-order logic is based on set theory. The semantic meaning of the predicate symbols is often defined with the help of a positivity domain which is determined by an interpretation of the logical language. The positivity domain of a unary predicate is a subset of a given universe. It contains those objects for which the predicate is said to be true. Predicates with higher arity can be defined similarly using some Cartesian product of the universe as base set [9]. Since an approximation space gives the ability to create the lower or upper approximation of sets, it can be used to approximate the positivity domain of predicates.

In a reasonable logical system, the positivity and negativity domains of the predicates must be disjoint. From this point of view, the use of the lower approximation can represent our certain knowledge (supposing that the lower approximation of a set S is a subset of S). In these circumstances, it is a legitimate expectation that the derived results in the approximated system, if there is any, must coincide with the results we could receive from the crisp (approximation free) world. These expectations can be satisfied by a three-valued logic system where the relationship between an object and a unary predicate can be the following:

– the object certainly belongs to the positivity domain of the predicate, or
– the object certainly belongs to the negativity domain of the predicate, or
– it cannot be determined whether the object belongs to the positivity or negativity domain (the object is it in the border).

The cases above are usually represented with the truth values 1, 0, $\frac{1}{2}$ respectively, so the result is a three-valued logic system [8]. The way how these systems extend the semantics of the logical connectives is crucial. A widely accepted principle to define the existential and the universal quantifiers so that they generalize the zero-order connectives: the disjunction and the conjunction. A partial approximation space requires partial logic system [10]. It gives us the ability to distinguish situations where we cannot say anything certain about the above-mentioned relationship between an object and a predicate:

– we do not know anything about the object (it is missing from the approximation),
– we do not know how the object is related to the predicate (the object is in the border).

The partiality causes the appearance of the truth value gap (usually denoted by 2 which extends the three-valued system). It is also widely accepted, that the connectives are defined so that the truth value gap is inherited.

The pessimistic scenario says that missing knowledge can refute the conclusions derived from our available knowledge. Keeping in mind how we defined the goal of the logic system, we have to adopt this pessimistic approach. In other words, from our viewpoint, it is better to say nothing than to say something unsure.

The disadvantage of the pessimistic approach is that, if we respect all the earlier mentioned widely accepted properties of the partial three-valued logic system, it makes the quantification useless in the case of partial approximation space. For example, to evaluate a universally quantified formula, we need to evaluate the subformula substituted all the objects of the universe in place of the bound variable (with the help of modified assignments). Since the approximation space is partial, at least one evaluation of the subformula will cause truth value gap. An often-used solution is to modify the semantics of the quantifiers so that truth value gap appears only if all subformula evaluations raise truth value gap [7] but it also voids the pessimistic approach. The second approach is to avoid partial approximation spaces.

6   Similarity-Based Rough Sets with Annotation

Sometimes it can happen that an object does not belong to a base set (non-singleton cluster) because the system could not consider it similar to any other objects based on the background information. This does not mean that this object is only similar to itself, but without proper information (maybe due to noisy data) the system could not insert it into any base set (non-singleton cluster) to decrease the number of conflicts. Correlation clustering is based on a tolerance relation that represents similarity. The degree of similarity is between -1 and 1. It can also happen that some relevant information is lost when we map the difference of two objects to $[-1, 1]$. In [13] we proposed a possible way to handle this situation. The users can use their knowledge to help the system by inserting the members of some singletons into base sets (non-singleton clusters). With the help of this manual annotation, the users can put their knowledge into the system. It also decreases the partiality by decreasing the number of singletons. One of the issues with this approach is that it assumes that the user has some background knowledge. It must also be performed manually, so it cannot be used in the case of a huge number of points because it requires too much time.

Artificial neural networks (ANN) are inspired by the biological neural networks in machine learning. They can be used to perform classification. The annotation process is a classification problem as we need to find a proper cluster to an object. Here, the cluster IDs can be treated as class labels. Given the specifics of an approximation space, the deep learning algorithms can help the user in the annotation process. Each layer of the neural network can identify the main properties of the base sets. Based on these characteristics, it can perform an automated comparison and offer options accordingly. In this way, annotation can be executed completely automatically which means it can also be used in the case of huge data. Naturally, during the annotation, not every object must be inserted into a base set. The neural network identifies if an object is an outlier and discards it. Figure 1 shows how the similarity-based rough sets approximation is constructed from the original data.

In a real-world application, it can happen that an attribute value of an object is missing. This means that it can be unknown, unassigned, or inapplicable (e.g. maiden name of a male). Handling these data is usually a difficult task. In many cases, these values are imputed. It is common to replace them with the mean or the most frequent value. Typically, this gives a rather good result in many situations. In early-stage diabetes, it is not unusual that the patient has only an elevated blood sugar level. If this value is missing for a patient, then it should not be replaced by the mean because the mean is usually the normal blood sugar level. After the substitution, this patient can be treated as healthy. This type of substitution does not consider the information of an object itself but the information of a collection of objects, therefore it can lead to a false conclusion. In this paper, we propose another method to handle missing data. If an object has a missing attribute value, then it cannot be treated as similar to any other objects, so this entity becomes a member of a singleton. As mentioned earlier, such a cluster cannot be treated as a base set. However, with the annotation, it can be placed into a base set. The neural network should only consider the non-missing attribute values and based on them it should find the appropriate base set.

In machine learning, it is very common to combine clustering with classification. Classification always requires class labels. After clustering the data, the cluster IDs can be treated as these class labels. In our approximation space, the

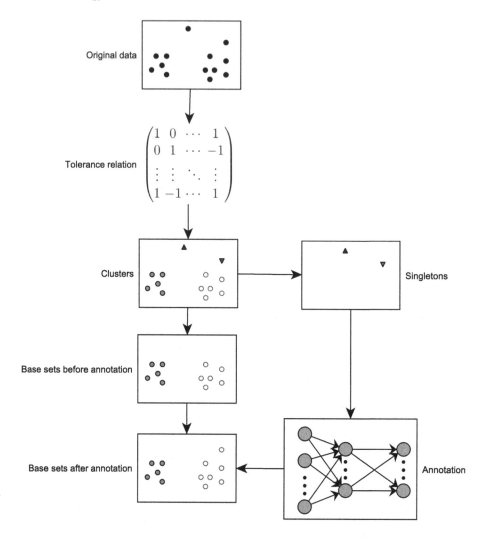

Fig. 1. The main steps of the similarity-based rough sets approximation space

non-singleton clusters are treated as base sets. If there is a new object for which we need to find the appropriate base set, then this object can be considered as a singleton. If it is a singleton, then the annotation can be applied to find the fitting base set.

7 Conclusion and Future Work

In [12] the authors introduced a partial approximation space relying on a tolerance relation that represents similarity. The novelty of this approximation space is that the systems of base sets are the result of correlation clustering. Thus the similarity is taken into consideration generally. Singleton clusters have no real information in the approximation process, these clusters cannot be taken as base sets, therefore the approximation space is partial. In the present paper, a new possibility is proposed to embed some information into the approximation space. A neural network may decide the status of a member of a singleton cluster. It can be put into a base set, and the approximation of a set changes according to the new system of base sets. This possibility is crucial in practical applications because it decreases the degree of partiality. Neural networks can also be used when we need to decide to which base set a new object belongs. In machine learning, it is common to combine clustering with classification. In our proposed system, a neural network decides and puts the new objects into the chosen base set. This is especially promising for a large number of new objects as we do not need to perform correlation clustering, which is an NP-hard problem, for each object.

Acknowledgement. This work was supported by the construction EFOP-3.6.3-VEKOP-16-2017-00002. The project was co-financed by the Hungarian Government and the European Social Fund.

References

1. Aigner, M.: Enumeration via ballot numbers. Discret. Math. **308**(12), 2544–2563 (2008). https://doi.org/10.1016/j.disc.2007.06.012. http://www.sciencedirect.com/science/article/pii/S0012365X07004542
2. Aszalós, L., Mihálydeák, T.: Rough clustering generated by correlation clustering. In: Ciucci, D., Inuiguchi, M., Yao, Y., Ślęzak, D., Wang, G. (eds.) RSFDGrC 2013. LNCS (LNAI), vol. 8170, pp. 315–324. Springer, Heidelberg (2013). https://doi.org/10.1007/978-3-642-41218-9_34
3. Aszalós, L., Mihálydeák, T.: Rough classification based on correlation clustering. In: Miao, D., Pedrycz, W., Ślęzak, D., Peters, G., Hu, Q., Wang, R. (eds.) RSKT 2014. LNCS (LNAI), vol. 8818, pp. 399–410. Springer, Cham (2014). https://doi.org/10.1007/978-3-319-11740-9_37
4. Aszalós, L., Mihálydeák, T.: Corrclation clustering by contraction. In: 2015 Federated Conference on Computer Science and Information Systems (FedCSIS), pp. 425–434. IEEE (2015)
5. Bansal, N., Blum, A., Chawla, S.: Correlation clustering. Mach. Learn. **56**(1–3), 89–113 (2004)
6. Becker, H.: A survey of correlation clustering. In: Advanced Topics in Computational Learning Theory, pp. 1–10 (2005)
7. Kádek, T., Mihálydeák, T.: Some fundamental laws of partial first-order logic based on set approximations. In: Cornelis, C., Kryszkiewicz, M., Ślęzak, D., Ruiz, E.M., Bello, R., Shang, L. (eds.) RSCTC 2014. LNCS (LNAI), vol. 8536, pp. 47–58. Springer, Cham (2014). https://doi.org/10.1007/978-3-319-08644-6_5
8. Mihálydeák, T.: First-order logic based on set approximation: a partial three-valued approach. In: 2014 IEEE 44th International Symposium on Multiple-Valued Logic, pp. 132–137, May 2014. https://doi.org/10.1109/ISMVL.2014.31
9. Mihálydeák, T.: Partial first-order logical semantics based on approximations of sets. Non-classical Modal and Predicate Logics, pp. 85–90 (2011)

10. Mihálydeák, T.: Partial first-order logic relying on optimistic, pessimistic and average partial membership functions. In: Pasi, G., Montero, J., Ciucci, D. (eds.) Proceedings of the 8th Conference of the European Society for Fuzzy Logic and Technology, pp. 334–339 (2013)
11. Mihálydeák, T.: Logic on similarity based rough sets. In: Nguyen, H.S., Ha, Q.-T., Li, T., Przybyła-Kasperek, M. (eds.) IJCRS 2018. LNCS (LNAI), vol. 11103, pp. 270–283. Springer, Cham (2018). https://doi.org/10.1007/978-3-319-99368-3_21
12. Nagy, D., Mihálydeák, T., Aszalós, L.: Similarity based rough sets. In: Polkowski, L., et al. (eds.) IJCRS 2017. LNCS (LNAI), vol. 10314, pp. 94–107. Springer, Cham (2017). https://doi.org/10.1007/978-3-319-60840-2_7
13. Nagy, D., Mihálydeák, T., Aszalós, L.: Similarity based rough sets with annotation. In: Nguyen, H.S., Ha, Q.-T., Li, T., Przybyła-Kasperek, M. (eds.) IJCRS 2018. LNCS (LNAI), vol. 11103, pp. 88–100. Springer, Cham (2018). https://doi.org/10.1007/978-3-319-99368-3_7
14. Pawlak, Z.: Rough sets. Int. J. Parallel Prog. 11(5), 341–356 (1982)
15. Pawlak, Z., Skowron, A.: Rudiments of rough sets. Inf. Sci. 177(1), 3–27 (2007)
16. Pawlak, Z., et al.: Rough sets: theoretical aspects of reasoning about data. In: System Theory, Knowledge Engineering and Problem Solving, vol. 9. Kluwer Academic Publishers, Dordrecht (1991)
17. Skowron, A., Stepaniuk, J.: Tolerance approximation spaces. Fund. Inform. 27(2), 245–253 (1996)
18. Zimek, A.: Correlation clustering. ACM SIGKDD Explor. Newsl. 11(1), 53–54 (2009)

P-T Probability Framework and Semantic Information G Theory Tested by Seven Difficult Tasks

Chenguang Lu[(⊠)] [iD]

College of Intelligence Engineering and Mathematics, Liaoning Technical University, Fuxin 123000, Liaoning, China
survival99@gmail.com

Abstract. To apply information theory to more areas, the author proposed semantic information G theory, which is a natural generalization of Shannon's information theory. This theory uses the P-T probability framework so that likelihood functions and truth functions (or membership functions), as well as sampling distributions, can be put into the semantic mutual information formula at the same time. Hence, we can connect statistics and (fuzzy) logic. Rate-distortion function $R(D)$ becomes rate-verisimilitude function $R(G)$ (G is the lower limit of the semantic mutual information) when the distortion function is replaced with the semantic information function. Seven difficult tasks are 1) clarifying the relationship between minimum information and maximum entropy in statistical mechanics, 2) compressing images according to visual discrimination, 3) multilabel learning for obtaining truth functions or membership functions from sampling distributions, 4) feature classifications with maximum mutual information criterion, 5) proving the convergence of the expectation-maximization algorithm for mixture models, 6) interpreting Popper's verisimilitude and reconciling the contradiction between the content approach and the likeness approach, and 7) providing practical confirmation measures and clarifying the raven paradox. This paper simply introduces the mathematical methods for these tasks and the conclusions. The P-T probability framework and the semantic information G theory should have survived the tests. They should have broader applications. Further studies are needed for combining them with neural networks for machine learning.

Keyword: Semantic information · Probability framework · Machine learning · Philosophy of science · Rate distortion · Boltzmann distribution · Maximum mutual information · Verisimilitude · Confirmation measure

1 Introduction

Although Shannon's information theory [1] has achieved great success since 1948, we cannot use it to measure semantic information. It is also not easy to apply this theory to machine learning because we cannot put likelihood functions in the information measure. According to this theory, we can only use the distortion criterion instead of

© IFIP International Federation for Information Processing 2021
Published by Springer Nature Switzerland AG 2021
Z. Shi et al. (Eds.): ICIS 2020, IFIP AICT 623, pp. 103–114, 2021.
https://doi.org/10.1007/978-3-030-74826-5_9

the information criterion to optimize detections and classifications. In 1949, Weaver first proposed to research semantic information [1]. In 1952, Carnap and Bar-Hillel [2] presented an outline of semantic information theory. There exist multiple different information theories relating to semantic information [3].

In 1993, the author proposed a generalized information theory [4, 5]. Now it is called semantic information G theory or G theory [3], where "G" means generalization. Early G theory was used mainly for semantic information evaluations and data compression [4, 5]. Recently, the author developed the P-T probability framework for G theory [6] so that likelihood functions and (fuzzy) truth functions (or membership functions) can be mutually converted by a pair of new Bayes' formulas. G theory now can resolve more problems with machine learning [3] and philosophy of science [6, 7].

This paper introduces the P-T probability framework and G theory; and uses them to complete seven difficult tasks for testing them.

2 P-T Probability Framework and Semantic Information G Theory

2.1 From Shannon's Probability Framework to P-T Probability Framework

The probability framework used by Shannon is defined as follows.

Definition 1. X is a discrete random variable taking a value $x \in U = \{x_1, x_2, ..., x_m\}$; $P(x_i)$ is the limit of the relative frequency of event $X = x_i$. Y is a discrete random variable taking a value $y \in V = \{y_1, y_2, ..., y_n\}$; $P(y_j) = P(Y = y_j)$. Shannon names $P(X)$ the source, $P(Y)$ the destination, and $P(Y|X)$ the channel. The latter consists of Transition Probability Functions (TPF): $P(y_j|x)$, $j = 1, 2, ..., n$.

The P–T probability framework is defined as follows.

Definition 2. The y_j is a label or a hypothesis, $y_j(x_i)$ is a proposition, and $y_j(x)$ is a predicate. The θ_j is a fuzzy subset of universe U, $y_j(x) = $ "$x \in \theta_j$" = "x is in θ_j". The θ_j is also treated as a model or a set of model parameters. A probability that is defined with " = ", such as $P(y_j) = P(Y = y_j)$, is a statistical probability. A probability that is defined with "$\in$", such as $P(X \in \theta_j)$, is a logical probability denoted by $T(y_j) = T(\theta_j) = P(X \in \theta_j)$. $T(\theta_j|x) = P(x \in \theta_j) = P(X \in \theta_j|X = x) \in [0, 1]$ is the truth function of y_j and the membership function of θ_j.

According to Davidson's truth condition semantics [8], the truth function of y_j ascertains the semantic meaning of y_j. Truth functions $T(\theta_j|x)$ ($j = 1, 2, ..., n$) form a semantic channel. Bayes, Shannon, and the author used three types of Bayes' Theorem [3]. Two asymmetrical formulas can express the third type:

$$P(x|\theta_j) = T(\theta_j|x)P(x)/T(\theta_j), \quad T(\theta_j) = \sum_i P(x_i)T(\theta_j|x_i), \tag{1}$$

$$T(\theta_j|x) = [P(x|\theta_j)/P(x)]/\max[P(x|\theta_j)/P(x)]. \tag{2}$$

2.2   From Shannon's Information Measure to Semantic Information Measure

Shannon's mutual information [1] is defined as:

$$I(X;Y) = \sum_j \sum_i P(x_i, y_j) \log \frac{P(x_i|y_j)}{P(x_i)}, \tag{3}$$

Replacing $P(x_i|y_j)$ with $P(x_i|\theta_j)$ after the log, we have semantic mutual information:

$$I(X;\Theta) = \sum_j \sum_i P(x_i, y_j) \log \frac{P(x_i|\theta_j)}{P(x_i)} = \sum_j \sum_i P(x_i, y_j) \log \frac{T(\theta_j|x_i)}{T(\theta_j)}. \tag{4}$$

When $Y = y_j$, we have generalized Kullback-Leibler (KL) information:

$$I(X;\theta_j) = \sum_i P(x_i|y_j) \log \frac{P(x_i|\theta_j)}{P(x_i)} = \sum_i P(x_i|y_j) \log \frac{T(\theta_j|x_i)}{T(\theta_j)}, \tag{5}$$

Further, if $X = x_i$, we have semantic information of y_j about x_i (illustrated in Fig. 1):

$$I(x_i;\theta_j) = \log \frac{P(x_i|\theta_j)}{P(x_i)} = \log \frac{T(\theta_j|x_i)}{T(\theta_j)}. \tag{6}$$

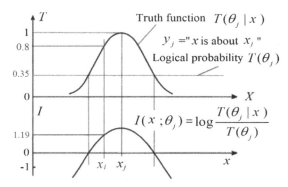

Fig. 1. Semantic information changes with x. When real x is x_i, the truth value is $T(\theta_j|x_i) = 0.8$; information $I(x_i;\theta_j)$ is 1.19 bits. If x exceeds a certain range, the information is negative.

If the truth function is a Gaussian truth function, i.e., $T(\theta_j|x) = \exp[-(x - x_j)^2/(2\sigma_j^2)]$, $I(X;\Theta)$ is equal to the generalized entropy minus the mean relative squared error:

$$I(X;\Theta) = -\sum_j P(y_j) \log T(\theta_j) - \sum_i \sum_j P(x_i, y_j)(x_i - y_j)^2/(2\sigma_j^2). \tag{7}$$

We can find that the maximum semantic mutual information criterion is like the Regularized Least Square (RLS) criterion that is getting popular in machine learning.

Shannon [9] proposed information rate-distortion function $R(D)$ for lossy data compression limit. D is the upper limit of the average of distortion $d_{ij} = d(x_i, y_j)$. Replacing d_{ij} with $I_{ij} = I(x_i, y_j)$ and let G be the lower limit of $I(X; \Theta)$, we obtain $R(G)$ function:

$$G(s) = \sum_i \sum_j P(x_i)P(y_j|x_i)I_{ij}, \quad R(s) = sG(s) - \sum_i P(x_i) \log \lambda_i, \quad (8)$$

$$P(y_j|x_i) = P(y_j)P(x_i|\theta_j)^s/\lambda_i, \quad \lambda_i = \sum_j P(y_j)P(x_i|\theta_j)^s. \quad (9)$$

Every $R(G)$ function has a point where information efficiency $G/R = 1$ (see Fig. 2).

3 To Complete Seven Difficult Tasks

3.1 Clarifying the Relationship Between Minimum Information and Maximum Entropy

Researchers found that Boltzmann's distribution can be used for machine learning [10] and is related to the rate-distortion function [11]. Using the P-T probability framework, we can explain this relationship better. The Boltzmann distribution is:

$$P(x_i|T) = \exp(-\frac{e_i}{kT})/Z, \quad Z = \sum_i \exp(-\frac{e_i}{kT}), \quad (10)$$

where $P(x_i|T)$ is the probability of a particle in the ith state x_i with energy e_i, T is the absolute temperature, k is the Boltzmann constant, and Z is the partition function. If x_i is the state with the ith energy, G_i is the number of states with e_i, and G is the total number of all states, then $P(x_i) = G_i/G$. Hence, the above formula becomes.

$$P(x_i|T) = P(x_i)\exp(-\frac{x_i}{kT})/Z', \quad Z' = \sum_i P(x_i)\exp(-\frac{x_i}{kT}). \quad (11)$$

Now, we can find that $\exp[-e_i/(kT)]$ can be treated as a truth function or a Distribution Constraint Function (DCF), Z' as a logical probability, and Eq. (11) as a Bayes' formula (see Eq. (1)). A DCF means that there should be $1 - P(x|y_j) \leq 1 - P(x|\theta_j)$. The author [4] has proved that for given $P(X)$ and a group of DCFs: $T(\theta_{xi}|y)$, $i = 1, 2, ..., m$, the minimum Shannon's mutual information $R(\Theta)$ is equal to the semantic mutual information $I(Y; \Theta)$, e.g. $R(\Theta) = R(D)$ [7]. From equation $F = E - TS$ (F is Helmholtz free energy, E is total energy, and S is entropy), we can derive [7]

$$R(\Theta) = -\sum_j P(y_j)\frac{e_j}{kT_j} - \sum_j P(y_j)\ln(Z_j/G) = \ln G - S/(kN) \quad (12)$$

This formula indicates the maximum entropy principle is equivalent to the minimum mutual information principle.

3.2 *R(G)* Function for Compressing Data According to Visual Discrimination

Suppose that x_i represents a gray level, a color, or a pixel with a certain position and color, y_j is the corresponding perception, which means "x is about x_j." The visual discrimination function is $T(\theta_j|x) = \exp[-(x-x_j)/(2d^2)]$. For given $P(x)$, the subjective information function is $I(x; \theta_j) = \log[T(\theta_j|x)/T(\theta_j)]$. Then we can obtain the relationship between R, G, and d, as shown in Fig. 2. Figure 2 reveals that the matching point ($R = G$) changes with the discrimination, and too high resolution is unnecessary. For more results, see [4, 5].

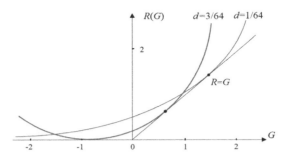

Fig. 2. Two $R(G)$ functions for different discrimination parameters d.

3.3 Multilabel Learning for Labels' Extensions or Truth Functions

We need to obtain truth functions, membership functions, or similarity functions from samples or sampling distributions in multilabel learning. Multilabel learning is difficult [12]; we can only use a pair of Logistic functions for two complementary labels' learning. However, with the P-T probability framework, multilabel learning is also easy. From the people ages' prior and posterior distributions $P(x)$, $P(x|$ "adult"), and $P(x|$ "elder") (see Fig. 3). Can we find the extensions or the truth functions of two labels?

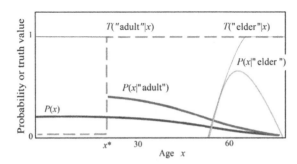

Fig. 3. Solving the truth functions of "adult" and "elder" using prior and posterior distributions. The human brain can guess T("adult" |x) and T("elder" |x).

Let **D** be a sample $\{(x(t), y(t))|t = 1 \text{ to } N; x(t) \in U; y(t) \in V\}$, where $(x(t), y(t))$ is an example. We can obtain sampling distribution $P(x|y_j)$ from **D**. According to Fisher's maximum likelihood estimation, the optimized $P(x|\theta_j)$ is $P^*(x|\theta_j) = P(x|y_j)$. According to the third type of Bayes' Theorem, we have the optimized truth functions:

$$T * (\theta_j|x) = [P * (x|\theta_j)/P(x)]/\max(P * (x|\theta_j)/P(x)) = [P(x|y_j)/P(x)]/\max(P(x|y_j)/P(x)) \quad (13)$$

We can use the above formula to solve two truth functions in Fig. 3. Further, we have.

$$T * (\theta_j|x) = P(y_j|x)/\max(P(y_j|x)), j = 1, 2, \ldots, n. \quad (14)$$

If samples are not big enough, we may use the generalized KL formula to obtain

$$T * (\theta_j|x) = \underset{T(\theta_j|x)}{\arg\max} \sum_i P(x_i|y_j) \log \frac{T(\theta_j|x_i)}{T(\theta_j)}. \quad (15)$$

We call the above method for the truth function Logical Bayesian Inference [3]. For classifications, we can use the maximum semantic information classifier:

$$y_j * = f(x) = \underset{y_j}{\arg\max} \log I(x; \theta_j) = \underset{y_j}{\arg\max} \log[T(\theta_j|x)/T(\theta_j)], \quad (16)$$

which is compatible with the maximum likelihood classifier.

3.4 The Channels Matching Algorithm for Maximum Mutual Information (MMI) Classifications

In Shannon's information theory, the distortion criterion instead of the information criterion is used to detect and classify. Without the classification, we cannot express mutual information, whereas we cannot optimize the classification without mutual information's expression. However, G theory can avoid this loop [3].

Assume that we classify every instance with unseen true label x according to its observed feature $z \in C$. That is to provide a classifier $y = f(z)$ to get a label y (see Fig. 4).

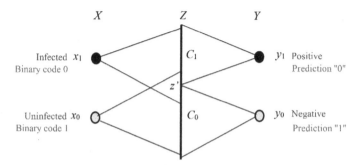

Fig. 4. Illustrating the medical test and the signal detection. We choose y_j according to $z \in C_j$.

Let C_j be a subset of C and $y_j = f(z|z \in C_j)$; hence $S = \{C_1, C_2, \ldots\}$ is a partition of C. Our aim is, for given $P(x, z)$, to find the optimized S:

$$S^* = \arg\max_S I(X; \theta|S) = \arg\max_S \sum_j \sum_i P(C_j)P(x_i|C_j) \log \frac{T(\theta_j|x_i)}{T(\theta_j)}. \tag{17}$$

Matching I: We obtain the Shannon channel for given S:

$$P(y_j|x) = \sum_{z_k \in C_j} P(z_k|x), \; j = 1, 2, \ldots, n \tag{18}$$

From this channel, we can obtain $T^*(\theta|x)$ (see Eq. (14)) and the semantic information $I(x_i; \theta_j)$. For given z, we have conditional information or reward functions:

$$I(X; \theta_j|z) = \sum_i P(x_i|z)I(x_i; \theta_j), \; j = 0, 1, \ldots, n. \tag{19}$$

Matching II: Let the Shannon channel match the semantic channel by the classifier:

$$y_j* = f(z) = \arg\max_{y_j} I(X_i; \theta_j|z), \; j = 0, 1, \ldots, n. \tag{20}$$

Repeat Matching I and II until S converges to S^*. Figure 5 shows an example.

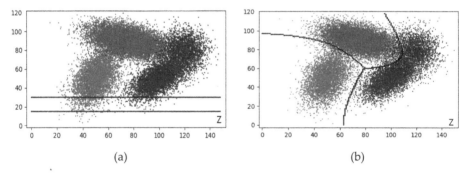

Fig. 5. The MMI classification. (a) Bad initial partition; (b) After two iterations.

After two iterations, the MMI is 1.0434 bits. The convergent MMI is 1.0435 bits. The result indicates that this algorithm is high-speed.

Using the $R(G)$ function, we can easily prove the above algorithm converges [3].

3.5 The Convergence Proof and the Improvement of the EM Algorithm for Mixture Models

If a probability distribution $P_\theta(x)$ comes from n likelihood functions' mixture, e. g.,

$$P_\theta(x) = \sum_{j=1}^n P(y_j)P(x|\theta_j), \tag{21}$$

then we call $P_\theta(x)$ a mixture model. If every predictive model $P(x|\theta_j)$ is a Gaussian function, then $P_\theta(x)$ is a Gaussian mixture model. Assume that sampling distribution $P(x)$ comes from the mixture of two true models with ratios $P^*(y_1)$ and $P^*(y_2) = 1 - P^*(y_1)$. That is $P(x) = P^*(y_1)P(x|\theta_1^*) + P^*(y_2)P(x|\theta_2^*)$. Our task is to find the true model parameters and mixture proportions θ_1^*, θ_2^*, and $P^*(y_1)$ from $P(x)$.

The EM algorithm includes two steps [13, 14]:

E-step: Write the conditional probability functions (e. g., the Shannon channel):

$$P(y_j|x) = P(y_j)P(x|\theta_j)/P_\theta(x), \quad P_\theta(x) = \sum_j P(y_j)P(x|\theta_j). \tag{22}$$

M-step: Improve $P(y)$ and θ to maximize the complete data log-likelihood:

$$Q = \sum_i \sum_j P(x_i)P(y_i|x_i) \log P(x_i, y_j|\theta)$$
$$= L_X(\theta) + \sum_i \sum_j P(x_i)P(y_j|x_i) \log P(y_j|x_i), \tag{23}$$

where $L_X(\theta) = \sum_i P(x_i)\log P_\theta(x_i)$ is the log-likelihood as the objective function. If Q cannot be improved further, then end iterations; otherwise, go to the E-step.

The M-step can be divided into two steps: the M1-step for optimizing $P(y)$ and the M2-step for maximizing semantic mutual information $I(X; \Theta)$ by letting $P(x|\theta_j) = P(x|y_j)$. The iterative method for the rate-distortion function reminds us that we can repeat Eq. (23) and $P^{+1}(y_j) = \sum_i P(x_i)P(y_j|x_i)$ until $P^{+1}(y_j) = P(y)$. The improved EM algorithm is called the Channels Matching EM (CM-EM) algorithm [3, 15], which repeats the M1-step many or several times so that $P^{+1}(y) \approx P(y)$.

Many researchers believe that Q and $L_X(\theta)$ are always positively correlated; we can achieve maximum $L_X(\theta)$ by maximizing Q. However, the author found that this is not true. We need reliable convergence proof to avoid blind improvements.

Using the semantic information method, we can derive [3]

$$H(P||P_\theta)=H(X) - H_\theta(X) = R - G + H(Y||Y^{+1}), \tag{24}$$

$$H(Y^{+1}||Y) = \sum_j P^{+1}(y_j) \log[P^{+1}(y_j)/P(y_j)]. \tag{25}$$

Using the variational and iterative methods that Shannon [9] and others used for analyzing the rate-distortion function $R(D)$, we can prove that both the E-step and the M1-step minimize $H(P||P_\theta)$ or maximize $L_X(\theta)$ [15].

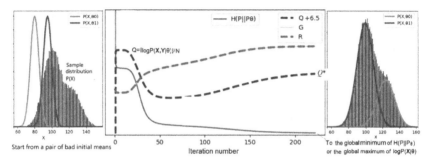

Fig. 6. Q and $H(P||P_\theta)$ change with iterations. The EM algorithm needs about 340 iterations, whereas the CM-EM algorithm needs about 240 iterations. The sample size is 50000.

Figure 6 shows an example used for the Deterministic Annealing EM (DAEM) algorithm in [14]. The true model is $(\mu_1^*, \mu_2^*, \sigma_1^*, \sigma_2^*, P^*(y_1)) = (80, 95, 5, 5, 0.5))$; the initial model is $(\mu_1, \mu_2, \sigma_1, \sigma_2, P^*(y_1)) = (125, 100, 10, 10, 0.7))$. This example indicates that Q does not always increase while $H(P||P_\theta)$ decreases or $L_X(\theta)$ increases; the CM-EM algorithm is better in this case. For the detailed convergence proof and the improvement, see [15].

3.6 To Resolve the Problem with Popper's Verisimilitude and Explain the RLS Criterion

To evaluate hypotheses, Popper [16] replace trueness with verisimilitude (or truthlikeness [17]). Researchers use two approaches to interpret verisimilitude: the content approach and the consequence approach [17]. The former emphasizes tests' severity, unlike the latter that emphasizes hypotheses' truth or closeness to truth. Some researchers think that the content approach and the likeness approach are irreconcilable.

The truth function $T(\theta_j|x)$ is also the confusion probability function; it reflects the likeness between x and x_j. The x_i (e. g., $X = x_i$) is the consequence, and the distance between x_i and x_j in the feature space reflects the likeness. The $\log[1/T(\theta_j)]$ represents the testing severity and potential information content. Using the formula for $I(x_i; \theta_j)$, we can easily explain an often-mentioned example: why "the sun has 9 satellites" (8 is true) has higher verisimilitude than "the sun has 100 satellites" [17].

Now, we can explain why the RLS criterion is getting popular. It is similar to the maximum mean verisimilitude criterion and the maximum semantic mutual information criterion.

3.7 To Provide Practical Confirmation Measures and Clarify the Raven Paradox

There have been many confirmation measures [7]. Researchers wish that 1) a confirmation measure can be used to evaluate tests and predictions like likelihood ratio; 2) it can be used to clarify the raven paradox.

We use the medical test, shown in Fig. 4, as an example to explain the two measures. Now x becomes h, and y becomes e. A major premise to be confirmed is "if e_1 then h_1"

(e.g., $e_1 \rightarrow h_1$). A confirmation measure is denoted by $c(e \rightarrow h)$. We can obtain four examples' numbers a, b, c, and d (see Table 1) to construct confirmation measures for a given classification. We wish that a confirmation measure $c(e_1 \rightarrow h_1)$ changes between -1 and 1 and possesses consequence symmetry: $c(e_1 \rightarrow h_1) = -c(e_1 \rightarrow h_1)$ [7].

Table 1. The numbers of four examples of confirmation measures.

	e_0 (negative)	e_1 (positive)
h_1 (infected)	b	a
h_0 (uninfected)	d	c

We regard the truth function of $e_1(h)$ as a believable part plus an unbelievable part, as shown in Fig. 7.

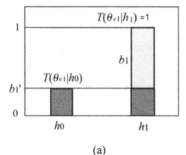 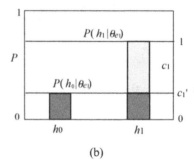

(a)   (b)

Fig. 7. The truth function (a) or the likelihood function (b) can be regarded as a believable part plus an unbelievable part.

The optimized b_1 is the degree of confirmation of major premise $e_1 \rightarrow h1$ [7]. It is

$$b_1* = b * (e_1 \rightarrow h_1) = \frac{P(e_1|h_1) - P(e_1|h_0)}{\max(P(e_1|h_1), P(e_1|h_0))} = \frac{ad - bc}{\max(a(c+d), c(a+b))} = \frac{LR^+ - 1}{\max(LR^+, 1)} \quad (26)$$

where LR^+ is the positive likelihood ratio. Since b_1* reflects how well the test serves as a means or a channel, we call $b*(e \rightarrow h)$ the channel confirmation measure, which is compatible with the likelihood ratio measure.

Suppose that likelihood function $P(h|\theta_{e1})$ includes a believable part and an unbelievable part, as shown in Fig. 7 (b), we derive the prediction confirmation measure

$$c_1* = c * (e_1 \rightarrow h_1) = \frac{P(h_1, e_1) - P(h_0, e_1)}{\max(P(h_1, e_1), P(h_0, e_1))} = \frac{a - c}{\max(a, c)} \quad (27)$$

which indicates how well the test serves as a prediction.

We can use measure $c*$ to clarify the Raven Paradox. Hemple [18] proposed the Raven Paradox. According to the Equivalence Condition (EC) in the classical logic, "if

x is a raven, then x is black" (Rule I) is equivalent to "if x is not black, then x is not a raven" (Rule II). A piece of white chalk supports Rule II; hence it also supports Rule I. However, according to the Nicod Criterion (NC), a black raven supports Rule I, a non-black raven undermines Rule I, and a non-raven thing, such as a black cat or a piece of white chalk, is irrelevant to Rule I. Hence, there exists a paradox between EC and NC.

Among all confirmation measures [7], only measure c^* supports the NC and objects the EC because c^* is only affected by a and c. For example, when $a = 6$, $c = 1$ and $b = d = 10$, $c^*(e_1 \rightarrow h1) = (6 - 1)/6 = 5/6$. The author has demonstrated that except for c^*, all popular confirmation measures cannot explain that a black raven can confirm "ravens are black" better than a piece of white chalk [7].

4 Summary

This paper has shown that using the P-T probability framework can obtain truth function or membership functions from sampling distributions and connect statistics and logic (fuzzy logic). It has been explained that the semantic information criterion is the verisimilitude criterion and similar to the RLS criterion. Using the semantic information criterion instead of the distortion criterion, we can solve the MMI classification better.

This paper has introduced how to apply the P-T probability framework and the G theory to semantic communication, machine learning, and philosophy of science for completing some challenging tasks. These applications reveal that the P-T probability framework and the G theory have great potential. We should be able to find more applications. We need further studies for combining the P-T probability framework and the G theory with neural networks for machine learning.

References

1. Shannon, C.E., Weaver, W.: The Mathematical Theory of Communication. The University of Illinois Press, Urbana (1949)
2. Bar-Hillel, Y., Carnap, R.: An outline of a theory of semantic information. Technical Report No. 247, Research Lab. of Electronics. MIT (1952)
3. Lu, C.: Semantic information G theory and logical Bayesian inference for machine learning. Information **10**(8), 261 (2019)
4. Lu, C.: A Generalized Information Theory. China Science and Technology University Press, Hefei (1993)
5. Lu, C.: A generalization of Shannon's information theory. Int. J. Gen. Syst. **28**(6), 453–490 (1999)
6. Lu, C.: The P-T probability framework for semantic communication, falsification, confirmation, and Bayesian reasoning. Philosophies **5**(4), 25 (2020)
7. Lu, C.: Channels' confirmation and predictions' confirmation: from the medical test to the raven paradox. Entropy **22**(4), 384 (2020)
8. Davidson, D.: Truth and meaning. Synthese **17**(3), 304–323 (1967)
9. Shannon, C.E.: Coding theorems for a discrete source with a fidelity criterion. IRE Nat. Conv. Rec. **4**, 142–163 (1959)
10. Ackley, D.H., Hinton, G.E., Sejnowski, T.J.: A learning algorithm for Boltzmann machines. Cognit. Sci. **9**(1), 147–169 (1985)

11. Li, Q., Chen, Y.: Rate distortion via restricted Boltzmann machines. In: 56th Annual Allerton Conference on Communication, Control, and Computing, Monticello, pp. 1052–1059 (2018)
12. Zhang, M.L., Zhou, Z.H.: A review on multilabel learning algorithm. IEEE Trans. Knowl. Data Eng. **26**(8), 1819–1837 (2014)
13. Dempster, A.P., Laird, N.M., Rubin, D.B.: Maximum likelihood from incomplete data via the EM algorithm. J. Roy. Stat. Soc. B **39**(1), 1–38 (1997)
14. Ueda, N., Nakano, R.: Deterministic annealing variant of the EM algorithm G. In: Tesauro, et al. (eds.) Advances in NIPS 7, pp. 545–552. MIT Press, Cambridge, MA (1995)
15. Lu, C.: From the EM Algorithm to the CM-EM Algorithm for Global Convergence of Mixture Models. https://arxiv.org/abs/1810.11227. Accessed 12 Oct 2020
16. Popper, K.: Conjectures and Refutations. Repr. Routledge, London and New York (1963/2005)
17. Oddie, G.: Truthlikeness, the Stanford Encyclopedia of Philosophy. https://plato.stanford.edu/entries/truthlikeness/. Accessed 12 Jan 2020
18. Hempel, C.G.: Studies in the Logic of Confirmation. Mind, **54**, 1–26, 97–121 (1945)

Trilevel Multi-criteria Decision Analysis Based on Three-Way Decision

Chengjun Shi[✉] and Yiyu Yao[✉]

Department of Computer Science, University of Regina,
Regina, SK S4S 0A2, Canada
{csn838,Yiyu.Yao}@uregina.ca

Abstract. Based on the principles of three-way decision as thinking in threes, this paper proposes a framework of trilevel multi-criteria decision analysis (3LMCDA). The main motivations and contributions are a new understanding of multi-criteria decision analysis at three levels: 1) single-criterion-level analysis with respect to individual criteria, 2) multi-criteria-level aggregation by pooling results from many criteria, and 3) multi-methods-level ensemble of different multi-criteria decision analysis methods. A full understanding of multi-criteria decision analysis requires understandings at the three levels. While the existing studies have been focused on the first two levels, new efforts at the third level may enable us to take advantages of many different multi-criteria decision analysis methods.

Keywords: Three-way decision · Multi-criteria decision-making · Trilevel thinking · Trilevel analysis

1 Introduction

Multiple criteria decision making (MCDM), or multiple criteria decision analysis (MCDA), deals with the evaluation and comparison of a set of decision alternatives with respect to a set of multiple, possibly conflicting, criteria. A basic task is to sort or rank the set of alternatives according to the set of criteria [7,9,16,17]. There are extensive studies on models and methods of MCDM, for example, Weighted Sum Model (WSM) [5], Weighted Product Model (WPM) [3,12], Analytic Hierarchy Process (AHP) [15], Technique for Order of Preference by Similarity to Ideal Solution (TOPSIS) [7], ELimination Et Choix Traduisant la REalité (ELECTRE) [2], and many more. A review of these studies brings our attention to two issues, which motivates the present study.

First, some studies (e.g., ELECTRE) start with constructing a set of rankings according to the evaluations of individual attributes and then build an aggregate ranking based on the set of rankings. On the other hand, the majority of studies (e.g., TOPSIS) directly construct a ranking based on an aggregate evaluation of the evaluations of the set of criteria. A lack of consideration of

© IFIP International Federation for Information Processing 2021
Published by Springer Nature Switzerland AG 2021
Z. Shi et al. (Eds.): ICIS 2020, IFIP AICT 623, pp. 115–124, 2021.
https://doi.org/10.1007/978-3-030-74826-5_10

rankings according to individual criteria perhaps needs a closer examination, as the individual rankings reveal some kinds of local properties that may be useful in MCDA. Second, many methods have been proposed and existing methods are continuously being refined in order to avoid some of their difficulties. Although there is no general agreement on the existence of the best or the most superior method, research efforts are constantly being made in search of such a method. It might be the time to think in a different direction. Instead of searching for new methods, we may search for ways to combine existing methods. Since each method produces one ranking, a set of methods produces a set of rankings. The set of rankings may be aggregated into one ranking, in the same way that the set of rankings obtained from individual criteria can be aggregated into one ranking.

The main objective of this paper is to propose a framework of MCDA to address the two issues. For this purpose, we make use of the principles and ideas of a theory of three-way decision (3WD). The theory of 3WD, proposed by Yao [20, 21, 23], is about thinking, working, and processing in threes. That is, we divide a whole into three parts and understand the whole by studying the three parts, as well as their interactions and dependencies. Trilevel thinking is a special mode of three-way decision [22], involving interpretations and understandings of a problem at three levels. To apply principles of trilevel thinking, we identify three levels in MCDA: 1) single-criterion analysis, 2) multi-criteria analysis, and 3) multi-methods analysis. This enables us to build a framework of trilevel multi-criteria decision analysis (3LMCDA), based on the results from three-way decision in general and trilevel analysis in specific.

There are several studies on combining three-way decision and MCDM. Jia and Liu [8] introduced a three-way decision model for solving MCDM problems. Based on the notions of loss functions, they provided algorithms for constructing three-way decision rules. Liu et al. [10] proposed a new three-way decision model with intuitionistic fuzzy numbers to solve MCDM problems, in which they used a different kind of loss functions. Liu and Wang [11] suggested a three-way decision model with intuitionistic uncertain linguistic variables to solve multi-attribute group decision-making problem, which takes the advantages of intuitionistic fuzzy sets and uncertain linguistic variables. Zhan et al. [25] applied three-way decision into outranking approach and built a new three-way multi-attribute decision making model. Wang et al. [19] combined three-way decision and MCDM in a hesitant fuzzy information environment. Cui et al. [4] discussed a method for deriving the attribute importance based on a combination of three-way decision and AHP method. In this paper, we take a different direction of research for applying three-way decision to MCDA by proposing a framework of trilevel MCDA (i.e., 3LMCDA). Instead of focusing on a particular method, we look at a conceptual model that enables us to unify existing studies in a common setting. In particular, we look at rankings generalized at three levels. The single-criterion level focuses on individual criteria locally, the multi-criteria level considers a set of criteria globally but from the angle of a particular MCDA method, the multi-methods level looks at MCDA from multiple angles given by different MCDA methods. The trilevel framework provides a much richer understanding than any of the three levels.

This paper is organized as follows. In Sect. 2, we review the basic concepts of three-way decision and trilevel thinking. In Sect. 3, we discuss a framework of trilevel multiple criteria decision analysis (3LMCDA). We briefly explain, through examples, some basic issues and tasks at each level. In Sect. 4, we summarize the key points of the new framework and point out future research topics.

2   Three-Way Decision and Trilevel Analysis

The fundamental philosophy of three-way decision is thinking in threes or triadic thinking [20–22]. A triad of three things is a basic pattern and structure used for understanding and problem solving [24]. Thinking in threes appears everywhere. It is flexible and widely applicable. We look at a problem based on its three components; we search for a full understanding of an information-processing system through three levels of understanding (i.e., theory, algorithms, and implementation); we build an explanation from three perspectives. By considering three things, we avoid the over-simplification of thinking in twos and, at the same time, the complication of thinking in fours or more.

Yao [21] proposed a trisecting-acting-outcome (TAO) model of three-way decision. The basic idea is to divide a whole into three parts and to approach the whole through the three parts. There are three related components of the TAO model. Trisecting is to construct a trisection from a whole, acting is to devise and apply strategies on the three parts of the trisection, and outcome is a desirable result from a combination of trisecting and acting. A useful class of trisections is the type of trisections consisting of three levels. It gives rise to thinking in three levels or trilevel thinking [22]. A triad of three levels normally consists of a top level, a middle level, and a bottom level. There are three basic tasks of trilevel thinking: (i) interpretation of the three levels, (ii) relationships between the three levels, and (iii) analysis to be done at each of the three levels, as well as at two adjacent levels and, simultaneously, at all three levels. Concrete examples of trilevel structures include three levels of abstraction, three levels of scope and scale, and three levels of complexity. In a trilevel abstraction, the top level is the most general level, the bottom level is the most detailed and specific one, and the middle level is between the two. Levels of scope and scale determine the different types of research questions. A lower level is at a narrower scope and scale and may focus on localized issue. A higher level deals with problems from a wider scope and scale and may focus on some global issues. Levels of complexity aims to decompose a problem into sub-problems with decreasing complexity. The solutions to or the results of a simpler sub-problem become supporting materials to the next higher level.

Typically, in a trilevel structure, a higher level guides and determines its lower level, and a lower level supports its higher level. There are three possible modes of processing of a trilevel, namely, top-down, bottom-up, and middle-out [22]. From the bottom to the top, details are extracted into abstract information. Redundant or less relevant information is removed. The problems become more general and more abstract. The results obtained at a lower level provide hints and support for the work at a higher level. In the top-down direction, the procedure is reversed. Goals or tasks at a higher level control the concrete ideas at a lower level. In the middle-out mode, the middle level is guided by the top and is supported by the bottom at the same time. That is, the middle level is developed in the context of both the top and bottom levels.

The support-and-control relationships of the levels are useful for learning, understanding, and processing. Trilevel thinking and analysis are generally applicable to a wide range of domains. For any problem, once we can identify and separate three levels, we can immediately apply the results of trilevel analysis. In the rest of the paper, we will demonstrate this method by considering the problem of multi-criteria decision analysis.

3   A Trilevel Framework of Multi-criteria Decision Making

We apply a trilevel thinking model to the problem of MCDM and construct a novel trilevel analysis framework in this section. The following three levels are reasonable to be considered: the bottom single-criterion level, the middle multi-criteria level, and the top multi-methods level. We start from the basic notions of MCDM problem, analyze it from three levels by reviewing existing studies, and discuss the significance and prospects of each level.

3.1   Trilevel Thinking in Multi-criteria Decision Analysis

A multi-criteria decision problem is a complex decision analysis problem. The main task is to find the optimal decision among a set of decision alternatives, or to rank the alternatives, by comparing their performances under a multiple criteria environment.

Definition 1. *A multi-criteria decision table is a triplet* $MCDT = \{A, C, p\}$, *where* A *is a finite and non-empty set of* n *alternatives* $A = \{a_1, a_2, \ldots, a_n\}$, C *is a finite and non-empty set of* m *criteria* $C = \{c_1, c_2, \ldots, c_m\}$, *and* p *is a real-valued mapping function, i.e.,* $p : A \times C \to \mathcal{R}$.

The performance function p assigns a real number to each alternative on every criterion. For example, $a_i \in A$, $p_j(a_i)$ is a real number and indicates a_i's performance on criterion c_j. An assumption is made that a larger real number reflects a better performance.

A trilevel framework (3LMCDA) is built as depicted in Fig. 1. From the bottom to the top, the three levels are explained as follows:

(i) The single-criterion level ranks alternatives on a single criterion. It provides detailed information of alternatives on every single criterion. The local ranking of alternatives is important at this level. Existing studies at this level include outranking relations, utility-based preference, etc.

(ii) The multi-criteria level ranks alternatives on a set of multiple criteria. It considers globally the ranking or the ordering of alternatives according to their total performance. Relevant methods are weighted models, AHP, TOPSIS, ELECTRE, and many more.

(iii) The multi-methods level ranks alternatives from a set of multiple MCDM methods. It considers strengths and drawbacks of existing approaches at the middle level. The ultimate goal is to construct a ranking based on an ensemble of results from multiple MCDM methods.

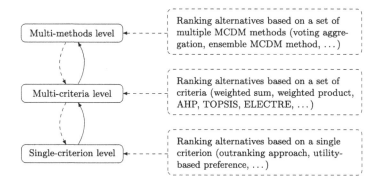

Fig. 1. Trilevel framework of MCDA (3LMCDA)

The connections between levels are clearly shown in Fig. 1. The bottom-up mode is denoted by the solid lines with arrows. The top-down mode is denoted by the dashed lines with arrows.

We use an example to explain the three levels in a bottom-up mode. Suppose that we have two alternatives $a_1, a_2 \in A$ and a set of criteria $C = \{c_1, c_2, c_3\}$, the problem is to rank them on the three criteria. A more specific sub-task is to compare them on each criterion individually. This is what the bottom single-criterion level focuses on. However, the results might not be consistent on the three criteria. A possible case is that a_1 is better than a_2 on $\{c_1, c_2\}$ but worse on c_3. A middle multi-criteria level serves to solve the conflicts by integrating results from the bottom level. This is usually done by a particular MCDM method. For example, ELECTRE is such a method that calculates several indices based on the comparison results at the single-criterion level. The role of the top multi-methods level is to rank based on an ensemble of results from a set of MCDM methods, taking into considerations of advantages and disadvantages of each method. The rankings produced by different MCDM methods are normally inconsistent. There is no agreement on which method produces the best ranking. Thus, multiple MCDM methods provide various angles to solve the ranking problem. To take advantages from a set of methods, an ensemble approach is needed.

The structure of the proposed trilevel analysis framework (3LMCDA) is consistent to the interpretations of trilevel thinking. It reflects three levels of abstraction, three levels of scope and scale, and three levels of complexity. In the following sections, we briefly outline several approaches with respect to the trilevel model.

3.2 Single-Criterion Level

The single-criterion level concentrates on ranking alternatives' performances locally based on one criterion. In the definition of a multi-criteria decision table, the assumption is that alternatives' performances are described in a numerical value. Given two alternatives $a_1, a_2 \in A$, there are three basic results from a comparison of $p_j(a_1)$ and $p_j(a_2)$: less than ($<$), equal to ($=$), and greater than ($>$).

The ranking of a_1 and a_2 is decided by the following rules. If $p_j(a_1) < p_j(a_2)$, then we say that a_1 is worse than a_2, denoted by $a_1 \prec_{c_j} a_2$; if $p_j(a_1) = p_j(a_2)$, then a_1 is considered as good as a_2 with respect to c_j, denoted by $a_1 \sim_{c_j} a_2$; if $p_j(a_1) > p_j(a_2)$, then we believe that a_1 performs better than a_2 on c_j, denoted by $a_1 \succ_{c_j} a_2$. For a better representation, we consider comparing $(p_j(a_1) - p_j(a_2))$ with 0 and the decision rules are given by:

$$\begin{cases} a_1 \prec_{c_j} a_2 \iff (p_j(a_1) - p_j(a_2)) < 0, \\ a_1 \sim_{c_j} a_2 \iff (p_j(a_1) - p_j(a_2)) = 0, \\ a_1 \succ_{c_j} a_2 \iff (p_j(a_1) - p_j(a_2)) > 0. \end{cases} \tag{1}$$

In a decision-theoretic setting, it is possible to interpret the values $p_j(a_i)$ as utilities. The relations may be viewed as preference relations and indifference relations [6].

Roy proposed an outranking approach [14] to decide the relationship between two alternatives at the single-criterion level. The meaning of outranking is "at least as good as" or "not worse than". Roy's approach is built by introducing two thresholds and four binary relations. The outranking relation can be considered as an union of other three relations, namely, indifferent with, strictly preferred to, and hesitation between the former two.

Definition 2. *Given a multi-criteria decision table, $c_j \in C$ is a criterion, $a_1, a_2 \in A$ are two alternatives. An outranking approach introduces two thresholds α_j and β_j ($0 < \alpha_j < \beta_j$). Four binary relations are defined as follows:*

Outranking relation S: $a_1 S_{c_j} a_2 \iff (p_j(a_1) - p_j(a_2)) \geq -\alpha_j$,
Indifference relation I: $a_1 I_{c_j} a_2 \iff |(p_j(a_1) - p_j(a_2))| \leq \alpha_j$,
Strict preference relation P: $a_1 P_{c_j} a_2 \iff (p_j(a_1) - p_j(a_2)) > \beta_j$,
Hesitation relation Q: $a_1 Q_{c_j} a_2 \iff \alpha_j < (p_j(a_1) - p_j(a_2)) < \beta_j$.

The binary relations in Definition 2 are defined based on one specific criterion and rely on the alternatives' values on the criterion. The outranking relation is equivalent to a union of the other three relations. That is, if $a_1 S_{c_j} a_2$ holds, then one and only one of the three statements $a_1 I_{c_j} a_2$, $a_1 P_{c_j} a_2$, or $a_1 Q_{c_j} a_2$ must hold.

The results of the single-criterion level will be aggregated and combined at the middle level.

3.3 Multi-criteria Level

The middle level sets up a task to aggregate results from the bottom level and generate a global ranking of alternatives according to a set of criteria. The individual comparisons on each criterion at the bottom level make contribution to the middle level.

The weighted sum model is a classical way to synthesize information from a set of criteria. It considers the importance of criteria by introducing a weight vector $W = \{w_1, w_2, \ldots, w_m\}$. The vector W denotes the relative importance of the criteria set. The following requirements should be satisfied: (i) there is no negative importance: $\forall w_j \in W$, $w_j \geq 0$, (ii) the summation of all weights equals to 1: $\sum_{j=1}^{m} w_j = 1$. To compare two alternatives a_1, a_2, the difference is calculated by:

$$\Delta^{\text{WSM}}(a_1, a_2) = \sum_{j=1}^{m} w_j \times (p_j(a_1) - p_j(a_2)). \tag{2}$$

In Eq. (2), $(p_j(a_1) - p_j(a_2))$ calculates the difference value between a_1 and a_2 on c_j, w_j is used to produce a weighted difference value. All of the weighted differences are added together. We have the following decision rules at the multi-criteria level:

$$\begin{cases} a_1 \prec_C a_2 \iff \Delta^{\text{WSM}}(a_1, a_2) < 0, \\ a_1 \sim_C a_2 \iff \Delta^{\text{WSM}}(a_1, a_2) = 0, \\ a_1 \succ_C a_2 \iff \Delta^{\text{WSM}}(a_1, a_2) > 0. \end{cases} \tag{3}$$

In general, we can also replace 0 by setting up a pair of thresholds.

We consider the ELECTRE method, as an example, to illustrate an aggregation method at the middle. ELECTRE calculates a concordance index and a discordance index to decide the ranking of a_1 and a_2 at the multi-criteria level. The concordance index measures how sufficiently the set of criteria supports "a_1 outranks a_2", denoted by $con(a_1 S a_2)$, while the discordance index is determined by a specific criterion indicates the degree of opposition to "a_1 outranks a_2", denoted by $disc(a_1 S a_2)$. When concordance degree is high enough and discordance degree is low enough at the same time, we say that the assertion $a_1 S a_2$ is validated.

In ELECTRE I, the concordance index is computed by the weights associated with the criteria on which a_1 outranks a_2 out of the total weights:

$$con(a_1 S a_2) = \frac{1}{\sum_{j=1}^{m} w_j} \sum_{j:a_1 S_{c_j} a_2} w_j. \tag{4}$$

If the weight vector W satisfies $\sum_{j=1}^{m} w_j = 1$, then Eq. (4) is equivalent to $con(a_1 S a_2) = \sum_{j:a_1 S_{c_j} a_2} w_j$. The discordance index $disc(a_1 S a_2)$ is defined as:

$$disc(a_1 S a_2) = \max_{j=1,2,\ldots,m} \{p_j(a_2) - p_j(a_1)\}. \tag{5}$$

It takes the greatest one of the difference values $(p_j(a_2) - p_j(a_1))$ on all c_j. The discordance index indicates the utmost degree of opposition.

To test the concordance degree and discordance degree, two thresholds δ and v are introduced. The threshold δ is a given concordance level and usually satisfies $\delta \in (0.5, 1]$. If $con(a_1 S a_2) \geq \delta$, we say that it passes the concordance test. Similarly, v is a given veto threshold. If $disc(a_1 S a_2) \leq v$, it passes the non-discordance test, because the discordant criterion is not powerful enough to override $a_1 S a_2$. The assertion $a_1 S a_2$ is valid if and only if both of the tests have been passed.

3.4   Multi-methods Level

Many MCDA methods have been proposed and applied in various domains and have been verified to be practically useful and effective. At the same time, most of them put their efforts at the middle level. Several studies [1, 16–18] reviewed the main ideas of common methods and summarized their relative advantages and disadvantages. These MCDA methods rank the set of alternatives in numerous and different ways. This may impose a difficulty for a decision maker when selecting a particular method among them.

At the top level, the idea of ensemble is a process of learning from a set of multiple methods and improving the model quality by considering multiple points of view. We simply discuss one of the possible ways. Given two alternatives a_1, a_2, we consider $a_1 \succ a_2$ to be a hypothesis and treat a MCDA method as an agent who votes on the hypothesis. A three-way voting decision is defined as follows.

Definition 3. *Given a multi-criteria decision table, a finite set of MCDM methods is denoted by T. In order to determine whether a_1 is ranked ahead of a_2, a hypothesis is written as $h : a_1 \succ a_2$. Each method $t_k \in T$ makes one and only one of the three voting decisions according to the rules:*

Support: t_k admits that $a_1 \succ a_2$,
Neutral: t_k neither admits $a_1 \succ a_2$ nor admits $a_1 \prec a_2$,
Oppose: t_k admits $a_1 \prec a_2$.

Aggregating the voting results from the set of methods may help us rank the given pairs of alternatives.

Mohammadi and Rezaei [13] considered the ensemble problem from another angle. They proposed a novel approach to producing an ensemble ranking by modeling the problem of aggregating rankings as a new multi-criteria decision-making. In their approach, each method is treated as a criterion and the weight of it is determined by the half-quadratic theory.

4   Conclusion and Future Work

Trilevel thinking, namely, a special mode of three-way decision, is a powerful idea for intelligent data analysis. Inspired by the principles of trilevel thinking, we have proposed a framework of trilevel multi-criteria decision analysis

(3LMCDA). The 3LMCDA consists of three levels: a single-criterion level, a multi-criteria level, and a multi-methods level. Each of the three levels considers a different type of questions and contributes uniquely to MCDA. The single-criterion level analysis ranks alternatives based on individual criteria, the multi-criteria level aggregates the set of rankings from individual criteria into one ranking (which is typically given by one MCDA), and multi-methods level is an ensemble of multiple different MCDA methods. As we move up, we gradually take a more global view, namely, single criterion, multiple criteria and single method, and multiple criteria and multiple methods. This new view, offered by 3LMCDA, may be worthy of further research efforts.

The conceptual 3LMCDA model offers many possible future research opportunities. By following the same idea, it is possible to look at other trilevel models. With the 3LMCDA model, one may unify results from existing studies. It may also be necessary to examine detailed interpretations of MCDA methods and experimental implementations within the 3LMCDA model, in order to demonstrate its practical values. New methods for combining three-way decision and multi-criteria decision-making may be investigated in light of 3LMCDA.

Acknowledgement. We would like to thank Professor Mihir Chakraborty and Professor Zhongzhi Shi for encouraging us to write this paper. We thank reviewers for their constructive and valuable comments. This work was partially supported by an NSERC (Canada) Discovery Grant.

References

1. Aruldoss, M., Lakshmi, T.M., Venkatesan, V.P.: A survey on multi criteria decision making methods and its applications. Am. J. Inf. Syst. **1**, 31–43 (2013)
2. Benayoun, R., Roy, B., Sussman, B.: ELECTRE: Une méthode pour guider le choix en présence de points de vue multiples. Note de travail 49, SEMA-METRA International, Direction Scientifique (1966)
3. Bridgman, P.W.: Dimensional Analysis. Yale University Press, New Heaven (1922)
4. Cui, X., Yao, J.T., Yao, Y.: Modeling use-oriented attribute importance with the three-way decision theory. In: Bello, R., Miao, D., Falcon, R., Nakata, M., Rosete, A., Ciucci, D. (eds.) IJCRS 2020. LNCS (LNAI), vol. 12179, pp. 122–136. Springer, Cham (2020). https://doi.org/10.1007/978-3-030-52705-1_9
5. Fishburn, P.C.: Additive utilities with incomplete product sets: Applications to priorities and assignments. Oper. Res. Soc. Am. **15**, 537–542 (1967)
6. Fishburn, P.C.: Nonlinear Preference and Utility Theory. Johns Hopkins University Press, Baltimore (1988)
7. Hwang, C.L., Yoon, K.: Methods for multiple attribute decision making. In: Hwang, C.L., Yoon, K. (eds.) Multiple Attribute Decision Making. LNEMS, vol. 186, pp. 58–191. Springer, Heidelberg (1981). https://doi.org/10.1007/978-3-642-48318-9_3
8. Jia, F., Liu, P.D.: A novel three-way decision model under multiple-criteria environment. Inf. Sci. **471**, 29–51 (2019)
9. Kaliszewski, I., Miroforidis, J., Podkopaev, D.: Interactive multiple criteria decision making based on preference driven evolutionary multiobjective optimization with controllable accuracy. Eur. J. Oper. Res. **216**, 188–199 (2012)

10. Liu, P.D., Wang, Y.M., Jia, F., Fujita, H.: A multiple attribute decision making three-way model for intuitionistic fuzzy numbers. Int. J. Approximate Reasoning **119**, 177–203 (2020)
11. Liu, P.D., Wang, H.Y.: Three-way decisions with intuitionistic uncertain linguistic decision-theoretic rough sets based on generalized Maclaurin symmetric mean operators. Int. J. Fuzzy Syst. **22**, 653–667 (2020)
12. Miller, D.W., Starr, M.K.: Executive Decisions and Operations Research. Prentic-Hall, Englewood Cliffs (1969)
13. Mohammadi, M., Rezaei, J.: Ensemble ranking: aggregation of rankings produced by different multi-criteria decision-making methods. Omega **96**, 102254 (2020)
14. Roy, B.: The outranking approach and the foundations of ELECTRE methods. In: Bana e Costa, C.A. (ed.) Readings in Multiple Criteria Decision Aid, pp. 155–183. Springer, Heidelberg (1990). https://doi.org/10.1007/978-3-642-75935-2_8
15. Saaty, T.L.: What is the analytic hierarchy process? In: Mitra, G., Greenberg, H.J., Lootsma, F.A., Rijkaert, M.J., Zimmermann, H.J. (eds.) Mathematical Models for Decision Support. NATO ASI Series (Series F: Computers and Systems Sciences), vol. 48, pp. 109–121. Springer, Heidelberg (1988). https://doi.org/10.1007/978-3-642-83555-1_5
16. Triantphyllou, E.: Multi-criteria Decision Making Methods: A Comparative Study. Springer, Boston (2000)
17. Umm-e-Habiba, Asghar, S.: A survey on multi-criteria decision making approaches. In: International Conference on Emerging Technologies, ICET, pp. 321–325 (2009)
18. Velasquez, M., Hester, P.T.: An analysis of multi-criteria decision making methods. Int. J. Oper. Res. **10**, 56–66 (2013)
19. Wang, J.J., Ma, X.L., Xu, Z.S., Zhan, J.M.: Three-way multi-attribute decision making under hesitant fuzzy environments. Inf. Sci. **552**, 328–351 (2021)
20. Yao, Y.Y.: Three-way decisions and cognitive computing. Cogn. Comput. **8**, 543–554 (2016)
21. Yao, Y.Y.: Three-way decision and granular computing. Int. J. Approximate Reasoning **103**, 107–123 (2018)
22. Yao, Y.: Tri-level thinking: models of three-way decision. Int. J. Mach. Learn. Cybern. **11**(5), 947–959 (2019). https://doi.org/10.1007/s13042-019-01040-2
23. Yao, Y.Y.: Set-theoretic models of three-way decision. Granular Comput. **6**, 133–148 (2020)
24. Yao, Y.: The geometry of three-way decision. Appl. Intell. (2021). https://doi.org/10.1007/s10489-020-02142-z
25. Zhan, J.M., Jiang, H.B., Yao, Y.Y.: Three-way multi-attribute decision-making based on outranking relations. IEEE Trans. Fuzzy Syst. (2020). https://doi.org/10.1109/TFUZZ.2020.3007423

Recurrent Self-evolving Takagi–Sugeno–Kan Fuzzy Neural Network (RST-FNN) Based Type-2 Diabetic Modeling

Quah En Zhe[1], Arif Ahmed Sekh[2(✉)], Chai Quek[1], and Dilip K. Prasad[2]

[1] Nanyang Technological University, Singapore, Singapore
ashcquek@ntu.edu.sg
[2] UiT The Arctic University of Norway, Tromsø, Norway
{arif.ahmed.sekh,dilip.prasad}@uit.no

Abstract. Diabetes mellitus affected an estimated 463 million people in the year 2019. The number of diabetic patients is projected to increase to an alarming figure of 700 million by the year 2045, out of which 90–95% of them are expected to be type 2 diabetes mellitus (T2DM) patients. The research presented an alternative way of state-of-the-art insulin therapy using manual insulin infusion. The T2DM model that simulates the body reaction of a T2DM patient has been developed using real human clinical data that uses insulin pump therapy. The proposed system uses a closed-loop control together with fuzzy gain scheduling and recurrent self-evolving Takagi–Sugeno–Kang fuzzy neural network (RST-FNN). Such a system will help the patient remove the need for manual insulin infusion. This proposed system will record the blood glucose level and predict the next iteration's blood glucose level. The change in blood glucose level will help detect the food intake (carbohydrates) with reference to the gain scheduler and the controller will communicate with the insulin pump to infuse the corresponding amount of insulin.

Keywords: Type-2 diabetes · Insulin therapy automation · Diabetes modeling

1 Introduction

Food that is consumed will be broken down into glucose, which makes its way into the bloodstream and is converted into energy to support cell function and growth [1]. For glucose to be transported into the cells, insulin, which is automatically produced by the pancreas, must be present. Thus, insufficient amount or poor quality insulin will result in the cells not getting the essential energy needed to function properly.

In a healthy person, the pancreas releases sufficient glucagon or insulin to regulate low blood glucose and high blood glucose respectively, thus maintaining

© IFIP International Federation for Information Processing 2021
Published by Springer Nature Switzerland AG 2021
Z. Shi et al. (Eds.): ICIS 2020, IFIP AICT 623, pp. 125–136, 2021.
https://doi.org/10.1007/978-3-030-74826-5_11

normoglycemia - a state of normal blood glucose level. When blood glucose level remains low over a long period of time, hypoglycemia occurs [2]. On the contrary, when blood glucose level remains high over a long period of time, hyperglycemia occurs. Either condition can put the patient in great danger and ultimately death, if not treated in time [3,4]. Figure 1 demonstrates how this balance is achieved.

Diabetes mellitus, or more commonly known as just diabetes, is the inability of the body to produce sufficient insulin to help glucose enter the body's cells, resulting in high glucose levels in the bloodstream that cannot be utilized by the cells. In Type-1 Diabetes Mellitus patients (T1DM), this is caused by the pancreas' inability to produce insulin due to $\beta - cell$ destruction. In Type-2 Diabetes Mellitus (T2DM) patients, the

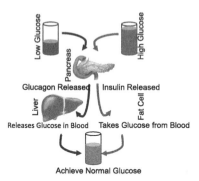

Fig. 1. Glucose balance procedure.

insulin produced by the pancreas is insufficient or of poor quality, and this is also known as insulin resistance [5]. T1DM affects approximately 5–10% of the diabetes patients and T2DM affects approximately 90–95% [6]. Apart from the direct risk diabetes patients face, they are exposed to increased risks of other diseases or ailments indirectly caused by the effects of diabetes [7]. Patients have a higher propensity of getting stroke, renal (kidney) failure, visual impairment and blindness, tuberculosis or the need to have their lower limbs amputated [8]. One of the most common and effective ways of treating diabetes is through insulin therapy. Insulin therapy consists of two different modes, subcutaneous injection and insulin pump. Insulin pump therapy has been gaining popularity as it helps patients to significantly improve on their glycemic control [9]. To further improve the lives of diabetic patients, automating the insulin pump to behave like an artificial pancreas by releasing the correct amount of insulin without the need for human intervention would be a major milestone. This would reduce or even remove the risk of poor carbohydrate estimation that often results in hypoglycemia or hyperglycemia in current patients.

The research motivations and objectives of the article are: (**i**) to design a computational model for T2DM that can mimic actual blood glucose concentration levels based on a real T2DM patient's data, (**ii**) design a neural network based system to predict blood glucose level, and (**iii**) design an automated system for insulin pump therapy.

To archive the research objectives, we have designed a computational model for T2DM that can mimic actual blood glucose concentration levels based on a real T2DM patient's data. The area of focus for the T2DM model would be the inclusion of the insulin profile that is used by the T2DM patient and finding the optimal constant glucose uptake value. The research also seeks to verify the new T2DM model through comparison with real patient data and design a closed-loop circuit control to remove human intervention regarding insulin input. In addition, a recurrent network will be introduced to allow accurate prediction of blood glucose levels to reduce the time taken in administration of insulin.

The article is organized as follows: Sect. 2 discuss the proposed pipeline and detail of the type-2 diabetes simulator, recurrent TSK fuzzy neural network, and the automated insulin control system. Section 3 discusses the results containing the dataset, simulation results, and results on clinical data. Finally, Sect. 4 presents the conclusion of the article.

2 Proposed Method

The proposed method consists of five modules as shown in Fig. 2. Firstly, a virtual type-2 diabetes patient simulator is used for generating training data. The parameters of the simulator are estimated using genetic algorithm (GA) [10]. Next, a Recurrent TSK Fuzzy Neural Network (RST-FNN) [11] is trained to predict the suitable insulin dose based on the health parameters of the patient. Finally, a control system is designed to deliver the suitable amount of insulin dose. Here we discuss each module in detail.

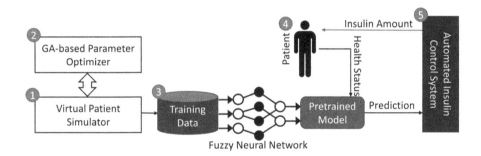

Fig. 2. Proposed type-2 diabetes modeling and insulin injection automation system.

2.1 Type-2 Diabetes Simulator (T2DS)

The proposed T2DS is based on GlucoSim [12], which is primarily used for Type-1 diabetes simulation. Here, we proposed a few modification of the system. We propose a submodel of total glucose utilization (TGU), that primarily consists of two parts: (A) insulin and glucose dependent utilization and (B) insulin and glucose independent utilization. Equation 1 and 2 indicate the calculations of the total glucose utilization rate:

$$TGU = k.I_A.G_B + CNU + (1 - e^{-G_B/kg.body.weight}) \tag{1}$$

$$\frac{dTGU}{dt} = -\frac{1}{T_{IA}(I_{BD} - I_A)}, I_A(t = 0) = I_{B_0} \tag{2}$$

where, I_A = effective insulin concentration or activity, I_{BD} = circulating blood insulin concentration delayed by $T_{D,TGU}$, $T_{D,TGU}$ = pure time delay for the total glucose utilization, T_{IA} = first order time constant for insulin action activation, G_B = circulating blood glucose concentration, k = rate constant for the total glucose utilization. CNU = glucose and insulin independent utilization rate, I_{B0} = initial blood insulin concentration. The parameter k is the most important parameter in the computational model for T2DS. It denotes the constant rate of total glucose uptake and also reflects the degree of insulin resistance (IR) which is a unique characteristic in T2DS. The value of k is positively correlated to IR and can be expressed as $k \propto \frac{1}{IR}$. This representation shows that if the T2DM patient has a high IR, he/she possesses a smaller glucose uptake rate, resulting in the need to utilize a larger amount of insulin to help to convert glucose into energy in the cells. In other words, the higher the IR value, the higher the k value. To find the optimal value of k, we have used GA. The objective is to maximize the rate of TGU, hence, Eq. 2 is used as the fitness function for the GA.

GA is used to find the optimal value of k. We have used 70% crossover rate, 0.1% mutation rate, 100 population size. We have used 1000 generations selecting the top 10 chromosome for next generation. The upper and lower bounds of each variable are $K = [0.00, 0.001]$, $I_A = [14000, 500]$, $G_B = [200, 30]$, $CNU = [2.0, 0.5]$, and $TGU = [500, 80]$. After running GA, optimal value of k is found as 0.000222.

Insulin Submodel-Based Simulation: To further improve the GlucoSim model, we propose an insulin submodel that simulates closer to reality. To do that, we must first understand the profile of Actrapid (Actrapid is a solution for injection that contains the active substance human insulin). This requires an understanding of the onset, peak, and duration of Actrapid. Here, we will use a fixed onset, peak and duration of 30 min, 90 min and 480 min. Next, we need a mathematical formula to calculate the amount of insulin being infused into the body at each interval. As it would be very computationally complex to model the full graph, it would be split into three parts (as shown (a), (b), (c) in Fig. 3(b)) to simplify the problem as presented in Fig. 3(a), for an example of 10 units of insulin bolus. Figure 3(b) illustrates the modelled insulin release.

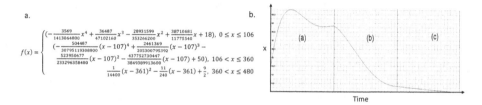

Fig. 3. (a) Function to calculate proposed insulin submodel. (b) Example of the division.

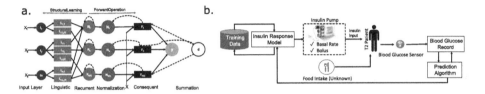

Fig. 4. (a) Proposed six layred Self-evolving TSK Fuzzy Neural Network. (b) Proposed type-2 diabetes modeling and insulin injection automation system.

This is noted that there is a need for different amounts of insulin infusion based on the requirements of the T2DM patient. Thus, we need to be able to make the equation (Fig. 3(a)), which is based on 10 units of insulin bolus, flexible to be able to handle this requirement. Here, we made the assumption that the difference between each interval of insulin infusion is constant. This need for different amounts of insulin infusion can be resolved through the Eq. 3.

$$insulin = \frac{bolus}{10} \times f(x) \tag{3}$$

where $f(x)$ s calculated from equation (Fig. 3(a)) with the proposed insulin sub-model. The optimal value of k is 0.000265, and used in experiments.

2.2   Recurrent Self-evolving TSK Fuzzy Neural Network (RST-FNN) for Prediction

Recurrent neural fuzzy networks are being used to solve temporal and dynamic problems. It brings the low-level learning and computational power of neural networks into fuzzy systems and provides the high-level human-like thinking and reasoning of fuzzy systems into neural networks. The main advantage of recurrent neural network over the feedforward neural network is the ability to involve dynamic elements in the form of feedback as internal memories to make the output history-sensitive [11].

Figure 4 illustrates the RST-FNN predictive model, which is a recurrent network. This model starts with an empty rule base with subsequent new rules added to be based on a novel fuzzy clustering algorithm which is completely data-driven. Membership functions that are highly overlapping are merged while obsolete rules are constantly pruned to maintain a compact fuzzy rule base that is relevant and accurate.

Most of the existing recurrent networks like Elman network [13] and Hopfield Network [14] do not have the capability to prune irrelevant rules by forgetting the obsolete data. Over time, these irrelevant rules may degrade the quality of the overall model, which may lead to poor prediction results. In contrast, the RST-FNN has the ability to forget/unlearn by implementing a rule pruning algorithm, which applies a 'gradual' forgetting approach and adopts the Hebbian learning mechanism behind the long-term potentiating phenomenon in the brain. RST-FNN does not require prior knowledge of the number of clusters or rules present in the training data set.

As shown in Fig. 4(a), the six-layered RST-FNN model defines a set of TSK-type IF THEN fuzzy rules. The fuzzy rules are incrementally constructed by presenting the training observations $\{..., (X(t), d(t)), ...\}$ one after another, where $X(t)$ and $d(t)$ denote the vectors containing the inputs and the corresponding desired outputs respectively at any given time, which is represented by t. Each fuzzy rule in RST-FNN has the form shown in Eq. 4:

$$R_k : IF(x_1 \ is \ L_{1,j_1^k})... \ AND \ (x_i \ is \ L_{i,j_i^k})$$
$$THEN \ f_k = b_{0k} + b_{1k}x_1 + ... + b_{ik}x_i + ... + b_{nk}x_n \tag{4}$$

where $X = [x_1, ..., x_i, ..., x_n]^T$ represents the numeric inputs of RST-FNN, $L_{i,j_i^k}(j_i = 1, ..., j_i(t), k = 1, ..., K(t))$ denotes the j_{th} linguistic label of the input x_i that is part of the antecedent of rule R_k. $j_i(t)$ is the number of fuzzy sets of $x : i$ at time t. $K(t)$ is the number of fuzzy rules at time t, f_k is the crisp output of rule R_k, n is the number of inputs, and $[b_{0k}, ..., b_{nk}]$ represents a set of consequent parameters of rule R_k. As seen in Fig. 4, the RST-FNN network is modelled as a multiple-input–single-output (MISO) network, requiring multiple inputs at Layer I and producing a single output at Layer VI with the intermediate layers being hidden from the user.

2.3   Automated Insulin Control System

The automated insulin control system proposed in [15] is an open-loop control system where no feedback is available to immediately inform the user whether the amount of input insulin, had achieved its task of reaching the desired output, normoglycemia. It is based on the estimated food intake and to infuse the corresponding amount of insulin. If the T2DM patient detects a high blood glucose level a period after the initial insulin infusion, the T2DM patient would then be required to infuse insulin again to bring down his/her blood glucose level as the initial infusion was insufficient. A closed-loop control (automated insulin infusion) with feedback is proposed here. The feedback from the blood glucose sensor would be able to help the system identify the amount of carbohydrate in the food intake and hence infuse the appropriate corresponding amount of insulin. Figure 4(b) illustrates the proposed closed-loop control system that is being presented in this article. The two main modules of the proposed system are Insulin Response Model (IRM) and the prediction algorithm that are connected via feedback.

To reduce the risk of incorrect amount of insulin infusion in the open-loop design reported in [15], closed-loop design is capable of detecting the amount of carbohydrates in unknown food intake based on the blood glucose records and matching the pattern against the training data. Using the predicted food intake, the insulin response model would inform the insulin pump to release the corresponding amount of insulin to maintain normoglycemia.

Insulin Response Model (IRM): There are two proposals for the implementation of the IRM. The first proposal is when a constant amount of insulin is infused into the body to neutralise the increasing blood glucose level. For example, from (t) to $(t + 5)$ there was an increase in $0.1\,\mathrm{mmol/L}$, where t is the time. Based on this increase, the IRM refers to a table or formula and pumps in the corresponding amount of insulin to neutralise this increase.

The second proposal is when the amount of insulin infused into the body is dependent on the prediction of food intake. Based on the same example stated above, the system is to predict the food intake by referencing the increase of blood glucose level to a lookup table or formula which indicates the food intake. The amount of insulin infused would then be decided based on the model reported in [16].

The initial risk of the first proposal is the long onset of 20–30 min with Actrapid. Without the aid of insulin for 30 min, the blood glucose level of the T2DM patient will potentially rise to a much higher level. When the insulin takes effect, it is insufficient to control the blood glucose level as the insulin dose is only sufficient for the initial increase in blood glucose level, which is lower as compared to 30 min later. This will result in the insulin consistently trying to keep on the increasing blood glucose level.

The subsequent risk is when the blood glucose level is nearing its peak; the blood glucose level is increasing at a decreasing rate, but the insulin reservoir is still high due to the onset. This insulin reservoir will then take effect when the blood glucose level is declining (30 min after infusion), resulting in a dangerous steep drop in blood glucose level. For example, the gradient of points $(t + 40)$ and $(t + 45)$ is 0.1 and the corresponding amount of insulin will be infused based on the lookup table or formula. At points $(t + 45)$ and $(t + 50)$, the gradient is -0.1. This implies that the peak is at $(t + 45)$ and no insulin should be infused. However, the insulin infused at $(t + 45)$ will only take effect after the onset at $(t + 75)$, causing a steep drop in blood glucose level. This has the potential to cause the T2DM patient to experience hypoglycemia.

The initial risk of the second proposal is similar to the initial risk of the first proposal due to the long onset of 20–30 min taken for insulin to take effect. However, instead of neutralising every increase in blood glucose level which results in the need for the insulin to constantly "catch up", the second proposal causes a later one-time insulin infusion due to the need to monitor the blood glucose level pattern and matching it against a lookup table or formula to predict the food intake. As a result, there will still be a risk of the T2DM patient going into hyperglycemia before insulin takes effect. With the predicted food intake, the corresponding insulin can be calculated using the using [16] and infused into the body. To reduce the time required to find the pattern, a prediction algorithm can be introduced to predict the following insulin level.

3 Results and Discussion

Training Data: Training data is obtained through simulations with known food intake using the model mentioned earlier. We use six different amounts of food intakes; 15 g, 30 g, 45 g, 60 g, 75 g and 90 g of carbohydrates. These six intakes are critical based on the clinical data but are not exhaustive for all cases. For each simulation, there will be two key attributes, carbohydrates (g) and blood glucose level (mmol/L). Based on these two key attributes, the initial two consecutive positive gradients and the difference of both gradients are recorded. $Gradient = \frac{y_i - y_{i-1}}{x_i - x_{i-1}}$ is used to calculate the gradient, where (x_i, y_i) are data points. Positive gradients represent an increasing blood glucose level which is caused by the absorption of food by the body, thus only positive gradients are recorded. Finding the difference in gradients is also important as we need to ensure that the blood glucose level is increasing at an increasing rate, which represents the start of food absorption. When it starts to increase at a decreasing rate, it represents food absorption reaching its peak as food consumption has stopped for some time and there is less food to be absorbed by the body. This information is also recorded at different blood glucose levels to create a more comprehensive training data. Here, we use three common points of blood glucose levels; at 3.5, 5.5 and 8.0 mmol/L respectively.

As it is computationally complex and expensive to formulate a nonlinear equation that represents the insulin reaction at different blood glucose levels and food intake, using a fuzzy gain scheduling would be able to reduce the complex problem into a simpler state which can be handled more easily. One method of implementing gain scheduling is the classification of the raw data and breaking it down to be handled locally. This will not only reduce the effort to product the complex formula, but also reduce the time taken during runtime. We have used a two-step method for the classification of the training data. The first step is breaking down of multiple blood glucose level, looking at each blood glucose level as an individual rather than a whole. By passing on the blood glucose level parameter, it helps to narrow down the search function to a much smaller pool of data, making it more efficient. We have three categories of blood glucose levels here (3.5, 5.5 and 8.0 mmol/L). The second step is arranging the actual data from each test for standardisation. It will be stored in the order of [(gradient1) (gradient2) (difference between gradients)] format, where (difference between gradients) is from (gradient2) – (gradient1). The IRM is used to predict the unknown food intake by passing in the parameter [(current blood glucose level) (gradient1) (gradient2) (difference between gradients)]. As it is impossible to find an exact match during runtime unless the training data contains all possible permutations, we need to introduce another technique that would be able to help us link the input parameters to the training data. We use a fuzzy rule based for the task. The rules are:

Step 1: (a). IF (current blood glucose level) exists THEN return the column

(b). IF (current blood glucose level) does not exists THEN return the column of the rounded-off value

Step 2: (a). IF (difference between gradients) is larger than 0.02 AND rounded-off to the nearest (difference between gradients) AND (gradient1) is less than 50% different, THEN return current level of (food intake)

(b). IF (difference between gradients) is larger than 0.02 AND rounded-off to the nearest (difference between gradients) AND (gradient1) is more than 50% different, THEN return upper level of (food intake)

(c). IF (difference between gradients) is smaller or equal to 0.02 THEN do nothing

Clinical Data: The clinical data used in this project was obtained from KK Women's and Children's Hospital, Singapore. The data was collected from a T2DM patient using a continuous glucose monitoring system (CGMS) over a period of four days. Throughout this period, blood glucose readings, carbohydrate intakes and insulin inputs were recorded. Insulin inputs consist of two types, insulin basal rate and insulin bolus. The basal rate is a constant fixed insulin infusion that is pumped into the body through the insulin pump at intervals of one minute. The bolus is a single insulin infusion into the body based on carbohydrate intake. Both are unique to individuals and prescribed by the doctor. This is to aid T2DM patients' blood glucose levels in maintaining normoglycemia similar to that of a healthy person's. The insulin used in this study is Actrapid, a short-acting insulin. It will begin absorption into the bloodstream after 20–30 min (onset), peak after 90–120 min and have duration of about 8–12 h. The information has been verified by experts. The T2DM patient involved in the study is obese, thus more glucose uptake is required to satisfy the daily requirements. In addition, obese people have higher risk of insulin resistance, hence making it necessary for more insulin consumption. The prescription given to the patients involved in this study is 1.5 units/h of insulin for basal rate and 2.5 units per 15 g of carbohydrate consumed. $Bolus = (\frac{2.5}{15} \times c)$, where $c = carbohydrate(g)$ is used to calculate the amount of insulin, in units needed based on the amount of carbohydrates consumed. To maintain the current blood glucose level, a pre-bolus of insulin would be recommended to mitigate the spike in blood glucose levels from the food consumed. With an onset of 20–30 min for Actrapid, a pre-bolus of insulin of a similar timing is recommended to allow the insulin to take effect together with the food consumed. Although the patient usually does the pre-bolus 5–10 min before food consumption, it is still within the acceptable range of time given for insulin infusion. Within this time frame, insulin can still serve its purpose of regulating the blood glucose level quickly enough not to cause any endangerment to the patient's life. Over these four days, the CGMS records the blood glucose level every five minutes. The normal range of blood glucose level is defined from 3.9–7.8 mmol/L. Blood glucose level above this range is labelled as hyperglycemia while blood glucose level below this range is considered as hypoglycemia. A prolonged period in either case is dangerous and would require immediate attention in returning the blood glucose level to normoglycemia.

Results of the Simulation: Here, we present the results of the proposed simulation engine. We have taken a real patient for a case study and put the similar value input to the simulator. In the best case, the simulator should produce a highly correlated observation. We have validated the simulator using the difference in peak and through. We also report the Pearson correlation among them. Figure 5(a) shows the simulation using this insulin submodel and Fig. 5(b) represents the comparisons made against clinical data. From Table (Fig. 5(b)), we can see that based on the peaks and troughs, the time difference between the data generated by the proposed insulin model and the clinical data are acceptable. The correlation is 0.57015.

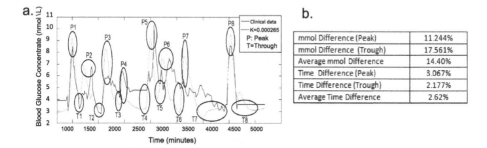

mmol Difference (Peak)	11.244%
mmol Difference (Trough)	17.561%
Average mmol Difference	14.40%
Time Difference (Peak)	3.067%
Time Difference (Trough)	2.177%
Average Time Difference	2.62%

Fig. 5. (a) Performance of the proposed T2DM patient simulation. (b) Difference of the simulation result and clinical data.

Results of the RST-FNN-Based Prediction: The proposed RSTSK is trained using the simulated data. Furthermore, based on the explanation given earlier, we split the simulated data into three parts and parts a and b are used as training data. The RST-FNN is configured using the parameter $[(t-20)(t-15)(t-10)(t-5)(t)(t+5)]$, where $(t+5)$ is the output of the prediction process and the rest are inputs. We have achieved Pearson's Correlation Coefficient 0.99 for the simulation and predicted values. Figure 6(a) shows the comparisons between simulations based on the study, the proposed design with RST-FNN and the ideal situation. From Fig. 6(a), we can see that with the introduction of RST-FNN, the results edge closer to the ideal situation. This is due to the earlier insulin infusion which makes it more similar to the recommended timing for insulin infusion. We found a correlation value of 0.99783 compared to the ideal situation (recommended). Figure 6(b) shows a case when the method fails to predict correctly. As we can see from Fig. 6(b), both graphs do not fit perfectly, but the general shape (peaks and troughs) show signs of similarity. The major differences can be spotted after each day's dinner (approximately time = 1700 and 3400). Based on the clinical data, the T2DM patient's blood glucose level fluctuates mainly around 3.9–7.8 mmol/L, which is normal, although occasionally it would dip below it.

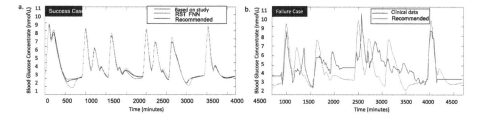

Fig. 6. (a) Results of proposed RST-FNN based prediction. (b) A case when the prediction fails.

4   Conclusion

The ultimate objective is to help T2DM patients manage and improve the disease as the implications of diabetes is not limited to the disease itself. This article is split into two major components, producing a relevant model that is capable of simulating a T2DM patient and designing a system that is capable of removing human intervention regarding insulin infusion. This allows to reduce human errors such as wrong estimation of food intake. The introduction of a closed-loop control helps provide feedback to the input, which allows the system to understand how different food intakes affect the blood glucose level. The system can then more accurately infuse the necessary amount of insulin to achieve the desired output. With the proposed design, we have managed to achieve a closer fit to the ideal situation as compared to manual infusion performed by the T2DM patient himself. The T2DM model and the proposed design provide a basis for future research on the modelling and automation of insulin infusion for T2DM patients. The design proposed in this article has illustrated that it is able to improve on the regulation of blood glucose level in T2DM patients.

References

1. Kullmann, S., et al.: Central nervous pathways of insulin action in the control of metabolism and food intake. Lancet Diab. Endocrinol. **8**(6), 524–534 (2020)
2. Deshmukh, H., et al.: Effect of flash glucose monitoring on glycemic control, hypoglycemia, diabetes-related distress, and resource utilization in the association of British clinical diabetologists (ABCD) nationwide audit. Diab. Care **43**(9), 2153–2160 (2020)
3. Li, S., et al.: Evaluating dasiglucagon as a treatment option for hypoglycemia in diabetes. Expert Opinion on Pharmacotherapy, pp. 1–8 (2020)
4. Nathan, D.M., et al.: Medical management of hyperglycemia in type 2 diabetes: a consensus algorithm for the initiation and adjustment of therapy: a consensus statement of the American diabetes association and the European association for the study of diabetes. Diabetes Care **32**(1), 193–203 (2009)
5. Petersen, M.C., Shulman, G.I.: Mechanisms of insulin action and insulin resistance. Physiol. Rev. **98**(4), 2133–2223 (2018)

6. WHO Definition: Diagnosis and classification of diabetes mellitus and its complications report of a who consultation part 1: Diagnosis and classification of diabetes mellitus. 1999. World Health Organization, Geneva, Switzerland. WHO/NCD/NCS/99.2. p, 59 (2017)

7. Zhu, B., Xiaomei, W., Bi, Y., Yang, Y.: Effect of bilirubin concentration on the risk of diabetic complications: a meta-analysis of epidemiologic studies. Sci. Rep. **7**(1), 1–15 (2017)

8. Ruiza, B., Ang, G., Yu, G.F.: The role of fructose in type 2 diabetes and other metabolic diseases. J. Nutr. Food Sci. **8**(1), 659 (2018)

9. Karges, B., et al.: Association of insulin pump therapy vs insulin injection therapy with severe hypoglycemia, ketoacidosis, and glycemic control among children, adolescents, and young adults with type 1 diabetes. Jama **318**(14), 1358–1366 (2017)

10. Li, H., Yuan, D., Ma, X., Cui, D., Cao, L.: Genetic algorithm for the optimization of features and neural networks in ECG signals classification. Sci. Rep. **7**, 41011 (2017)

11. Aziz Khater, A., El-Nagar, A.M., El-Bardini, M., El-Rabaie, N.M.: Online learning based on adaptive learning rate for a class of recurrent fuzzy neural network. Neural Comput. Appl. **32**(12), 8691–8710 (2020)

12. Erzen, F.C., Birol, G., Cinar, A.: Glucosim: a simulator for education on the dynamics of diabetes mellitus. In: 2001 Conference Proceedings of the 23rd Annual International Conference of the IEEE Engineering in Medicine and Biology Society, vol. 4, pp. 3163–3166. IEEE (2001)

13. Ghosh, S., Tudu, B., Bhattacharyya, N., Bandyopadhyay, R.: A recurrent elman network in conjunction with an electronic nose for fast prediction of optimum fermentation time of black tea. Neural Comput. Appl. **31**(2), 1165–1171 (2019)

14. Yang, J., Wang, L., Wang, Y., Guo, T.: A novel memristive hopfield neural network with application in associative memory. Neurocomputing **227**, 142–148 (2017)

15. Zaharieva, D.P., McGaugh, S., Pooni, R., Vienneau, T., Ly, T., Riddell, M.C.: Improved open-loop glucose control with basal insulin reduction 90 minutes before aerobic exercise in patients with type 1 diabetes on continuous subcutaneous insulin infusion. Diab. Care **42**(5), 824–831 (2019)

16. Puckett, W.R.: Dynamic modelling of diabetes mellitus (1993)

Granulated Tables with Frequency by Discretization and Their Application

Hiroshi Sakai$^{(\boxtimes)}$ and Zhiwen Jian

Graduate School of Engineering, Kyushu Institute of Technology,
Tobata, Kitakyushu 804-8550, Japan
sakai@mns.kyutech.ac.jp, zhiwen.jian389@mail.kyutech.jp

Abstract. We have coped with rule generation from tables with discrete attribute values and extended the Apriori algorithm to the DIS-Apriori algorithm and the NIS-Apriori algorithm. Two algorithms use table data characteristics, and the NIS-Apriori generates rules from tables with uncertainty. In this paper, we handle tables with continuous attribute values. We usually employ continuous data discretization, and we often had such a property that the different objects came to have the same attribute values. We define a *granulated table with frequency* by discretization and adjust the above two algorithms to granulated tables due to this property. The adjusted algorithms toward big data analysis improved the performance of rule generation. The obtained rules are also applied to rule-based reasoning, which gives one solution to the black-box problem in AI.

Keywords: Rule generation · The Apriori algorithm · Rule-based reasoning · Big data analysis and machine learning

1 Introduction

We are applying rough sets [10,13,19] to rule generation from table data sets and are adjusting the Apriori algorithm for transaction data sets [1,2] to table data sets. We proposed a framework termed "NIS-Apriori" [16–18] based on the combination of equivalence classes in rough sets and the effective enumeration of the candidates of rules in the Apriori algorithm.

We term such a table in Table 1 as a *Deterministic Information System* (DIS). Several rough-set based rule generation methods are proposed [5,10,13,15,19,21] in DISs. In Table 1, we have an implication τ: [P1,c]$\Rightarrow$ [Dec,d1] from object $x1$. It occurs 1 time for 5 objects, namely $support(\tau)$ (a ratio of occurrence) $=1/5$. It occurs 1 time for 2 objects with [P1,c], namely $accuracy(\tau)$ (a ratio of consistency)$=1/2$. We term such formulas like [P1,c] and [Dec,d1] *descriptors*. We usually specify constraints $support(\tau) \geq \alpha$ and $accuracy(\tau) \geq \beta$ for $0 < \alpha, \beta \leq 1$, and we see each implication satisfying the constraints as a rule.

© IFIP International Federation for Information Processing 2021
Published by Springer Nature Switzerland AG 2021
Z. Shi et al. (Eds.): ICIS 2020, IFIP AICT 623, pp. 137–146, 2021.
https://doi.org/10.1007/978-3-030-74826-5_12

Table 1. An exemplary DIS ψ.

Object	P1	P2	P3	Dec
$x1$	c	1	b	d1
$x2$	b	2	b	d1
$x3$	a	2	b	d2
$x4$	a	3	c	d2
$x5$	c	2	c	d3

Table 2. An exemplary DIS ψ with missing values.

Object	P1	P2	P3	Dec
$x1$	c	?	b	d1
$x2$	b	?	b	d1
$x3$	?	2	b	d2
$x4$	a	3	c	?
$x5$	c	2	?	d3

Table 3. An exemplary NIS Φ. Each ? is changed to a set of all possible values.

Object	P1	P2	P3	Dec
$x1$	c	{1, 2, 3}	b	d1
$x2$	b	{1, 2, 3}	b	d1
$x3$	{a, b, c}	2	b	d2
$x4$	a	3	c	{d1, d2, d3}
$x5$	c	2	{a, b, c}	d3

To cope with information incompleteness in DISs, *missing values '?'* [6] (Table 2) and a *Non-deterministic Information System* (NIS) [11,12] (Table 3) were also investigated. In NIS, some attribute values are given as a set of possible attribute values. In Table 3, we interpret $\{1, 2, 3\}$ in $x2$ as that *one of 1, 2, and 3 is the actual value, but there is not enough information to decide it* due to information incompleteness.

If we replace each set of attributes in NIS Φ with one element of the set, we have one possible DIS. In Table 3, there are 243 ($=3^5$) possible DISs and let $DD(\Phi)$ denote a set of possible DISs. We considered the certain rules and the possible rules from NIS below:

- An implication τ is a certain rule, if τ is a rule in each $\psi \in DD(\Phi)$.
- An implication τ is a possible rule, if τ is a rule in at least one $\psi \in DD(\Phi)$.

This definition seems natural, but the $|DD(\Phi)|$ increases exponentially. However, we proved some properties and developed the NIS-Apriori algorithm, which does not depend upon the number of $|DD(\Phi)|$ [17,18].

This paper extends the previous frameworks and considers rule generation from tables with continuous attribute values. Furthermore, this paper applies the obtained rules to decision making. The reasoning based on the obtained rules will recover the black-box problem in AI.

This paper's organization is as follows: Sect. 2 clarifies the discretization of the continuous attribute values and considers an example of the Iris data set [4]. Section 3 proposes granulated tables Γ with frequency by discretization and extends the previous algorithms for NIS to that for Γ. Section 4 applies the extended algorithms to some data sets and shows the improvement of rule generation performance. An application of the obtained rules to decision making is considered. Section 5 concludes this paper.

2   Tables with Continuous Attribute Values

This section briefly examines the discretization of continuous attribute values and considers the case of the Iris data set [4].

2.1   Rules and Discretization of Continuous Attribute Values

To generate rules from tables with continuous attribute values, we need to discretize tables because there may be too many descriptors. There seem to be several methods for discretization [7]. Most of them focus on the optimal discretization specified by the constraints like minimal entropy, equal-interval width, equal-interval frequency, etc. In these researches on discretization, descriptors seem to be obtained as a side effect.

However, our research purpose is to generate rules by specified descriptors. We at first specify descriptors and their intervals, then generate rules by them. Thus, we can have our rules for our specifications.

2.2   An Example of the Iris Data Set

The Iris data set consists of 150 objects, four condition attributes $\{spl, spw, pel, pew\}$ (each attribute value of them is continuous), one decision attribute *class* whose attribute value is one of *setosa*, *versicolor*, and *virginica*. Figure 1 is a part of the Iris data set.

	A	B	C	D	E	F
1	object	spl	spw	pel	pew	class
2	1	5.1	3.5	1.4	0.2	setosa
3	2	4.9	3	1.4	0.2	setosa
100	99	5.1	2.5	3	1.1	versicolor
101	100	5.7	2.8	4.1	1.3	versicolor
150	149	6.2	3.4	5.4	2.3	virginica
151	150	5.9	3	5.1	1.8	virginica

	small	medium	large
spl	5.5<	5.5<= & < 6.7	6.7<=
spw	3<	3<= & < 4	4<=
pel	2.5<	2.5<= & < 5	5<=
pew	1<	1<= & < 2	2<=

Fig. 1. A part of the Iris data set.

Fig. 2. A definition of the discretization of continuous attribute values.

Figure 2 shows the discretization specification; namely, we want to generate rules using the attribute values *small*, *medium*, and *large* for each attribute. Here, every object is identified as one element of the Cartesian product $\{small, medium, large\}^4 \times \{setosa, versicolor, virginica\}$, whose number of

```
[1,(('spl','m'),('spw','s'),('pel','l'),('pew','m'),('class','versicolor')),1]
[2,(('spl','m'),('spw','m'),('pel','l'),('pew','m'),('class','virginica')),3]
[3,(('spl','l'),('spw','s'),('pel','l'),('pew','l'),('class','virginica')),2]
[4,(('spl','s'),('spw','l'),('pel','s'),('pew','s'),('class','setosa')),1]
[5,(('spl','l'),('spw','s'),('pel','l'),('pew','l'),('class','virginica')),3]
[6,(('spl','m'),('spw','l'),('pel','s'),('pew','s'),('class','setosa')),3]
[7,(('spl','m'),('spw','s'),('pel','m'),('pew','l'),('class','versicolor')),27]
[8,(('spl','m'),('spw','m'),('pel','l'),('pew','l'),('class','virginica')),7]
[9,(('spl','m'),('spw','m'),('pel','m'),('pew','m'),('class','virginica')),2]
[10,(('spl','s'),('spw','m'),('pel','m'),('pew','s'),('class','setosa')),42]
[11,(('spl','m'),('spw','s'),('pel','m'),('pew','l'),('class','virginica')),1]
[12,(('spl','m'),('spw','s'),('pel','m'),('pew','m'),('class','virginica')),2]
[13,(('spl','l'),('spw','s'),('pel','m'),('pew','m'),('class','versicolor')),1]
[14,(('spl','s'),('spw','s'),('pel','s'),('pew','s'),('class','setosa')),2]
[15,(('spl','m'),('spw','m'),('pel','m'),('pew','m'),('class','versicolor')),10]
[16,(('spl','l'),('spw','m'),('pel','l'),('pew','m'),('class','virginica')),2]
[17,(('spl','s'),('spw','m'),('pel','m'),('pew','m'),('class','versicolor')),1]
[18,(('spl','l'),('spw','m'),('pel','l'),('pew','l'),('class','virginica')),15]
[19,(('spl','m'),('spw','s'),('pel','l'),('pew','m'),('class','virginica')),8]
[20,(('spl','l'),('spw','m'),('pel','l'),('pew','m'),('class','versicolor')),1]
[21,(('spl','s'),('spw','s'),('pel','m'),('pew','m'),('class','versicolor')),5]
[22,(('spl','m'),('spw','s'),('pel','l'),('pew','l'),('class','virginica')),4]
[23,(('spl','m'),('spw','m'),('pel','s'),('pew','s'),('class','setosa')),2]
[24,(('spl','s'),('spw','s'),('pel','m'),('pew','m'),('class','virginica')),1]
[25,(('spl','l'),('spw','m'),('pel','m'),('pew','m'),('class','versicolor')),4]
```

Fig. 3. Discretized table with frequency.

Table 4. A relationship between data sets and the discretized data sets. Here, type I: tables with continuous values, type II: tables with discrete values, type III: tables with missing values, #object: the number of objects, #con: the number of condition attributes, #Cartesian: the number of elements of the Cartesian product, #object_discre: the number of objects after discretization.

data sets	type	#object	#con	#Cartesian	#object_discre
Iris	I	150	4	243 $(=3^4 \times 3)$	25
Wine quality [4]	I	4898	11	1240029 $(=3^{11} \times 7)$	564
Htru2	I	17898	8	13122 $(=3^8 \times 2)$	134
Phishing [4]	II	1353	9	59049 $(=3^9 \times 3)$	724
Car Evaluation [4]	II	1728	6	27648 $(=4^4 \times 3^3)$	1728
Suspicious Network [3]	II	39427	51	–	39427
Mammographic [4]	III	961	5	3200 $(=4^3 \times 5^2 \times 2)$	301
Congress Voting [4]	III	435	16	131072 $(=2^{17})$	342

elements is 243 $(=3^5)$. On the other hand, the number of objects is 150. Thus, there exist many such elements of the Cartesian product that do not correspond to objects.

We examined the relationship between the Iris data set and the elements of the Cartesian product. Figure 3 shows lists of the sequential number, the Cartesian product element, and the duplicated number of objects. For example, the 10th list represents 42 objects for 150 objects. We may say that 42 objects are granulated to one granule represented by the 10th Cartesian product element. Thus, we have a discretized table with re-numbered 25 objects from a table with 150 objects. This phenomenon seems interesting because we can handle re-numbered 25 objects for a total of 150 objects.

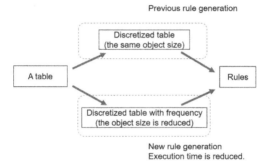

Previous rule generation

Fig. 4. Previous rule generation and new rule generation.

We dealt with this phenomenon for other data sets with continuous attribute values. Table 4 shows the results. For example, there are 17898 objects in the Htru2 data set [4]. We employed a similar discretization, which divides each attribute into three classes, as Fig. 2. Then, each object is identified with an element of the Cartesian product with 13122 elements. Furthermore, 17898 objects are granulated to 134 elements in the Cartesian product.

Remark 1. In tables with continuous attribute values, we specify descriptors for discretization. If the number of discretized objects is much smaller than that of the original data sets, we will reduce the execution time of rule generation. Of course, we have the same rules as that of the original data set. It will be meaningful to consider the new rule generation in Fig. 4.

The strategy in Fig. 4 will be related to researches on rough sets (coarse classes) [13, 19], the Infobright technology [20], granular computing [14], and the zoom out operation [21]. As for the Car Evaluation and Suspicious Network data sets (type II) in Table 4, each tuple was different, and we cannot reduce the number of objects. Remark 1 will be applied to tables with continuous attribute values, and it may not be useful to tables with discrete attribute values.

3   Rule Generation from Discretized Tables with Frequency

This section defines a granulated table Γ with frequency.

Definition 1. *A discretized table Γ with frequency (from a DIS ψ) consists of the following:*

1. *A finite set AT of attributes,*
2. *A finite set $DESC_A$ of descriptors for $A \in AT$,*
3. *A pair $(LIST, freq)$ $(LIST \in \Pi_{A \in AT} DESC_A$: the Cartesian product of sets of descriptors, and $freq (> 0)$: the number of objects in ψ satisfying $LIST$),*

Input: Table data set DIS ψ, decision attribute Dec, threshold values α, β.
Output: A set $Rule(\psi)$ of minimal rules.
1: $Rule(\psi) \leftarrow \{\}$; $i \leftarrow 1$;
2: create a set CAN_1 of candidates of rules with 1 condition;
3: **while** $(|CAN_i| \geq 1)$ **do**
4:     $Rest_i \leftarrow \{\}$; $Rule_i \leftarrow \{\}$;
5:     **for all** $\tau_{i,j} \in CAN_i$ **do**
6:         **if** $support(\tau_{i,j}) \geq \alpha$ **then**
7:             **if** $accuracy(\tau_{i,j}) \geq \beta$ **then** add $\tau_{i,j}$ to $Rule_i$; **else** add $\tau_{i,j}$ to $Rest_i$;
8:             **end if**
9:         **end if**
10:     **end for**
11:     $i \leftarrow i + 1$;
12:     create CAN_i (candidates of rules with i-th conditions) from $Rest_{i-1}$ and $Rest_1$;
13: **end while**
14: **return** $Rule(\psi){=}\cup_{k<i}Rule_k$

Fig. 5. The DIS-Apriori algorithm adjusted to table data set DIS ψ [9].

4. We assign a number num to each pair and see a tuple $(num, LIST, freq)$ as an object in Γ.

Figure 3 is an example of Γ. Now, we consider rule generation from Γ. Previously, we have investigated rules and rule generators [17,18] in the following.

1. Rules in DIS ψ and the DIS-Apriori rule generator,
2. Certain rules and possible rules in NIS Φ and the NIS-Apriori rule generator,
3. Decision-making tool based on the obtained rules.

We identify each descriptor [attribute, value] as an item and adjusted the Apriori algorithm for transaction data sets to that of table data sets. The overview of the adjusted DIS-Apriori algorithm is in Fig. 5. There is usually one decision attribute in every table, and we can see one itemset defines one implication. Furthermore, in the line 12 in Fig. 5, we have proved that CAN_i can be generated from $Rest_{i-1}$ and $Rest_1$ [9]. Due to these characteristics, we reduced the number of candidates of rules and the execution time of rule generation.

Remark 2. The following properties are related to the DIS-Apriori algorithm.

1. We replace DIS ψ with NIS Φ, *support* and *accuracy* values with *minsupp* and *minacc* values [17,18], respectively. Then, this algorithm generates all minimal certain rules.
2. We replace DIS ψ with NIS Φ, *support* and *accuracy* values with *maxsupp* and *maxacc* values [17,18], respectively. Then, this algorithm generates all minimal possible rules.
3. We term the above two algorithms the NIS-Apriori algorithm.
4. Both DIS-Apriori and NIS-Apriori algorithms are logically sound and complete for rules. They generate rules without excess and deficiency.

To handle a discretized table Γ with frequency, we revise the calculation of *support* and *accuracy* values in Fig. 5. To calculate them, we are handling equivalence classes defined by the concept of rough sets. For example, in Table 1, we have equivalence classes $\{x1, x5\}$ for [P1,c] and $\{x1, x2\}$ for [Dec,d1], respectively. We can easily know that τ: [P1,c] $\Rightarrow$ [Dec,d1] is supported by $\{x1\}(= \{x1, x5\} \cap \{x1, x2\})$. Here, the occurrence of τ is 1, however in Γ, we need to count the frequency $freq(x1)$ of $x1$. Due to this consideration, we have the next Remark 3 for handling Γ.

Remark 3. In a DIS ψ, if an implication τ is supported by an equivalence class $\{x1, x2, \cdots, x_n\}$, τ is supported by n objects. However, in Γ, τ is supported by $freq(x1) + freq(x2) + \cdots + freq(x_n)$ objects (Here, $freq(x_i)$ means the frequency of x_i in Γ). If we replace the number of occurrence (i.e., 1) of one object x_i with $freq(x_i)$, we can have the DIS-Apriori algorithm for handling Γ.

4 An Apriori-Based Rule Generator for Γ and Some Experiments

We revised rule generation programs, the DIS-Apriori algorithm for Γ and the NIS-Apriori algorithm for Γ, in Python based on Remarks 2 and 3. For simplicity, we omit the details of the NIS-Apriori algorithm for Γ and show the execution time in Table 5.

Table 5. A Comparison of the execution time: the original tables and the granulated tables. As for Iris, we coped with four cases. We duplicated the original table by ten times, 100 times, and 1000 times.

Table	Support	Accuracy	Original		Granulated	
			#object	Exec (sec)	#object	Exec (sec)
Iris	0.01	0.9	150	0.022	25	0.018
Iris1500	0.01	0.9	1500	0.030	25	0.020
Iris15000	0.01	0.9	15000	0.206	25	0.022
Iris150000	0.01	0.9	150000	2.162	25	0.022
Wine quality	0.001	0.5	4898	13.365	564	10.962
Htru2	0	0.7	17898	3.167	134	0.175
Mammographic						
(certain rule)	0	0.8	961	0.343	299	0.297
(possible rule)	0	0.8	961	0.375	299	0.349
Congress Voting						
(certain rule)	0.1	0.7	435	0.434	342	0.417
(possible rule)	0.1	0.7	435	0.448	342	0.419

Due to Table 5, we know the new rule generator is more effective in three cases
of Iris15000, Iris150000, and Htru2. In the granulated tables Γ from Iris15000 and
Iris150000, the number of objects is the same, and only the frequency is changed.
The rule generation process is the same as that of Iris, and the calculation
of *support* and *accuracy* is slightly changed. Thus, the execution time of rule
generation is almost constant. We think that to consider Γ will be meaningful
for big data analysis. Figure 6 shows the obtained rules (the left hand side) and
$Rest_1$, $Rest_2$ (the right hand side). By using $Rest_1$, $Rest_2$, $\cdots$, we can reduce
the number of candidates of rules. This reduction causes to reduce the execution
time of rule generation.

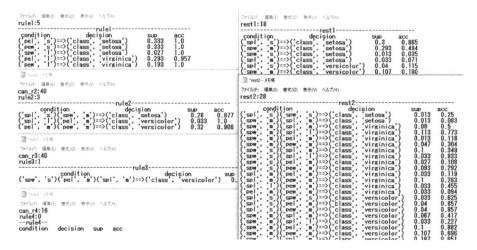

Fig. 6. Obtained rules $(support(\tau) \geq 0.01, accuracy(\tau) \geq 0.9)$ and $Rest_1$, $Rest_2$ from
Iris150000.

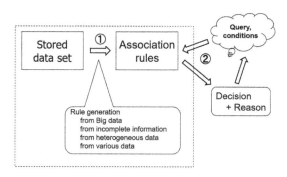

Fig. 7. A chart of rule-based reasoning.

Now, we consider the application of the obtained rules. We quickly think the
rule-based decision making in Fig. 7. We have coped with ① and have developed

the environment of rule generation. We also need to manage ②, i.e., decision making by the obtained rules. The decision making in Fig. 7 is based on the applied rules, and the applied rule supports the reasoning. Thus, the reason is apparent. This strategy will recover the black-box problem in AI. We need to clarify the following subjects:

1. How do we select one rule if there are some applicable rules?
2. How do we have a decision if there is no candidate for a rule?
3. How do we think that the different results may be concluded?

For the Suspicious Network data set in Table 4, we employed the $lift$ value for selecting one rule. We employed three-cross validation for 39427 objects and applied the obtained rules. This procedure is based on Fig. 7, and the averaged 94% correct estimation was obtained [8]. The research on Fig. 7 is in progress now.

5   Concluding Remarks

We considered discretization for tables with continuous attribute values and proposed granulated tables with frequency. In some cases, we can reduce the number of objects. This property causes to reduce the execution time of rule generation. We implemented a new rule generator and examined that the adjusted algorithms for big data analysis improved rule generation performance. The obtained rules are also applied to rule-based reasoning, which gives one solution to the black-box problem in AI. We need to improve much more comparative analysis and sensitive analysis comprehensively.

Acknowledgment. The authors would be grateful to the anonymous referees for their useful comments. This work is supported by JSPS (Japan Society for the Promotion of Science) KAKENHI Grant Number JP20K11954.

References

1. Agrawal, R., Srikant, R.: Fast algorithms for mining association rules in large databases. In: Proceedings of the VLDB 1994, pp. 487–499, Morgan Kaufmann (1994)
2. Agrawal, R., Mannila, H., Srikant, R., Toivonen, H., Verkamo, A.I.: Fast discovery of association rules. In: Advances in Knowledge Discovery and Data Mining, pp. 307–328. AAAI/MIT Press (1996)
3. Bigdata Challenge. https://knowledgepit.ml/. Accessed 14 July 2019
4. Frank, A., Asuncion, A.: UCI machine learning repository. University of California, School of Information and Computer Science, Irvine, CA (2010). http://mlearn.ics.uci.edu/MLRepository.html. Accessed 10 July 2019
5. Greco, S., Matarazzo, B., Słowiński, R.: Granular computing and data mining for ordered data: the dominance-based rough set approach. In: Meyers, R.A. (ed.) Encyclopedia of Complexity and Systems Science, pp. 4283–4305. Springer, New York (2009). https://doi.org/10.1007/978-0-387-30440-3_251

6. Grzymała-Busse, J.W., Werbrouck, P.: On the best search method in the LEM1 and LEM2 algorithms. Incomplete Information: Rough Set Analysis, Studies in Fuzziness and Soft Computing, vol. 13, pp. 75–91. Springer, Heidelberg (1998). https://doi.org/10.1007/978-3-7908-1888-8_4

7. Grzymała-Busse, J.W., Stefanowski, J.: Three discretization methods for rule induction. Int. J. Intell. Syst. **16**, 29–38 (2001)

8. Jian, Z., Sakai, H., Watada, J., Roy, A., Hassan, M.B.: An Apriori-based data analysis on suspicious network event recognition. Proc. IEEE Big Data **2019**, 5888–5896 (2019)

9. Jian, Z., Sakai, H., et. al.: An adjusted Apriori algorithm to itemsets defined by tables and an improved rule generator with three-way decisions. In: Bello, R. et al. (eds.) IJCRS 2020, Springer LNCS, vol. 12179, pp. 95–110 (2020). https://doi.org/10.1007/978-3-030-52705-1_7

10. Komorowski, J., Pawlak, Z., Polkowski, L., Skowron, A.: Rough sets: a tutorial. In: Pal, S.K., Skowron, A. (eds.) Rough Fuzzy Hybridization: A New Method for Decision Making, pp. 3–98. Springer, Cham (1999)

11. Lipski, W.: On databases with incomplete information. J. ACM **28**(1), 41–70 (1981)

12. Orłowska, E., Pawlak, Z.: Representation of nondeterministic information. Theor. Comput. Sci. **29**(1–2), 27–39 (1984)

13. Pawlak, Z.: Rough sets. Int. J. Comput. Inf. Sci. **11**(5), 341–356 (1982)

14. Pedrycz, W.: Granular computing for data analytics: a manifesto of human-centric computing. IEEE/CAA J. Automatica Sinica **5**(6), 1025–1034 (2018)

15. Riza, L.S., et al.: Implementing algorithms of rough set theory and fuzzy rough set theory in the R package RoughSets. Inf. Sci. **287**(10), 68–89 (2014)

16. Sakai, H.: Execution logs by RNIA software tools. http://www.mns.kyutech.ac.jp/~sakai/RNIA. Accessed 10 July 2019

17. Sakai, H., Nakata, M.: Rough set-based rule generation and Apriori-based rule generation from table data sets: a survey and a combination. CAAI Trans. Intell. Technol **4**(4), 203–213 (2019)

18. Sakai, H., Nakata, M., Watada, J.: NIS-Apriori-based rule generation with three-way decisions and its application system in SQL. Inf. Sci. **507**, 755–771 (2020)

19. Skowron, A., Rauszer, C.: The discernibility matrices and functions in information systems. In: Słowiński, R. (ed.) Intelligent Decision Support - Handbook of Advances and Applications of the Rough Set Theory, pp. 331–362. Kluwer Academic Publishers (1992)

20. Ślęzak, D., Eastwood, V.: Data warehouse technology by Infobright. In: Proceedings of ACM SIGMOD 2009, pp. 841–846 (2009)

21. Yao, Y.Y., Liau, C., Zhong, N.: Granular computing based on rough sets, quotient space theory, and belief functions. In: Proceedings of ISMIS 2003, LNCS, vol. 2871, pp. 152–159. Springer, Hiedelberg (2003). https://doi.org/10.1007/978-3-540-39592-8_21

Data Intelligence

Person Authentication by Gait Data from Smartphone Sensors Using Convolutional Autoencoder

Ashika Kothamachu Ramesh[1], Kavya Sree Gajjala[1], Kotaro Nakano[2], and Basabi Chakraborty[3(✉)]

[1] Graduate School of Software and Information Science, Iwate Prefectural University, Takizawa, Iwate, Japan
[2] Research and Regional Co-operative Divison, Iwate Prefectural University, Takizawa, Iwate, Japan
[3] Faculty of Software and Information Science, Iwate Prefectural University, Takizawa, Iwate, Japan
basabi@iwate-pu.ac.jp

Abstract. Biometric authentication is a security process that relies on the unique biological characteristics of an individual to verify who he or she is. Human gait serves as an important non invasive biometric modality for an authentication tool in various security applications. Recently due to increased use of smartphones and easy capturing of human gait characteristics by embedded smartphone sensors, human gait related activities can be utilized to develop user authentication model. In this work, a new method for user authentication from smartphone sensor data by a hybrid deep network model named convolutional autoencoder has been proposed and the performance of the model is compared with other machine learning including deep learning based techniques by simulation experiments with bench mark data sets. It is found that our proposed authetication method from smartphone sensor data with convolutional autoencoder reduces the time for authentication and also produces fair authentication accuracy and EER. It can be potentially used for person authentication in real time.

Keywords: Biometric authentication · Smartphone sensors · Human gait · Convolutional autoencoder

1 Introduction

With increasing demands of user identification or authentication [1] for secured processing of today's big data and artificial intelligence based applications, biometric techniques play a major role. The biometric traits that are commonly used in different systems are face [2], fingerprints, palm print, handwriting, iris, gait, voice etc. Among them, gait recognition is a relatively new behavioral biometric technique which aims to recognize the person by the way they walk without intruding the persons privacy. It can be used in applications for criminal

Z. Shi et al. (Eds.): ICIS 2020, IFIP AICT 623, pp. 149–158, 2021.
https://doi.org/10.1007/978-3-030-74826-5_13

detection or for health monitoring to diagnose the abnormal walking that led to health issues. Every individual has unique walking style by which a person can be differentiated and identifed [3]. Authentication by gait can be carried out by extracting some salient properties related to the coordinated cyclic motions.

Smart devices are now equipped with top quality sensors, fast processing and communication power. User authentication based on gait characteristics captured by embedded smartphone sensors is one of the most active mode of biometrics for smartphone based security applications. The rapidly evolving field of wireless communication allows us to record the time series data from smartphone sensors without any hassle to the user [4]. Human activity recognition from smartphone sensor data is also an active research area because of its importance in health care and assisted living. Activity dependent authentication framework based on human gait characteristics from smartphone sensor data has been proposed by one of the authors in earlier works [17,25].

In recent years, machine learning including deep learning techniques has evolved a lot and increasingly applied to classification problems. Deep neural networks (DNN), especially Convolutional Neural Networks (CNN) have been proposed in many research works for user authentication. In this work, we have proposed an authentication approach using a new hybrid DNN model which is a combination of CNN and autoencoders (convolutional autoencoders) to reduce the time of authentication as well as to increase accuracy of authentication. We have performed simulation experiments with the proposed model by a few benchmark datasets and found that our model works better when compared with traditional machine learning or other deep learning models. In the next section brief description of related works on person authentication is presented followed by the outline of the proposed method in the following section. The next section contains simulation experiments and results and the last section is summarization and conclusion.

2   Related Work

In this modern world, almost everyone is dependent on smartphone for daily activities like connecting friends and relatives across the globe or managing personal obligations from monitoring health to paying bills online. Smartphones have in built motion sensors like accelerometer, gyroscope, magnetometer. Accelerometers can measure any movement of the phone while gyroscope can capture current orientation of the phone in all three axes (X, Y, and Z). Three dimensional time series data are generated from accelerometer and gyroscope. Time series is a sequence of data that describes the change of the observed phenomenon over time. Data from motion sensors of smartphone capture gait characteristics of the user carrying the phone. Sensors attached to fixed body positions are also capable of capturing gait characteristics but the popularity of smartphones motivates development of smartphone based authentication applications using gait characteristics [5,21]. Inertia based gait recognition is popular because it can analyze the details of movement characteristics [6]. An efficient higher order statistical analysis based gait person authentication which is able to operate on multichannel and multisensor data by combining feature-level and sensor-level fusion is explained in [7].

Recently deep neural networks (DNN) are found to be very efficient at delivering high quality results in pattern classification problems. A review of research works on human activity recognition by motion data from inertial sensors using deep learning is found in [8]. Among DNN models, CNN produce good results for many classification problems. Gait classification and person authentication using CNN is presented in [9]. Another deep recurrent network model, Long Short Term Memory (LSTM) is also suitable for analysis of sensor data for recognition. Person authentication from gait data during human activity with smartphone sensors using LSTM is explained in [10]. A good comprehensive study of the research work on authentication of smartphone users with behavioral biometric can be found in [11]. Different kinds of user authentication techniques are also explained in [12]. A comparative analysis of hybrid deep learning models for human activity recognition is considered in [13]. Unobtrusive user authentication on mobile phones using biometric gait recognition is presented in [14]. Authentication of smartphone user using behavioral biometrics is explained in [15]. The use of autoencoders for gait-based person authentication is found in [16]. For lowering computational cost of classifier, knowledge distillation is used as an approach for designing low cost deep neural network based biometric authentication model for smartphone user in [17].

3 Proposed Method and Comparative Study

In this work, person authentication approach by gait characteristics captured from smartphone sensor data is proposed with Convolutional Autoencoder (CAE), a new hybrid deep network model to reduce the time of authentication as well as to increase the accuracy. Convolutional autoencoders are usually popular in computer vision or in image analysis. Convolutional autoencoder along with LSTM has also been used in [18] for time series prediction. In authentication problem, CAE used to extract features for finger vein verification problem along with SVM (Support Vector Machine) for classification in [19]. The use of CAE in radar based classification of human activities shows promising improvement over SVM classifiers in [20]. Deployment of CAE for smartphone sensor based data analysis in user authentication has not been addressed yet.

This work aims to exploit CAE for analysing smartphone sensor data for person authentication. The performance of convolutional autoencoder based authentication method is evaluated by simulation experiments with bench mark data sets and a comparative study with other popular deep network models for person authentication has also been done. A brief introduction of CAE and other DNN models used in our study is presented in the next subsections.

3.1 Convolutional Autoencoder (CAE)

Convolutional Autoencoder is a combination of auto encoder and convolutional neural network. Autoencoders are neural networks that can be easily trained on any kind of input data. Generally encoders compress the given input into fixed dimension and decoder transforms the code into original input. In convolutional autoencoder, encoding and decoding use convolution and deconvolution, the encoder use the convolutional layer, batch normalization layer, an activation function and at last, a maxpooling function which reduces the dimensions of the feature maps. When encoder is complete, the feature maps are flattened and a dense layer is used for latent space representation and the deconvolution is used for upsampling of the incoming feature maps followed by batch normalization and activation function.

3.2   Deep Neural Network Models Used for Comparison

Convolutional Neural Networks (CNN). CNN is the most popular deep neural network model which has become dominant in various computer vision tasks. CNN is composed of multiple building blocks, such as convolution layers, pooling layers, and fully connected layers, and it is designed to automatically and adaptively learn spatial hierarchies of features.

Long Short Term Memory (LSTM) and Bi Directional Long Short Term Memory (BiLSTM). Long short term memory (LSTM) is a special kind of Recurrent Neural Network (RNN), capable of learning spatial dependencies. They have internal mechanisms called gates (input gate, output gate, forget gate) that can regulate flow of information and can learn data in a sequence. Another core concept of LSTM is the cell state that acts like memory of the network that transfers relative information all the way in the sequence chain. Bidirectional LSTM is an extension of LSTM that can improve model performance on classification problems. BiLSTMs are combination of two LSTMs one fed with data sequence in normal time order and other fed in reverse time order. The outputs of the two networks are then concatenated at each time step.

Gated Recurrent Unit (GRU) and Bidirectional Gated Recurrent Unit (BiGRU). GRU is an improved version of standard recurrent neural network similar to LSTM that aims to solve vanishing gradient problem. Main advantage of GRU is that it can be trained to keep the information from long back without removing it through time or information which is irrelevant to the prediction. As they have few operations, they are little speedier to train. BiGRU is combination of two GRUs one working on normal time order and the other one on reverse time order.

4   Simulation Experiments and Results

Simulation experiments with several bench mark datasets have been done to evaluate the eficiency of convolutional autoencoder in person authentication from smartphone sensor data. For comparison, several other deep learning methods have been used along with some popular traditional machine learning (ML) techniques such as k-nearest neighbour (KNN), Naive Bayes (NB), Support Vector Machine (SVM) and Linear Discriminant Analysis (LDA). In the next subsections, the data sets and simulation results are presented.

4.1   Datasets

1. WISDM (Wireless Sensor Data Mining)
   Smartphone is used to collect the data of 6 activities WALKING, JOGGING, UPSTAIRS, DOWNSTAIRS, SITTING, and STANDING from 36 subjects carrying their mobile device in front leg pocket. There are 1,098,207 samples available in the dataset which are sampled 20 Hz, 46 statistical measure like standard deviation, average absolute difference etc. are used. The details can be found in [22].
2. UCI-HAR data set
   Data of 30 volunteers within age limit of 19–48 are considered, each person performed six activities (WALKING, WALKINGUPSTAIRS, WALKING-DOWNSTAIRS, SITTING, STANDING, LAYING) wearing a smartphone on the waist. It consists of 748406 samples captured at the rate of 50 HZ. The sensor signals were preprocessed applying noise filters and sampled in 2.56 s fixed sliding window and 50% overlap and all the features are normalized and bounded within $[-1,1]$. The details are found in [23].
3. Motion sensor data set
   Data generated by accelerometer and gyroscope sensors with an iPhone 6s kept in the front pocket is collected from 24 persons of 6 activities in 15 trials in the same environment conditions (DOWNSTAIRS, UPSTAIRS, WALKING, JOGGING, SITTING, AND STANDING). This data set is obtained from Queen Mary University of London's repository. The details are in [24].

4.2   Simulation Results

In this section the performance of all the models are evaluated from the simulation results. Classification accuracy, elapsed time, true positive rate (TPR), false positive rate (FPR) and equal error rate (EER) are used to analyze the performance of the models.

Table 1 represents authentication accuracies of different machine learning models for different datasets. It seems that all the deep network based models perform better than traditional ML methods. Figure 1 and Fig. 2 present the comparative results of authentication accuracy for all the data sets by traditional ML methods and DNN based methods respectively. From the figures it can be seen that RF produces the best accuracy among ML methods for all the data sets while proposed CAE based authentication method produces the best accuracy among deep networks based models though CNN and BiGRU also produce high accuracy. As the input data is utilized twice for training in BiLSTM, it has additional training capability and it outperformed LSTM regarding accuracy though it takes longer time. LSTM is comparatively fast but each hidden

Table 1. Authentication accuracies of various models

Accuracies of models (in percentage)			
Classifiers	WISDM	UCIHAR	Motion Sense
KNN	0.44	0.65	0.56
Naive Bayes	0.36	0.59	0.62
SVM	0.46	0.68	0.49
LDA	0.53	0.78	0.74
RF	0.90	0.86	0.80
LSTM	0.66	0.88	0.77
BiLSTM	0.93	0.89	0.81
GRU	0.63	0.80	0.80
BiGRU	0.96	0.90	0.90
CNN	0.98	0.90	0.94
CONV AUTOENCODER	**0.995**	**0.936**	**0.971**

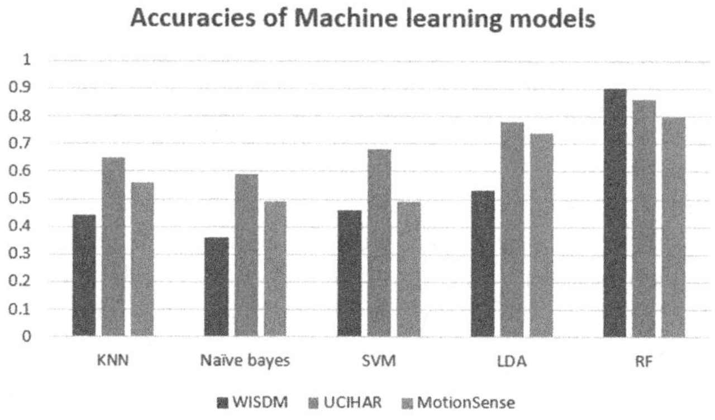

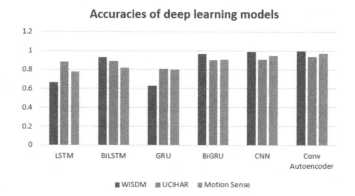

Fig. 1. Authentication accuracies of machine learning models.

Fig. 2. Authentication accuracies of deep learning models.

state has been computed until the previous hidden state computation is complete, training takes a lot of resources and it impacts the accuracy of the model. BiGRU also have same ability to keep memory from previous activations like LSTM but it performs both input and forget gates operation together with its reset gate, so it has fewer tensor operations and it is speedier than LSTM and BiLSTM while producing better accuracy for authentication.

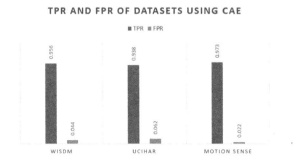

Fig. 3. TPR and FPR of datasets by using CAE

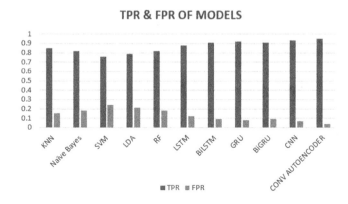

Fig. 4. TPR & FPR of different authentication models

Figure 3 represents TPR and FPR of our proposed CAE based authentication results for all the data sets while Fig. 4 represents TPR, FPR of all other models for WISDM dataset. It is found that deep network based models perform comparatively better than other models with high TPR and low FPR for WISDM data sets. Other data sets also produce similar results. Among all the models, CAE seems to be the best and CNN is the second best in terms of TPR and FPR.

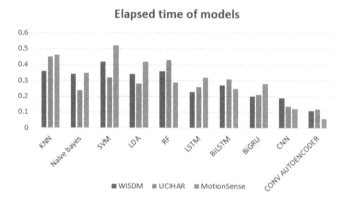

Fig. 5. Time taken for authentication in secs

For better authentication of a person, processing time also plays a key role. Less processing time represents, it can be potentially used for real time authentication of a person. Figure 5 presents authentication time taken by each model for each of the data sets. It is found that the proposed convolutional autoencoder based model is the fastest compared to the other models for all the data sets. Time taken for authentication by CAE seems sufficiently low to implement in real time on smartphone. EER of deep learning models are shown in Table 2. It is seen that CNN and GRU gives good EER values for WISDM and UCIHAR dataset. Low EER values represents that particular model authenticates person efficiently. Convolutional Autoencoder based proposed model produces the best EER values for all the data sets.

Table 2. Equal Error Rate (EER) of deep learning models

EER of deep learning models			
Classifiers	WISDM	UCIHAR	Motion Sense
LSTM	2.74	3.24	2.50
BiLSTM	2.14	3.31	3.82
GRU	1.80	2.45	2.78
BiGRU	4.41	4.25	3.12
CNN	1.92	2.95	2.10
CONV AUTOENCODER	**1.12**	**2.14**	**1.74**

5   Summarization and Conclusion

Person authentication with smartphone sensor data utilizing human gait characteristics by deep neural network model has been studied in this work. The capability of a convolutional autoencoder, a hybrid deep network model previously used in computer vision and image analysis, has been examined and proposed as a suitable candidate for person authentication by gait characteristics captured by smartphone sensor data. The performance of the proposed authentication method has been evaluated by simulation experiments with benchmark datasets and also compared with several traditional machine learning approaches as well as popular deep neural network based methods. It is found that all the deep network models perform better than traditional machine learning classifiers as they are capable of extracting proper features implicitly.

Proposed convolutional autoencoder based model gives the best accuracy in less processing time for all the data sets among all deep neural network based models. It is also found that the proposed CAE based model has the potential for development of real time smartphone based person authentication application. There are many limitations for gait authentication of a person like dressing style or if a person met with an accident walking style of the person changes and also depends on the environment persons walking style changes. So in the future work we will consider all the limitations and apply these techniques for accurate authentication in real time. As convolutional autoencoder yields good results for authentication, by applying continuous authentication techniques and transfer learning approach on this model, it can be a good candidate for developing smartphone based continuous person authentication for health care applications for elderly people.

References

1. Kataria, A.N., Adhyaru, D.M., Sharma, A.K., Zaveri, T.H.: A survey of automated biometric authentication techniques. In: Nirma University International Conference on Engineering (NUiCONE), Ahmedabad, pp. 1–6 (2013). https://doi.org/10.1109/NUiCONE.2013.6780190
2. Zulfiqar, M., Syed, F., Khan, M.J., Khurshid, K.: Deep face recognition for biometric authentication. In: International Conference on Electrical, Communication, and Computer Engineering (ICECCE), Swat, Pakistan, pp. 1–6 (2019). https://doi.org/10.1109/ICECCE47252.2019.8940725
3. Lu, J., Wang, G., Moulin, P.: Human identity and gender recognition from gait sequences with arbitrary walking directions. IEEE Trans. Inf. Forensics Secur. **9**(1), 51–61 (2014)
4. Singha, T.B., Nath, R.K., Nasimhadhan, A.V.: Person Recognition using Smartphones' Accelerometer data (2017). https://arxiv.org/pdf/1711.04689
5. Nixon, M., Tan, T., Chellappa, R.: Human Identification Based on Gait. Springer, Boston (2006). https://doi.org/10.1007/978-0-387-29488-9
6. Sprager, S., Juric, M.B.: Inertial sensor-based gait recognition: a review. Sensors (Basel) **15**(9), 22089-127 (2015). https://doi.org/10.3390/s150922089
7. Sprager, S., Juric, M.B.: An efficient HOS based gait authentication of accelerometer data. IEEE Trans. Inf. Forensics Secur. **10**(7), 1486–1498 (2015)
8. Chen, K., Zhang, D., Yao, L., et al.: Deep learning for sensor-based human activity recognition: overview. Challenges Opportunities (2018). https://arxiv.org/abs/2001.07416

9. Yuan, W., Zhang, L.: Gait classification and identity authentication using CNN. In: Li, L., Hasegawa, K., Tanaka, S. (eds.) AsiaSim 2018. CCIS, vol. 946, pp. 119–128. Springer, Singapore (2018). https://doi.org/10.1007/978-981-13-2853-4_10

10. Zhang, M.: Gait activity authentication using LSTM neural networks with smartphone sensors. In: 15th International Conference on Mobile Ad-Hoc and Sensor Networks (MSN), Shenzhen, China, pp. 456–461 (2019). https://doi.org/10.1109/MSN48538.2019.00092

11. Mahfouza, A., Mahmoud, T.M., Eldin, A.S.: A survey on Behavioural Biometric Authentication on Smartphones. J. Inf. Secur. Appl. **37**, 28–37 (2017)

12. Muhammad, E.U.H., Awais, A.M., Jonathan, L., et al.: Authentication of smartphone users based on activity recognition and mobile sensing. Sensors **17**(9), 2043–2074 (2017)

13. Abbaspour, S., Fotouhi, F., Sedaghatbaf, A., Fotouhi, H., Vahabi, M., Linden, M.: A comparative analysis of hybrid deep learning models for human activity recognition. Sensors **20**(19) (2020)

14. Derawi, M.O., Nickely, C., Bours, P., Busch, C.: Unobtrusive user authentication on mobile phones using Biometric gait recognition. In: Proceedings of the 6th International Conference on Intelligent Information Hiding and Multimedia Signal Processing, Germany, pp. 306–311, October 2010

15. Alzubaidi, A., Kalita, J.: Authentication of smartphone users using behavioral biometrics. J. IEEE Commun. Surv. Tutor. **8**(3), 1998–2026 (2015)

16. Cheheb, I., Al-Maadeed, N., Al-Madeed, S., Bouridane, A.: Investigating the Use of Autoencoders for Gait-based Person Recognition. In: 2018 NASA/ESA Conference on Adaptive Hardware and Systems (AHS), Edinburgh, pp. 148–151 (2018)

17. Chakraborty, B., Nakano, K., Tokoi, Y., Hashimoto, T.: An approach for designing low cost deep neural network based biometric authentication model for smartphone user. In: TENCON 2019–2019 IEEE Region 10 Conference (TENCON), Kochi, India, pp. 772–777 (2019)

18. Zhao, X., Han, X., Su, W., Yan, Z.: Time series prediction method based on Convolutional Autoencoder and LSTM. In: Chinese Automation Congress (CAC), Hangzhou, China, pp. 5790–5793 (2019)

19. Hou, B., Yan, R.: Convolutional autoencoder model for finger-vein verification. IEEE Trans. Instrum. Meas. **69**(5), 2067–2074 (2020)

20. Seyfioğlu, M.S., Özbayoğlu, A.M., Gürbüz, S.Z.: Deep convolutional autoencoder for radar-based classification of similar aided and unaided human activities. IEEE Trans. Aerospace Electron. Syst. **54**(4), 1709–1723 (2018)

21. Gafurov, D., Helkala, K., Sondrol, T.: Biometric gait authentication using accelerometer sensor. J. Comput. **1**(7), 51–59 (2006)

22. Wisdm dataset is publicly. https://www.cis.fordham.edu/wisdm/dataset.php

23. https://archive.ics.uci.edu/ml/datasets/human+activity+recognition+using+smartphones

24. https://www.kaggle.com/malekzadeh/motionsense-dataset

25. Chakraborty, B.: Gait related activity based person authentication with smartphone sensors. In: Proceedings of the 2018 12th International Conference on sensing Technology (ICST), pp. 208–212 (2018)

Research on Personal Credit Risk Assessment Model Based on Instance-Based Transfer Learning

Maoguang Wang and Hang Yang[✉]

School of Information, Central University of Finance and Economics, Beijing, China

Abstract. Personal credit risk assessment is an important part of the development of financial enterprises. Big data credit investigation is an inevitable trend of personal credit risk assessment, but some data are missing and the amount of data is small, so it is difficult to train. At the same time, for different financial platforms, we need to use different models to train according to the characteristics of the current samples, which is time-consuming. In view of these two problems, this paper uses the idea of transfer learning to build a transferable personal credit risk model based on Instance-based Transfer Learning (Instance-based TL). The model balances the weight of the samples in the source domain, and migrates the existing large dataset samples to the target domain of small samples, and finds out the commonness between them. At the same time, we have done a lot of experiments on the selection of base learners, including traditional machine learning algorithms and ensemble learning algorithms, such as decision tree, logistic regression, xgboost and so on. The datasets are from P2P platform and bank, the results show that The AUC value of Instance-based TL is 24% higher than that of the traditional machine learning model, which fully proves that the model in this paper has good application value. The model's evaluation uses AUC, prediction, recall, F1. These criteria prove that this model has good application value from many aspects. At present, we are trying to apply this model to more fields to improve the robustness and applicability of the model; on the other hand, we are trying to do more in-depth research on domain adaptation to enrich the model.

Keywords: Personal credit risk · Big data credit investigation · Instance-based transfer learning

1 Introduction

Personal credit risk is a part that both government and enterprises attach great importance to. A good personal credit risk assessment will not only help government to improve the credit system but also make some enterprises avoid risk effectively. The development of personal credit risk assessment model is from traditional credit assessment model to data mining credit risk assessment model. It has gone through the process from traditional credit assessment model to big data credit assessment model. Traditional credit

© IFIP International Federation for Information Processing 2021
Published by Springer Nature Switzerland AG 2021
Z. Shi et al. (Eds.): ICIS 2020, IFIP AICT 623, pp. 159–169, 2021.
https://doi.org/10.1007/978-3-030-74826-5_14

assessment model often uses discriminant analysis, liner regression, logistic regression, while data mining credit risk assessment model often use decision tree, neural network, support vector machine and other methods to evaluate credit [1].

At present, the existing data mining credit risk assessment models have relatively high accuracy, but only limited to the case of sufficient data and less missing values. When the data volume is small or the data is seriously missing, the prediction effect of the model is often poor. Based on this, we introduces Instance-based Transfer Learning, which migrates the existing large data set samples to the target field of small samples, finding out the commonness between them, and realizing the training of the target domain dataset.

In the other parts, the second section introduces the related works. The third section constructs the personal credit risk assessment model based on the idea of Instance-based transfer. The fourth section introduces the specific experimental process and the comparative analysis of the results. The fifth section is the summary of the full paper.

2 Related Works

The concept of transfer learning was first proposed by a psychologist. Its essence is knowledge transfer and reuse. Actually, it is to extract useful knowledge from one or more source domain tasks and apply it to new target task, so as to realize "renovation and utilization" of old data and achieve high reliability and accuracy. The emergence of transfer learning solves the contradiction between "big data and less tagging" and "big data and weak computing" in machine learning.

In terms of the classification of transfer learning, Pan, S. J. and Yang, Q. [2] summarized the concept of transfer learning and divided transfer learning in 2010 according to learning methods, which can be divided into Instance-based Transfer Learning, Feature based Transfer Learning, Model based Transfer Learning and Relation based Transfer Learning. According to the characteristic attributes, transfer learning can also be divided into Homogeneous Transfer Learning and Heterogeneous Transfer Learning [3]. According to the offline and online learning system, transfer learning can also be divided into Offline Transfer Learning and Online Transfer Learning. Among these classification, Instance-based Transfer Learning is the most commonly used model. The Instance-based Transfer Learning generate rules based on certain weights and reuse the samples. Dia et al. [4] proposed the classic TrAdaboost method, which is to apply the AdaBoost idea into transfer learning. It is used to increase the weight beneficial to the target classification task and reduce the weight harmful to the classification task, so as to find out the commonness between the target domain and the source domain and realize the migration.

At present, transfer learning has a large number of application, but mainly concentrated in Text Classification, Text Aggregation, Emotion Classification, Collaborative Filtering, Artificial Intelligence Planning, Image processing, Time Series, medical and health fields. [5–7] Dia et al. applied Feature based Transfer Learning to the field of text classification and achieved good results [8]. Zhu et al. [9] proposed a Heterogeneous Transfer Learning method in the field of image classification. Pan et al. [10] applied transfer learning algorithms to Collaborative Filtering. Some scholars also use transfer learning framework to solve problems in the financial fields. Zhu et al. introduced

TrBagg which can integrate internal and external information of the system, in order to solve the category imbalance caused by the scarcity of a few samples in customer credit risk [11]. Zheng, Lutao et al. improved TrAdaBoost algorithm to study the relationship between user behavior and credit card fraud [12]. Wang Xu et al. applied the concept of migration learning to quantitative stock selection [13]. But generally speaking, transfer learning is seldom used in the financial field, especially in the field of personal credit risk.

3   The Construction of Personal Credit Risk Assessment Model

3.1   The Build of Instance-Based Transfer Learning

Traditional machine learning assumes that training samples are sufficient and that training and test sets of the data are distributed independently. However, in most areas, especially in the field of financial investigation, these two situations are difficult to meet, data sets in some domains have not only small data volume but also a large number of missing, which leads to the traditional machine learning method can not train very good results. If other data sets are introduced to assist training, it will be unable to train because of the different distribution of the two data sets. In order to solve this problem, we introduces Transfer Learning. In transfer learning, we call the existing knowledge or source domain, and the new knowledge to be learned as the target domain. And Instance-based Transfer Learning, to make maximum use of the effective information in the source domain data to solve the problem of poor training results caused by the small sample size of the target domain data set.

In order to ensure the maturity of the transfer learning framework, we innovatively introduce the classic algorithm of Instance-based Transfer Learning, the tradabost algorithm, to apply to the data in the field of financial credit reference [4]. The TrAdaBoost algorithm comes from the Ensemble Learning-AdaBoost algorithm, which is essentially similar to the AdaBoost algorithm. First of all, it gives weight to all samples, and if a sample in the source data set is misclassified during the calculation process, we think that the contribution of this sample to the destination domain data is small, thus reducing the proportion of the sample in the classifier. Conversely, if the sample in the destination domain is misclassified, we think it is difficult to classify this sample, so we can increase the weight of the sample. The sample migration model built in this paper is based on the tradaboost framework, which is divided into two parts: one is the construction of the tradaboost framework [4, 14], the other is the selection of the relevant base Learners [15]. The following figure shows specific process (Fig. 1).

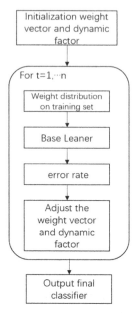

Fig. 1. Instance-based TL

Where we mark the source domain data as T_a, the destination domain data is marked T_b. Take 50% of all the source domain data and the target domain data as the training set T, take 50% of the target domain data as the test set, recorded as S, from which it is not difficult to find that T_b and S are same distribution.

Step 1. Normalized training set $(T_a \cup T_b)$ And each data weight in the test set (S) to make it a distribution.

Step 2. For t = 1,…,N

(1) Set and call the Base Learner.
(2) Calculate the error rate, and calculate the error rate on the training set S.
(3) Calculates the rate of weight adjustment.
(4) Update the weight. If the target domain sample is classified incorrectly, increase the sample weight; if the source domain sample is classified incorrectly, reduce the sample weight.

Step 3. Output final classifier.

3.2 Base Learner Selection

In general, for personal credit risk assessment, the commonly used algorithms include logistic regression, decision tree and other machine learning algorithms, as well as xgboost and other Ensemble Learning algorithms. When the dataset is sufficient, the application of machine learning algorithm on the dataset can achieve good results. Therefore, we can learn from these mature algorithms in the selection of Base Learner, and migrate the algorithm from the source domain to the target domain, so as to achieve better results in the target domain.

Learners are generally divided into weak learners and strong learners. At present, most researches choose weak learners, and then through many iterations to achieve better results. However, we find that in the field of credit risk, some scholars have applied xgboost algorithm and achieved good results [15]. Therefore, according to the characteristics of data in the field of credit risk, this paper selects the strong learner-xgboost algorithm as the Base Learner, which is also convenient for model parameter adjustment and optimization.

XGBoost (extreme gradient boosting) is a kind of Ensemble Learning algorithm, which can be used in classification and regression problems, based on decision tree. The core is to generate a weak classifier through multiple iterations, and each classifier is trained on the basis of the residual of the previous round. In terms of prediction value, XGBoost's prediction value is different from other machine learning algorithms. It sums the results of trees as the final prediction value.

$$\hat{y}_i = \varnothing(x_i) = \sum_{k=1}^{K} f_k(x_i), \ f_k \in F \tag{1}$$

Suppose that a given sample set has n samples and m features, which is defined as

$$D = \left\{ x_i^-, y_i \right\} |D| = n, x_i \in R^m, y_i^- \in R \tag{2}$$

For x_i, y_i, The space of CART tree is F. As follows:

$$F = \left\{ f(x) = w_{q(x)} \right\} \left(q : R^m \to T, w \in R^T \right) \tag{3}$$

Where q is the model of the tree, $w_{q(x)}$ is the set of scores of all leaf nodes of tree q; T is the number of leaf nodes of tree q. The goal of XGBoost is to learn such k-tree model f (x). Therefore, the objective function of XGBoost can be expressed as [13]:

$$obj^t = \sum_{i=1}^{n} l\left(y_j, \hat{y}^{(t)}\right) + \sum_{k=1}^{t} \Omega(f_t) \quad where \ \Omega(f) = \Upsilon T + \frac{1}{2}\lambda\|w\|^2 \tag{4}$$

4   Compare Experiments and Results Analysis

4.1   The Source of the Dataset

The source domain dataset and target domain dataset are from the Prosper online P2P lending website and a bank's April-September 2005, respectively. The data sets of both source domain and target domain data have data missing and high correlation among features. There are only 9000 pieces of data in the destination domain, and the source domain dataset contains more redundant fields. Therefore, it is necessary to fill in the missing values and select features by information divergence.

4.2 Missing Values Processing

There are several common missing value handling methods:

(1) Filling fixed values according to data characteristics;
(2) Fill the median/median/majority;
(3) Fill in the KNN data;
(4) Fill the predicted value of the model;

4.3 Feature Selection

The characteristics of the data will have a positive or negative impact on the experimental results. In particular, the amount of features in the source domains of this paper is huge, including many redundant features and highly relevant features. Firstly, delete redundant features according to the meaning of the features. The following table is the feature dictionary after the features are deleted (Table 1).

Table 1. Deleted characteristic values.

Common ground	Features	Meaning
redundant features	ListingKey	Unique key for each listing, same value as the 'key' used in the listing object in the API
	ListingNumber	The number that uniquely identifies the listing to the public as displayed on the website
	LoanNumber	Unique numeric value associated with the loan
	LenderYield	The Lender yield on the loan. Lender yield is equal to the interest rate on the loan less the servicing fee
	LoanKey	Unique key for each loan. This is the same key that is used in the API
Characteristics related only to investors	LP_InterestandFees	Cumulative collection fees paid by the investors who have invested in the loan
	LP_CollectionFees	Cumulative collection fees paid by the investors who have invested in the loan
	LP_GrossPrincipalLoss	The gross charged off amount of the loan
	LP_NetPrincipalLoss	The principal that remains uncollected after any recoveries
	PercentFunded	Percent the listing was funded
	InvestmentFromFriendsCount	Number of friends that made an investment in the loan
	InvestmentFromFriendsAmount	Dollar amount of investments that were made by friends

We chooses the method of Information divergence to select other features. Information divergence is often used to measure the contribution of a feature to the whole, which also can select features. The basis of Information divergence is entropy, a measure of the uncertainty of random variables. Entropy can be subdivided into information entropy and conditional entropy. The computational formula is shown in Table 2 [16].

Table 2. Calculation formula of entropy.

Information entropy	$H(S) = -\sum_{i=1}^{C} p_i log_2(p_i)$
Conditional entropy	$H(C|T) = P(t)H(C|t) + P(\bar{t})H(C|\bar{t})$

The calculation of Information divergence is based on information entropy and conditional entropy. The computational formula is as follows.

$$IG(T) = H(C) - H(C|T) \tag{5}$$

Using the python program, the entropy of the overall dataset and Information divergence of each feature can be obtained. At the same time, the greater the value of Information divergence, the greater the contribution of the feature to the overall dataset. Since there are many useless features in the source domain data in this paper, Information divergence of each feature is calculated and shown in Fig. 2. In order to keep consistent with the target domain, this paper selects the first 23 features with greater Information divergence to simplify the subsequent calculation process.

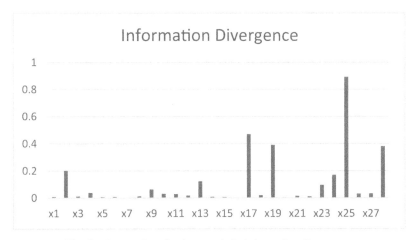

Fig. 2. Source domain characteristic information divergence

4.4 Experimental Results and Comparative Analysis Results

Firstly, apply the XGBoost algorithm to training T_a and T_b. The training results are as follows (Table 3).

Table 3. Training results.

Dataset	T_a	T_b
AUC	0.97	0.56

It is observed that training T_b alone cannot get a better performance. However, using the XGBoost algorithm to train T_a can get a higher AUC value, which proves that it is feasible to use the XGBoost experimental method as a Base Learner.

In the aspect of base learner selection, we have done a lot of experiments, including traditional machine learning algorithm and ensemble learning algorithm. In this paper, we choose xgboost as the base learner to construct the tradapoost (xgboost). The following table shows the experimental results. Figure 3 Shows the AUC value of the tradapoost (xgboost) (Table 4).

Table 4. The results of tradapoost (xgboost)

	AUC	Prediction	Recall	F1
TrAdaBoost(XGBoost)	0.80	0.79	0.65	0.71

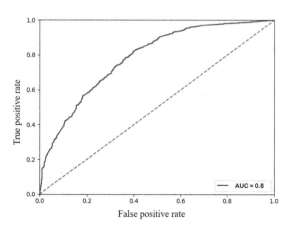

Fig. 3. The AUC of the tradapoost (xgboost)

It can be seen that the accuracy of T_b after transfer is significantly higher than that of training using only XGBoost algorithm.

The experiment is compared from two aspects: (1) Choose different Base Learners, and compare it from the transfer learning dimension.(2) Compare transfer learning with machine learning algorithms.

In the dimension of transfer learning, this paper adds the decision tree as the Base Learner of TrAdaBoost to predict data. Denote the algorithm using decision tree as the base learner as TrAdaBoost (DT). At the same time, Denote the algorithm using

XGBoost as the base learner as TrAdaBoost (XGBoost). Now, we input into the Base Learner using decision tree and XGBoost as TrAdaBoost construction separately to predict the target data. The AUC value is selected as the criterion of result evaluation. The models' evaluation uses AUC, prediction, recall, F1. Table 5 and Fig. 4 show the results.

Table 5. Comparison results-1.

TL VS TL	TrAdaBoost(DT)	TrAdaBoost(XGBoost)
AUC	0.62	0.80
Prediction	0.64	0.79
Recall	0.61	0.65
F1	0.63	0.71

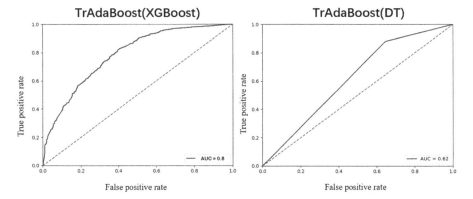

Fig. 4. Comparison results-1

It can be seen from the experimental results that using the Ensemble Learning algorithm XGBoost as the Base Learner increases the AUC value of the base learner by 18% compared with the simple algorithm decision tree as the Base Learner. Therefore, it reveals that the choice of Base Learner has an important influence on the final result.

To demonstrates the superiority of transfer learning algorithm, we also select decision tree, XGBoost, Logistic regression algorithm to predict the target domain respectively. Observe the results of training using only the target domain data and the models in this paper. The results are shown in Table 6 and Fig. 5.

Table 6. Comparison results-2.

TL VS ML	TrAdaBoost(XGBoost)	XGBoost	Decision Tree	Logistic
AUC	0.80	0.56	0.61	0.64
Prediction	0.79	0.59	0.61	0.62
Recall	0.65	0.59	0.64	0.67
F1	0.71	0.59	0.61	0.64

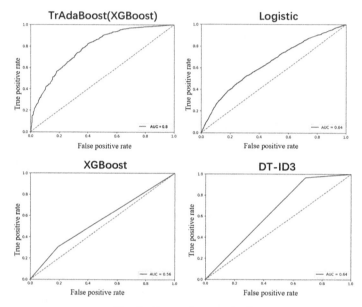

Fig. 5. Comparison results-2

From this, it is clear that using transfer learning algorithms to train the target domain has a higher AUC, prediction, recall and F1 than traditional machine learning. It also further verifies that transfer learning algorithms can better solve the prediction of small samples problem.

5 Conclusion

We constructs person Credit Evaluating Model based on Instance-based Transfer Learning, and focuses on the choice of Basic Learners in the design. The model shows better classification and forecasting capabilities, and can help banking and P2P financial institutions to avoid risks to a certain extent. Besides, the model uses Information divergence to select features with greater contribution to reduce computational complexity. We do a lot of experiments to select the Base Learner and improve the accuracy of the model. The TrAdaBoost (XGBoost) model makes full use of the source domain information to successfully complete the training of the target domain information, and solves the predicament that the data set cannot be trained due to the lack of samples and significant missing values. This article achieves the transfer of samples in the field of personal credit risk, which has certain reference value for the financial field. The model based on TrAdaBoost sample transfer proposed in this paper adds the XGBoost Ensemble Learning algorithm, which improves the accuracy of the model, enhances the performance of the model, and has good generalization capabilities.

References

1. He, S., Liu, Z., Ma, X.: A comparative review of credit scoring models——comparison between traditional methods and data mining. Credit Ref. **037**(002), 57–61 (2019)
2. Pan, S.J., Yang, Q.: A survey on transfer learning. IEEE TKDE **22**(10), 1345–1359 (2010)
3. Weiss, K., Khoshgoftaar, T.M., Wang, D.: A survey of transfer learning. J. Big Data **3**(1), 1–40 (2016). https://doi.org/10.1186/s40537-016-0043-6
4. Dai, W., Yang, Q., Xue, G.-R., Yu, Y.: Boosting for transfer learning. In: ICML, pp. 193–200. ACM (2007)
5. Cook, D., Feuz, K.D., Krishnan, N.C.: Transfer learning for activity recognition: a survey. Knowl. Inf. Syst. **36**(3), 537–556 (2013)
6. Kermany, D.S., et al.: Identifying medical diagnoses and treatable diseases by image-based deep learning. Cell **172**(5), 1122–1131 (2018)
7. Zhuang, F.-Z., Luo, P., He, Q., Shi, Z.-Z.: Survey on transfer learning research. Ruan Jian Xue Bao/J. Softw. **26**(1), 26–39 (2015). https://www.jos.org.cn/1000-9825/4631.html
8. Xiulong, H.: User credit rating modeling based on XGBoost. Comput. Knowl. Technol. **5** (2018)
9. Zhu, Y., et al.: Heterogeneous transfer learning for image classification. In: Burgard, W., Roth, D. (eds.) Proceedings of the AAAI, pp. 1304–1309. AAAI Press (2011)
10. Pan, W., Xiang, E.W., Yang, Q.: Transfer learning in collaborative filtering with uncertain ratings. In: Hoffmann, J., Selman, B. (eds.) Proceedings of the AAAI, pp. 662–668. AAAI Press (2012)
11. Zhu, B., He, C.-Z., Li, H.-Y.: Research on credit scoring model based on transfer learning. Oper. Res. Manage. Sci. **24**(002), 201–207 (2015)
12. Zheng, L., et al.: Improved TrAdaBoost and its application to transaction fraud detection. IEEE Trans. Comput. Soc. Syst. **99**, 1–13 (2020)
13. Xu, W.: Application of transfer learning in quantitative stock selection. Diss. (2019)
14. Zhongzhong, Y.: Ensemble transfer learning algorithm for imbalanced sample classification Acta Electronica Sinica **40**(007), 1358–1363 (2012)
15. Chen, T., Guestrin, C.: XGBoost: a Scalable Tree Boosting System (2016)
16. Liu, J.C., Jiang, X.H., Wu, J.: Realization of a knowledge inference rule induction system. Syst. Eng. **21**(3), 108–110 (2003)

Language Cognition

From Texts to Classification Knowledge

Shusaku Tsumoto[1]([⊠]), Tomohiro Kimura[2], and Shoji Hirano[1]

[1] Department of Medical Informatics, Faculty of Medicine, Shimane University,
Matsue, Japan
{tsumoto,hirano}@med.shimane-u.ac.jp
[2] Medical Services Division, Faculty of Medicine, Shimane University, Matsue, Japan
t-kimura@med.shimane-u.ac.jp

Abstract. Hospital information system stores all clinical information, whose major part is electronic patient records written by doctors, nurses and other medical staff. Since records are described by medical experts, they are rich in knowledge about medical decision making. This paper proposes an approach to extract clinical knowledge from the texts of clinical records. The method consists of the following three steps. First, discharge summaries, which include all clinical processes during the hospitalization, are extracted from hospital information system. Second, morphological and correspondence analysis generates a term matrix from text data. Then, finally, machine learning methods are applied to a term matrix in order to acquire classification knowledge. We compared several machine learning methods by using discharge summaries stored in hospital information system. The experimental results show that random forest is the best classifier, compared with deep learning, SVM and decision tree. Furthermore, random forest gains more than 90% classification accuracy.

Keywords: Discharge summary · Hospital information system · Text mining · Classification learning · Rough sets

1 Introduction

Computerization of patient records enables to store "big unstructured text data" in a Hospital Information System (HIS). For example, our system in Shimane university hospital, where about 1000 patients visit outpatient and about 600 patients stays inpatient, additionally stores about 200 GB text data per year, including patient records, discharge summaries and reports of radiology and pathology. Application of text mining to these resources is very important to discover useful knowledge from text data, which can be viewed as a new type of support of clinical actions, researches and hospital management.

This research is supported by Grant-in-Aid for Scientific Research (B) 18H03289 from Japan Society for the Promotion of Science (JSPS).

This paper is a first step to acquire clinical knowledge from texts: We focus on discharge summaries, which includes all clinical processes during the hospitalization, whose written style is formal, compared with regular records. Thus, conventional text mining approach can be used to extract enough keywords from each text. Then, we propose a method for construction of classifiers of diseases codes from discharge summaries, which consists of the following five steps. First, discharge summaries are extracted from hospital information system. Second, morphological analysis is applied to a set of summaries and a term matrix is generated. Second, correspond analysis is applied to the term matrix with the class labels, and two dimensional coordinates are assigned to each keyword. By measuring the distance between categories and the assigned points, ranking of key words for each category will be generated. Then, keywords are selected as attributes according to the rank, and training example for classifiers will be generated. Finally learning methods are applied to the training examples. We conduct experimental validation with four methods, random forest, deep learning (multi-layer perceptron), SVM and decision tree induction method. The results show that random forest achieved the best performance and the second best was the deep learner with a small difference, but decision tree methods with many keywords performed only a little worse than neural network or deep learning methods. The paper is organized as follows. Section 2 explains our motivation. Section 3 gives a proposed mining process. Section 4 shows the experimental results. Section 5 discusses the results obtained. Finally, Sect. 6 concludes this paper.

2   Motivation

Knowledge acquisition from medical experts is a very classical problem, but still a bottleneck for developing medical expert systems [8,10].

The reason why knowledge acquisition is difficult is that medical experts had difficulties in reformatting their knowledge into if-then rules. Thus, although deeper reasoning for experts emerged in the research of artificial intelligence in medicine in order to solve such bottle neck problems [1,9]. it is still a difficult problem.

However, medical staff write down the process of clinical decision making as texts: it suggests that such knowledge about decision making can be represented as language texts, but not as rule-based. Thus, if we can analyze texts written by medical staff, we will be able to get knowledge about medical decision making.

3   Methods

3.1   Discharge Summary

In this paper, we focus on discharge summaries. A discharge summary includes all clinical processes during the hospitalization, whose written style is formal, compared with regular records. Thus, conventional text mining approach can be used to extract enough keywords from each text. We apply ordinary text mining process as preprocessing.

3.2   Motivation for Feature Selection

It is well known that feature selection is important even for deep learners [6, 7]. Although deep learners gain good performance in image analysis, in other cases, differences between deep learners and other classification method are very small. It may be due to that we have not yet suitable network structure and we may not use suitable features for classification. The empirical fact that deep learners are good at recognition of images suggests that some kind of topological relations should be explicitly embedded into training data. So, here we propose a new feature selection method based on correspondence analysis, which calculates mapping attributes to points of multi-dimensional coordinates. The method can embed the topological relations between keywords and concepts into the data.

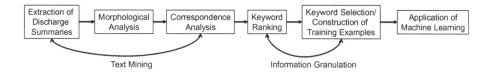

Fig. 1. Mining process

3.3   Mining Process

Figure 1 shows the proposed total mining process pipeline, composed of the following five steps. The first three methods can be called text mining process and the next two methods can be called information granulation process, which determines the granularity of knowledge for classification. Please note that these five methods are automated: after the users impose some conditions for text retrieval, such as collection period, and select the classification model, the process will start and the classification model will be generated.

Extraction of Discharge Summaries. First, discharge summaries are extracted from the hospital information system as texts.

Morphological Analysis. Next, morphological analysis (MeCab [2]), is applied, which outputs a term matrix, that is, a contingency table for keywords and concepts are generated.

Correspondence Analysis. Thirdly, correspondence analysis is applied to a term matrix. Although high dimensional coordinates can be selected, since a very large table is obtained, we focus on two dimensional analysis which can be easily used for visualization. To each key word and concept, two dimensional coordinate is assigned.

Ranking. The coordinates of a concept and a keyword will be used to calculate the euclidean distance between them. We use the distance values for ranking of keywords to each concept: the smaller the distance is, the higher its ranking is.

Table 1. DPC top twenty diseases (fiscal year: 2015)

No	DPC	Cases
1	Cataract (lateral)	445
2	Cataract (bilateral)	152
3	Type II Diabetes Mellitus (except for keto-acidosis)	145
4	Lung Cancer (with surgical operation)	131
5	Uterus Cancer (without surgical operation) 121	
6	Lung Cancer (without surgical operation, chemotherapy)	
7	Uterus benign tumor	111
8	Lung Cancer (without surgical operation, with chemotherapy)	110
9	Shortage of Pregnancy	110
10	Injury of Elbow and Knee	99
11	Autoimmune Disease	96
12	Non-Hodgkin Disease	94
13	Pneumonia	86
14	Lung Tumor (without surgical operation nor chemotherapy)	85
15	Chronic Nephritis	83
16	Liver Cancer	82
17	Gallbladder Stone	82
18	Cerebral Infarction	80
19	Retinal Detachment	75
20	Fetal Abnormalities	75

Keyword Selection. We assume the number of keywords selected before the analysis, say 100. Usually, since the lower rank keywords give information about specific cases, the number of selected keywords will determine the granularity of induced knowledge. Then, keywords whose ranking are up to 100 are selected keywords for classification. Some keywords may overlap, so such overlapped keywords will be deleted. Then, training examples with a classification label and the value of selected keywords (binary attributes) are constructed.

Classification. Finally, classification learning methods will be applied. In this paper, we selected random forest [5], deep learning (multi-layer perceptron) (darch), Support Vector Machine (SVM) [3], Backpropagation Neural Network (BNN) [14] and decision tree (rpart) [11] for comparison.

4   Experimental Evaluation

We selected top 20 frequent DPC codes in fiscal year of 2015 extracted discharge summaries from HIS in Shimane University Hospital. Table 1 shows their statistics and includes the averaged number of characters used in summaries.

Except for extraction from data from HIS, all the processes are implemented on R 3.5.0.

4.1   Mining Process

Correspondence Analysis. For morphological analysis, RMeCab [2] is used and the bag of keywords was generated. From the bag of keywords, a contingency table for these summaries are obtained. Then, correspondence analysis which is implemented in MASS package on R3.5.0 was applied to the table and two dimensional coordinate was assigned to each keyword and each class.[1] Figure 2 shows the two-dimensional plot of correspondence analysis.

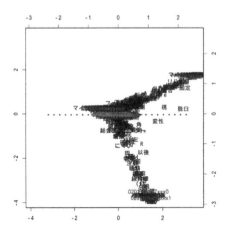

Fig. 2. Correspondence analysis

Ranking. Next, the distances between the coordinate of a keyword and that of a class is calculated, and the ranking of keywords for each class was obtained. By using the ranking, a given number of keywords were selected to generate a classification table.

---

[1] The method can also generate $p(p \geq 3)$-dimensional coordinates. However, higher dimensional coordinates did not give better performance that the experiments below.

Table 2. Number of selected keywords and actually used keywords

#Keyword	Selected keywords	
	DPC Top 10	DPC Top 20
1	10	19
2	19	37
3	27	54
4	36	71
5	44	88
10	88	167
20	115	309
30	247	449
40	334	597
50	406	718
100	724	1125
150	1000	1472
200	1192	1782
250	1382	1932
300	1547	2031
350	1676	2113
400	1797	2192
450	1929	2273
500	2028	2364
750	2304	2808
1000	2545	3000
ALL	13944	20417

Keyword Selection. In the case of 250 keywords, 250 keywords were selected for each DPC code and 5000 keywords in total. But, since the overlapped words should be removed, so only 1932 keywords were used for classification. Thus, some important keywords may be deleted due to the overlap if such words are frequently used at least in two diseases (Table 2).

Classification. Finally, decision tree (package: rpart [11]), random forest (package: randomForest [5], SVM (kernlab [3]), BNN(package: nnet [14]) and Deep Learner (multi-layer perceptron) (darch[2]) were applied to the generated table. For parameters of Darch, the number of intermediate neurons are 10, (10,5), (40,10) and (100,10), whose epoch was 100. For all other packages, the default settings of parameters were used.

---

[2] Darch was removed from R package. Please check the githb: https://github.com/maddin79/darch.

Evaluation Process. Evaluation process was based on repeated 2-fold cross validation [4][3].

First, a given dataset was randomly split into training examples and test samples half in half. Then, training examples was used for construction of a classifier, and the derived classifier was evaluated by using remaining test samples. The above procedures were repeated for 100 times in this experiment, and the averaged accuracy was calculated.

The number of keywords varied from 1 to 1000, selected according to the rank given by correspondence analysis. For analysis, two units of HP Proliant ML110 Gen9 (Xeon E5-2640 v3.2 2.6 GHz 8Core, 64GBDRAM) was used.

Table 3. Experimental results (averaged accuracies)

#keywords	Darch one layer (20)	Darch two layers (40, 20)	Darch two layers (80, 20)	SVM	Rpart	Random Forest	BNN
1	**0.247**	0.236	0.237	0.233	0.202	0.239	0.264
2	**0.442**	0.407	0.43	0.24	0.218	0.288	0.429
3	0.569	0.581	**0.584**	0.324	0.145	0.295	0.541
4	0.632	0.628	0.633	0.424	0.254	**0.676**	0.582
5	0.662	0.655	0.657	0.295	0.315	**0.714**	0.597
10	0.716	0.704	0.71	0.323	0.315	**0.767**	0.633
20	0.786	0.772	0.778	0.664	0.598	**0.826**	0.698
30	0.804	0.792	0.796	0.694	0.652	**0.841**	0.718
40	0.821	0.809	0.814	0.739	0.656	**0.855**	0.74
50	0.823	0.813	0.818	0.742	0.673	**0.855**	0.748
100	0.849	0.841	0.851	0.749	0.577	**0.875**	0.785
150	0.864	0.857	0.867	0.778	0.747	**0.896**	0.806
200	0.865	0.855	0.864	0.784	0.741	**0.907**	0.805
250	0.868	0.862	0.867	0.783	0.744	**0.906**	0.807
300	0.82	0.814	0.821	0.77	0.768	**0.907**	0.798
350	0.826	0.815	0.824	0.767	0.761	**0.907**	0.799
400	0.825	0.818	0.826	0.771	0.764	**0.908**	0.808
450	0.825	0.819	0.83	0.77	0.767	**0.908**	0.802
500	0.832	0.821	0.831	0.77	0.768	**0.908**	0.804
750	0.836	0.831	0.841	0.757	0.782	**0.907**	0.81
1000	0.846	0.836	0.845	0.753	0.79	**0.909**	0.82

All results are obtained by repeated two-fold cross validation (100 repetitions).
Layer(s) denote the number of intermediate layers.
and (a, b) shows the numbers of neurons for intermediate layers.

---

[3] The reason why 2-fold is selected is that the estimator of 2-fold cross-validation will give the lowest estimate of parameters, such as accuracy and the estimation of bias will be minimized.

4.2 Classification Results

Table 3 shows the evaluation results of the top 20 diseases. For four or fewer keywords, all the classifiers showed an accuracy of about 70%. At five or more keywords, however, SVM showed a decrease in accuracy, whereas the other methods showed monotonic increases in accuracy, with the latter plateauing at 200 keywords. The Random Forest method performed better than the other classifiers, followed by Darch deep learning. If more than 250 keywords were selected, the performance of Darch decreased, whereas the performances of random forest and decision trees increased monotonically. Although BNN showed poorer accuracy than Darch (default setting) with 5 to 100 selected keywords, the accuracy of BNN approached that of Darch classifiers with a larger number of keywords. Interestingly, the accuracy of the decision tree method increased monotonically, becoming maximal when all the keywords were used for analysis.

5 Discussion

5.1 Misclassified Cases

Figures 3 and 4 show confusion matrices of random forest and darch (multilayer perceptron), where DPC codes are set in order which means that similar diseases are assigned to similar codes.[4] Yellowed region indicates misclassified

True \ Prediction		010060x099030x brain infarction	020110xx97xxx0 Cataract (Uni)	020110xx97xxx1 Cataract (Bil)	020160xx97xxx0 Retina Detachment (Uni)	040040xx97x0xx Lung Cancer (with Surgical Operation)	040040xx9900xx Lung Cancer (without Surgery)	040040xx9904xx Lung Cancer (with Broncho-scopy)	040040xx99100x Lung Cancer (with Chemo theory)	040080x099x0xx Pneu-monia	060050xx97x0xx Liver Cancer (with Surgical Operation)	060340xx03x00x Gold-bladder Stone & Infla-mmation	070560xx99x0xx Severe Auto-immune Disease	100070xxxxxxxx Type II DB	110280xx99000x Chronic Nephritis	12002xxx99x40x Uterus Cancer	120060xx02xxxx Uterus Benign Tumor	120180xx01xxxx Fetal Ab-normalities	130030xx99x40x Non-hodgkin	140010x199x00x Pregnancy (early)	160620xx01xxxx Injury of Knee or Elbow	Total
010060x099030x	brain infarction	17																				17
020110xx97xxx0	Cataract (Uni)		20	5																		25
020110xx97xxx1	Cataract (Bil)		33	34																		67
020160xx97xxx0	Retina Detachment (Uni)		1		28																	29
040040xx97x0xx	Lung Cancer (with Surgical Operation)					6		1		1												8
040040xx9900xx	Lung Cancer (without Surgery)						16	10	3	2					2							33
040040xx9904xx	Lung Cancer (with Bronchoscopy)						1	16														17
040040xx99100x	Lung Cancer (with Chemotheory)							2	29	1												32
040080x099x0xx	Pneumonia	1	3		1		1	1		17		2		1	1							28
060050xx97x0xx	Liver Cancer (with Surgical Operation)								1	1	11	4		1								18
060340xx03x00x	Goldbladder Stone & Inflammation		5								1	9										15
070560xx99x0xx	Severe Autoimmune Disease								2	1			13		1							17
100070xxxxxxxx	Type II DB										1			15								16
110280xx99000x	Chronic Nephritis	2	2							1			1	4	14							24
12002xxx99x40x	Uterus Cancer															6						6
120060xx02xxxx	Uterus Benign Tumor		1														4					5
120180xx01xxxx	Fetal Abnormalities																	7				7
130030xx99x40x	Nonhodgikin								1										21			22
140010x199x00x	Pregnancy (early)																			3		3
160620xx01xxxx	Injury of Knee or Elbow														1							1
	Total	20	65	39	29	6	18	30	36	24	13	15	14	21	19	6	4	7	21	3	0	390

Fig. 3. Confusion matrix of random forest

[4] DPC codes are three-level hierarchical system and each DPC code is defined as a tree. The first-level denotes the type of a disease, the second-level gives the primary selected therapy and the third-level shows the additional therapy. Thus, in the tables, characteristics of codes are used to represent similarities.

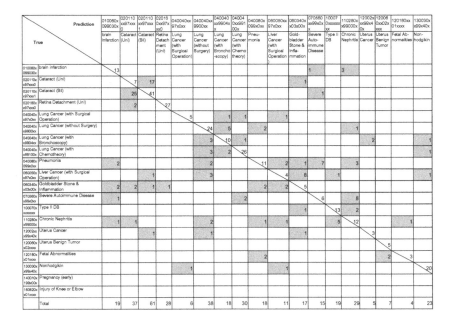

Fig. 4. Confusion matrix of deep learner

cases. It can be observed that whereas errors in random forest are located near the diagonal, errors in darch are more scattered. This suggests that random forest almost correctly classifies a case if similar dpc codes are grouped into one generalized class, while Darch has unexpected errors.

5.2 Execution Time

Table 4 shows an empirical comparison of repeated two-fold cross validations (100 trials). The times need for Random Forest and SVM were 183 and 156 min for 250 keywords, whereas Darch (20) required 672 min. For 1000 keywords, the times needed for Random Forest, SVM and Darch (20) were 261, 288, and 1101 min, respectively. The times required by random forest and BNN methods were close to those of Deep Learners. In the case of Darch, the number of intermediate layers resulted in greater computation times, although the growth rate was smaller than that of BNN. Although the performance of BNN and Deep Learners were high, the problem is that they need more time for computation.

Table 4. Times required for construction of classifications for the top 20 diseases

#keyword	Darch one layer (20)	Darch two layers (40, 20)	Darch two layers (80, 20)	SVM	Rpart	Random Forest	BNN
1	172	226	230	8	0	3	3
2	175	231	247	10	0	6	4
3	177	239	257	12	1	9	4
4	182	245	276	12	0	10	5
5	186	254	288	14	1	11	6
10	201	277	368	21	1	20	11
20	231	362	532	31	2	35	42
30	269	435	729	40	3	48	95
40	302	516	891	48	4	61	171
50	331	575	1063	57	4	76	239
100	453	752	1574	90	7	115	577
150	540	922	1948	114	9	152	943
200	635	1099	2431	142	12	172	1429
250	672	1281	2902	156	13	183	1679
300	751	1263	2701	164	13	198	1850
350	789	1507	3085	172	14	204	2010
400	817	1500	3114	179	15	202	2136
450	833	1550	3311	194	15	222	2296
500	883	1650	3381	201	17	223	2509
750	1027	2177	4802	244	19	271	3542
1000	1110	2504	5030	261	22	288	4062

All results are obtained by repeated two-fold cross validation (100 repetitions).
Layer(s) denote the number of intermediate layers
and (a,b) shows the numbers of neurons for intermediate layers.

5.3   Next Step

Experimental Validation shows that discharge summaries, as natural language texsts, include enough information for classification of disease codes. And it also shows that 200 keywords for each disease are enough for classification. Thus, the next step is to apply automated knowledge acquisition process [12,13] based on rough sets to the term matrix generated before application of learning methods (Fig. 5).

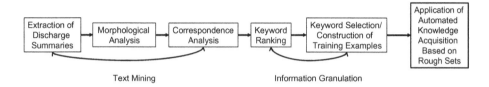

Fig. 5. Mining process for diagnostic rules induction

6   Conclusion

Knowledge acquisition from medical experts is a very classical problem, but still a bottleneck for developing medical expert systems.

The reason why knowledge acquisition is difficult is that medical experts had difficulties in reformatting their knowledge into if-then rules. Thus, although deeper reasoning for experts emerged in the research of artificial intelligence in medicine in order to solve such bottle neck problems: it is still a difficult problem.

However, medical staff write down the process of clinical decision making as texts: it suggests that such knowledge about decision making can be represented as language texts, but not as rule-based. Thus, if we can analyze texts written by medical staff, we will be able to get knowledge about medical decision making.

We propose a method for construction of classifiers of disease codes from discharge summaries as follows. First, discharge summaries are extracted from hospital information system. Then, second, morphological analysis is applied to a set of summaries and a term matrix is generated. Third, correspond analysis is applied to the classification labels and the term matrix and generates two dimensional coordinates. By measuring the distances between categories and the assigned points, ranking of key words will be generated. Then, keywords are selected as attributes according to the rank, and training example for classifiers will be generated. Finally learning methods are applied to the training examples. Experimental validation was conducted by discharge summaries stored in Shimane University Hospital in fiscal year of 2015. The results shows that random forest achieved the best performance around 93% classification accuracy. The second best was deep learners with a small difference, around 91%. Thus, they suggests that discharge summaries includes enough information of clinical processes for each disease code. It will be our future work to acquire more structured knowledge from texts, such as diagnostic rules, rules for therapies, rules for prediction of outcome and so on.

References

1. Amisha, P.M., Pathania, M., Rathaur, V.K.: Overview of artificial intelligence in medicine. J. Family Med. Primary Care **8**(7), 2328–2331 (2019)
2. Ishida, M.: Rmecab. http://rmecab.jp/wiki/index.php?RMeCabFunctions (2016)

3. Karatzoglou, A., Smola, A., Hornik, K., Zeileis, A.: Kernlab - an S4 package for kernel methods in R. J. Stat. Softw. **11**(9), 1–20 (2004). http://www.jstatsoft.org/v11/i09/

4. Kim, J.H.: Estimating classification error rate: Repeated cross-validation, repeated hold-out and bootstrap. Comput. Stat. Data Anal. **53**(11), 3735–3745 (2009). https://doi.org/10.1016/j.csda.2009.04.009

5. Liaw, A., Wiener, M.: Classification and regression by randomforest. R News **2**(3), 18–22 (2002). http://CRAN.R-project.org/doc/Rnews/

6. Mares, M.A., Wang, S., Guo, Y.: Combining multiple feature selection methods and deep learning for high-dimensional data. Trans. Mach. Learn. Data Mining **9**, 27–45 (2016)

7. Nezhad, M.Z., Zhu, D., Li, X., Yang, K., Levy, P.: SAFS: a deep feature selection approach for precision medicine. CoRR abs/1704.05960 (2017). http://arxiv.org/abs/1704.05960

8. Persidis, A., Persidis, A.: Medical expert systems: an overview. J. Manage. Med. **5**(3), 27–34 (1991). https://doi.org/10.1108/EUM0000000001316

9. Riaño, D., Wilk, S., ten Teije, A. (eds.): AIME 2019. LNCS (LNAI), vol. 11526. Springer, Cham (2019). https://doi.org/10.1007/978-3-030-21642-9

10. Shortliffe, E.: Medical expert systems-knowledge tools for physicians. W. J. Med. **145**(6), 830–839 (1986)

11. Therneau, T.M., Atkinson, E.J.: An Introduction to Recursive Partitioning Using the RPART Routines (2015). https://cran.r-project.org/web/packages/rpart/vignettes/longintro.pdf

12. Tsumoto, S.: Automated induction of medical expert system rules from clinical databases based on rough set theory. Inf. Sci. **112**, 67–84 (1998)

13. Tsumoto, S., Hirano, S.: Incremental induction of medical diagnostic rules based on incremental sampling scheme and subrule layers. Fundam. Informaticae **127**(1–4), 209–223 (2013). https://doi.org/10.3233/FI-2013-905

14. Venables, W.N., Ripley, B.D.: Modern Applied Statistics with S, 4th edn. Springer, New York (2002). http://www.stats.ox.ac.uk/pub/MASS4, ISBN 0-387-95457-0

ANAS: Sentence Similarity Calculation Based on Automatic Neural Architecture Search

Dong-Sheng Wang[1]([⊠]), Cui-Ping Zhao[1], Qi Wang[1], Kwame Dwomoh Ofori[1],
Bin Han[1], Ga-Qiong Liu[1], Shi Wang[2]([⊠]), Hua-Yu Wang[3], and Jing Zhang[1]

[1] School of Computer Science, Jiangsu University of Science of Technology, Zhenjiang 212001, Jiangsu, China
[2] Key Laboratory of Intelligent Information Processing, Institute of Computing Technology, Chinese Academy of Sciences, Beijing 100190, China
wangshi@ict.ac.cn
[3] School of Army Engineering University, Nanjing 210007, Jiangsu, China

Abstract. Sentence similarity calculation is one of several research topics in natural language processing. It has been widely used in information retrieval, intelligent question answering and other fields. Traditional machine learning methods generally extract sentence features through manually defined feature templates and then perform similarity calculations. This type of method usually requires more human intervention and constantly has to deal with domain migration problems. Automatically extracting sentence features using certain deep learning algorithms helps to solve domain migration problems. The training data is of the domain is required to complete the extraction process. However, the neural network structure design in the deep learning model usually requires experienced experts to carry out multiple rounds of the tuning design phase. Using an automatic neural architecture search (ANAS) technology deals with all the network structure design problems. This paper proposes a sentence similarity calculation method based on neural architecture search. This method uses a combination of grid search and random search. The method in this paper is tested on the Quora "Question-Pairs" data set and has an accuracy of 81.8%. The experimental results show that the method proposed in this paper can efficiently and automatically learn the network structure of the deep learning model to achieve high accuracy.

Keywords: Natural language processing · Deep learning · Neural architecture search · Sentence similarity calculation

1 Introduction

Sentence similarity calculation is a branch of artificial intelligence in natural language processing. Traditional machine and deep learning methods have proven to be highly efficient when dealing with problems encountered in sentence similarity calculation. The neural network structure design in a deep learning model usually requires experienced experts to carry out multiple rounds of tuning and design. Adjusting the network structure to adapt to the characteristics of the field then becomes necessary.

Z. Shi et al. (Eds.): ICIS 2020, IFIP AICT 623, pp. 185–195, 2021.
https://doi.org/10.1007/978-3-030-74826-5_16

At present, research done in automatic searches under neural architecture search have produced promising results. This paper proposes a sentence similarity calculation model based on neural architecture search. The model uses padding and label encoding technology to convert text inputs into a real number vector. It applies the GloVe model to get the word embedding representation next and finally returns a numerical output between 0 and 1. Adam algorithm is then used for intensive training. This algorithm ascertains the similarity between the two sentences according to the numerical results output by the model. The proposed model uses neural architecture search technology to initiate the model generation process and strategy. It then applies the random search method to iteratively find a better model combination. The main purpose of this paper is to design a deep neural network model using an automatic neural architecture search with minimal human intervention. This way, the domain migration ability increases hence the efficiency of the model in general is improved.

2 Related Work

Many researchers have conducted extensive research in sentence similarity calculation. Zhai Sheping [12] fully considered sentence structure and semantic information. He proposed a method for calculating semantic similarity of sentences based on multi-feature fusion of semantic distance. Liu Jiming et al. [13] combined smooth inverse frequency (SIF) with dependent syntax to calculate sentence similarity. Ji Mingyu et al. [14] calculated sentence similarity by reducing and normalizing the difference features between sentence vectors. He Yinggang [15] proposed a sentence similarity calculation method based on word vectors and LSTM… There are also numerous studies done on neural architecture search. Google [16] proposed neural architecture search for convolutional neural networks. Qi Fei et al. fitted a multivariate Gaussian process function to construct an acquisition function for optimized search based on the trained neural network architecture map and the corresponding evaluation value… Although the above method was highly efficient, there are recurring problems such as low accuracy and time-consuming search process. Therefore, this paper proposes an automated neural architecture search technology which greatly improves efficiency.

3 Our Approach

This paper can be divided into two aspects: the first part is the research and realization of automated neural architecture search technology; the second is the implementation of a deep learning model for sentence similarity calculation based on neural architecture search.

3.1 Neural Architecture Search Technology

With the increasing demand of deep learning models, designing a complex network structure requires more time and effort. Therefore, neural architecture search is used to automate the construction of neural networks. This section describes the three key components of ANAS: search space, search strategy and search performance evaluation strategy.

Search Space. There are two common search space types in the current research results which are the chain structure neural network space and the unit search space. The NAS search architecture used in this article refers to the idea of unit search space. By pre-defining some neural network structural units and then using NAS to search the stacking mode of these network units to build a complete neural network. The basic network block is created as the main search object and ultimately generates good models.

Search Strategy. Although the search space is theoretically limited, it requires a certain search strategy to balance exploration and utilization: on one hand, it is necessary to search for a good-performance architecture rapidly, on the other hand, it should be avoided due to the repercussion of early convergence.

Many different search strategies like Bayesian optimization and evolutionary algorithms are used to explore the neural architecture space. Restricted by experimental conditions, this paper uses a search strategy that combines grid search and random search commonly used in machine learning. Firstly, we use grid search on the framework parameter and then perform random search on other parameters. Next, we combine them into a complete model for training and evaluation. Finally, we keep the model parameters with the highest accuracy on the test set.

The ANAS search strategy algorithm used in this paper is as follows:

Algorithm ANAS Strategy Based on Random Search of Network Blocks algorithm
Input: training data set, search space
Output: good model in search space
Begin
Define search space, including learning rate space, batch size space, framework space, dense space, dropout space, random space.
(2)For i=1 to length(framework space) do
(3)   For j=1 to length(random space) do
(3)   Get model block parameters in search space randomly.
(4)   Generate a complete model by model block parameters in (3).
(5)   Training model in data set, get the result
(6)   If the results are better now than last time.
then record and cover model block parameters.
Else
continue.
(7)   End;
(8)   End;
(9) Get good model parameters.
(10)End.
End.

The input of the algorithm is the training set and the search space. The output is a set of network parameters. Using this set of parameters, a complete model can be generated. The model performance outperforms other models with same functions.

In the above analysis, the data set used is a subset of the target data set. The specific determination of the search space depends on the outcome of search space experiments carried out. The search space parameters used in this paper are shown in Table 1 below:

Table 1. Margins and print area specifications.

ANAS parameters	Value range
Learning rate space	(0.0001, 0.1)
Batch size space	(50, 10000)
Framework space	(0, 20)
Dense space	(10, 500)
Dropout space	(0, 0.5)
Random search times	100

The parameters in Table 1 are mainly defined based on the experience of designing deep learning models in the past. The upper limit of Random search times is the number of all permutations and combinations in the search space, but subject to the computing power of the experimental platform. We set this value to 100 and eventually generate 2000 models. The model was trained and tested, and it took more than 60 consecutive hours to get the final result.

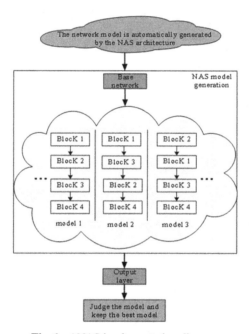

Fig. 1. ANAS implementation diagram

The neural architecture search technology proposed in this paper is based on the backbone network. It is also based on the network block stacking and internal parameter changes. It randomly searches for network combinations that perform well. The ANAS implementation is shown in Fig. 1.

Performance Evaluation Strategy. The goal of ANAS is to find a model architecture that achieves high predictive performance on unknown data, Therefore, effective performance evaluation strategies need to be developed. Considering the computational cost, only standard training and verification of 2000 generated models were performed. Under the premise of GPU acceleration, it took more than 60 h. In the subsequent research, measures need to be taken to improve the efficiency of ANAS search.

3.2   Model Implementation and Evaluation Optimization

Model Construction. This paper uses TensorFlow and Keras to implement the construction, training, and prediction of deep learning models. The deep learning model structure used in this paper is shown in Fig. 2:

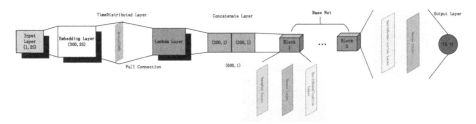

Fig. 2. Discriminant model neural network architecture

The input sentence is vectorized using the Glove word embedding of feature engineering and transformed into a real number matrix of shape (1, 25), which respectively represents two sentences to be differentiated. After the Embedding layer is processed, the matrix becomes (300, 25) and each vocabulary is transformed into a word embedding vector with a length of 300 dimensions. After calculation, a result of (1, 1) size is output and the value ranges between 0–1. The next task is to determine if the intent of the two sentences are the same according to the magnitude of the value. The model is implemented using TensorFlow-based Keras. Embedding is the advanced neural network layer that comes with Keras. The input layer serves as the initial occupancy layer of the Keras network and is used to agree on the matrix shape of the input data. The Embedding layer implements the mapping of data to the word embedding vector. The weight of the word embedding vector is set through the weight's parameter. The trainable parameter is usually set to False so that the training can proceed normally. The TimeDistributed layer is an advanced encapsulator. The application in this model is equivalent to independently applying a fully connected layer with 300 neurons on each vocabulary output by the Embedding layer. The Lambda layer is used to encapsulate any expression into the Layer object of the network model. In this model, the Lambda layer selects the word embedding vector with the largest activation value of each word in the output of the TimeDistributed layer. This is done to improve the nonlinear learning ability of the model. The Concatenate layer is used to connect two matrices together. In this model, the (300, 1) matrix output by the two Lambda layers is connected into a (600, 1) matrix. The

Dropout layer receives dropout parameters between 0 and 1 and randomly discards the output results of the previous layer according to the parameter size. This process only takes effect during network training which then reduces the overfitting of the model. The dense layer is a fully connected layer. After setting the number of neurons, Keras will automatically establish weight connections between all neurons in this layer and all outputs of the previous layer. BatchNormlization will normalize the calculation data of the current batch which reduces the over-fitting of the model.

Model Training. The Adam algorithm is used to train the model with the help of the fit method of the Keras framework. Using Adam algorithm can speed up the training speed, making the model converge in less epoch training. TensorFlow and Keras are used directly in the training, "*model.fit([x1, x2], y, epochs = 50, validation_split = 0.1,batch_size = 4096)*" can be executed to achieve model training.

Part of the information during the model training process is shown in Table 2 below:

Table 2. Part of the information during model training

epoch	time cost	loss	acc	val_loss	val_acc
1/50	6 s 18 us	0.6028	0.6781	0.5384	0.7247
2/50	4 s 11 us	0.5047	0.7484	0.5259	0.7264
3/50	4 s 11 us	0.4644	0.7728	0.4968	0.7537
4/50	4 s 11 us	0.4332	0.7913	0.4567	0.7797
5/50	4 s 11 us	0.4042	0.8080	0.4461	0.7877
6/50	4 s 11 us	0.3810	0.8219	0.4299	0.7937

In Table 2, val_loss and val_acc respectively represent the loss function value and accuracy of the current training batch of the model in the validation set.

Model Evaluation and Optimization. There are many ways to optimize models. This article only considers optimizing the accuracy of the model.

Underfitting. The word embedding of the GloVe model and TimeDistribute layer are used to improve the network structure. After embedding is output, a fully connected layer of the same 300 dimensions is applied to the representation of each word to make up for the insufficiency of the embedding layer trainable = False setting and improve the accuracy of the model. In order to deal with under-fitting, the neural network structure is deepened and the basic network. After the block, the neural architecture search mechanism determines how deep the network needs to be stacked.

Overfitting. By reducing the number of network parameters or the Dropout method, the Dropout layer of Keras receives the parameter rate and its value should be set between 0 and 1. This indicates the proportion of discarding. Use BatchNormlization layer to reduce model overfitting and speed up training.

4 Experiment

4.1 Datasets

The training set data uses a data set published by Quora in the form of a competition on the Kaggle platform with a size of about 55MB and a total of about 400,000 training samples. The format of the original data set is a text format with tabs separating different columns and line breaks separating different rows and each row represents a labeled sample.

Table 3. Example of training data

Index	Question1	Question2	is_duplicate
1	What is the step by step guide to invest in sh...	What is the step by step guide to invest in sh...	0
2	What does manipulation mean?	What does manipulation mean?	1
3	Why are so many Quora users posting questions...	Why do people ask Quora questions which can be...	1

As shown in Table 3, mark 0 means that the intent of question 1 and question 2 are inconsistent, and mark 1 means that the schematic diagram is consistent.

To train and evaluate the model, the labeled dataset is divided into three subsets: training set, validation set and test set. There is no strict basis for the division ratio of the data set. In this work, we divide the data set into 3 parts. The test set (10%), the validation set (5%) and the training set (85%).

4.2 Evaluation Index

The focus of this paper is to calculate the accuracy of the model but also to consider other evaluation indicators.

Model Accuracy. The research focuses on the accuracy of the model. The following is the accuracy calculation formula:

$$A = \frac{R}{S}$$

A is the accuracy rate, R is the number of samples the model predicts correctly in the data set and S is the number of all samples in the data set.

The following figure shows the changes in accuracy on training set and validation set during the training process of the model (Fig. 3):

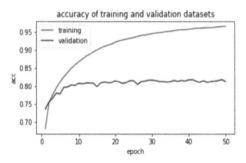

Fig. 3. Changes in the accuracy of the model in the training set

Table 4 shows the accuracy results of the three data sets after the model training is completed.

Table 4. Accuracy result after model training is completed

	Training set	Validation set	Test set
Accuracy	0.9640	0.8185	0.8180

Training Time and Prediction Time-Consuming. Model training time-consuming is mainly related to model structure, optimization algorithm, experimental software and hardware environment. GPU acceleration is used in the experimental environment of this paper. Table 5 shows whether GPU acceleration is used to train the model influences:

Table 5. The impact of GPU acceleration on training time

	Use GPU	Without GPU
1 training batch time	4.04 s	54.70 s
Total training time	202 s	2735 s
rate	1	13.54

Table 5 shows that in the experimental environment of this article, keeping other configurations unchanged and using GPU when training the model can be 13.54 times faster than not using GPU. In actual application, how long the model needs to be used to obtain the prediction result is also a factor that needs to be considered. This is slightly different from the factors that affect the time-consuming model training. It is related to the structure of the model and the experimental software and hardware environment, but is related to the use The optimization algorithm has nothing to do.

Table 6. The impact of GPU on model prediction time

	Use GPU	Without GPU
Time-consuming prediction of 1 sample	5 ms	17 ms
Time-consuming prediction of 10 samples	5 ms	17 ms
Time-consuming prediction of 100 samples	11 ms	39 ms

It is reflected from Table 6 that the prediction time is very short, in the order of milliseconds. However, the time consumption of predicting 1 sample and predicting 10 samples or even predicting 50 samples has not changed much in the two cases of using GPU acceleration and not using GPU acceleration. The reason is the design of the model in this article and the characteristics of TensorFlow. It is decided that the model can calculate multiple samples at the same time, which is the result of automatically calling multi-threaded calculations behind TensorFlow. Therefore, within a certain range of predictions, the prediction results can be quickly obtained without being affected by the number of samples.

4.3   Experimental Results and Analysis

This chapter compares and analyzes the training results of the deep learning model proposed mainly considering the accuracy and time-consuming aspects of the model. The models participating in the comparison include random forest algorithm, gradient boosting tree algorithm (GBDT), and support vector machine (SVM) model.

Model Accuracy. Table 7 shows the accuracy of each model.

Table 7. Accuracy performance of each model

	Acc. on the training set	Acc. on the test set
Random forest	0.9459	0.7450
GBDT	0.9843	0.7706
SVM	0.7875	0.6799
ANAS	0.9640	0.8180

The ANAS in Table 7 refers to the model used in this paper. The results in the table cannot directly judge the pros and cons of different model algorithms. Because this paper does not perform feature engineering on the data suitable for the specific algorithm, the data results does not reflect the best performance of the model on this data set.

Comparison of Prediction Time-Consuming Models. Table 8 shows the prediction time-consuming results of each model.

Table 8. Forecast time-consuming results of each model

	Prediction 1 sample	Prediction 10 samples	Prediction 100 samples
Random forest	110 ms	115 ms	270 ms
GBDT	3 ms	3 ms	9 ms
SVM	95 ms	98 ms	250 ms
ANAS	5 ms	5 ms	11 ms

As shown in Table 8, the prediction time is greatly affected by the implementation framework. This paper uses the random forest algorithm and support vector machine implemented by Sklearn. It also uses the LightGBM provided by Microsoft to implement the gradient boosting tree algorithm. A deep learning model implemented by Tensor-Flow and Keras is also used. Sklearn is mainly used in the data preprocessing stage and did not do calculation optimization work. Microsoft's LightGBM and Google's Tensor-Flow are commercial open source applications that have made a lot of optimizations on computing performance. Therefore, the time-consuming is significantly better than the model implemented by Sklearn.

5   Conclusion

This paper mainly focuses on the problem of sentence similarity calculation. In order to use the deep learning method, the deep learning model and neural architecture search are discussed and studied. Feature engineering chooses GloVe word embedding and this choice is also confirmed by the results. The research of neural architecture search mainly realizes the search based on the unit search space, creating the backbone network and the basic network block. We combine grid search and random search strategies using network generators and neural architecture search engines. Finally, a relatively good deep learning model is formed. The key points of the deep learning model are evaluated and analyzed. The basic structure and function principle and optimization direction of the deep learning model are described and summarized. The model obtained in this paper scores an accuracy of 81.8% on the test set. The results indicate that using the ensemble learning techniques commonly used in machine learning competitions, combined with multiple model methods should be further improved.

Acknowledgement. This paper is funded by the National Natural Science Foundation of China "Research on Automatic Learning Method of Constrained Semantic Grammar for Short Text Understanding" (61702234).

References

1. Angeli, G., Manning, C.D.: Naturalli: natural logic inference for common sense reasoning. In: EMNLP (2014)

2. Loshchilov, I., Hutter, F.: SGDR: stochastic gradient descent with warm restarts. In: International Conference on Learning Representations (2017)
3. Bahdanau, D., Cho, K., Bengio, Y.: Neural machine translation by jointly learning to align and translate. In: ICLR (2015)
4. Dagan, I., Glickman, O., Magnini, B.: The PASCAL recognising textual entailment challenge. In: Quiñonero-Candela, J., Dagan, I., Magnini, B., d'Alché-Buc, F. (eds.) MLCW 2005. LNCS (LNAI), vol. 3944, pp. 177–190. Springer, Heidelberg (2006). https://doi.org/10.1007/117367 90_9
5. Cubuk, E.D., Zoph, B., Mane, D., Vasudevan, V., Le, Q.V.: AutoAugment: learning augmentation policies from data. arXiv:1805.09501, May 2018
6. Hermann, K.M., et al.: Teaching machines to read and comprehend. In: NIPS (2015)
7. Vinyals, O., Fortunato, M., Jaitly, N.: Pointer networks. In: NIPS (2015)
8. Ahmed, K., Torresani, L.: MaskConnect: connectivity learning by gradient descent. In: Ferrari, V., Hebert, M., Sminchisescu, C., Weiss, Y. (eds.) ECCV 2018. LNCS, vol. 11209, pp. 362–378. Springer, Cham (2018). https://doi.org/10.1007/978-3-030-01228-1_22
9. Huang, G., Liu, Z., Weinberger, K.Q.: Densely connected convolutional networks. In: Conference on Computer Vision and Pattern Recognition (2017)
10. Yu, F., Koltun, V.: Multi-scale context aggregation by dilated convolutions (2016)
11. Chen, T., Goodfellow, I.J., Shlens, J.: Net2net: accelerating learning via knowledge transfer. In: International Conference on Learning Representations (2016)
12. Zhai, S., Li, Z., Duan, H., et al.: Sentence semantic similarity method based on multifeatured fusion. Comput. Eng. Design (2019). (In Chinese)
13. Liu, J., Tan, Y., Yuan, Y.: Calculation method of sentence similarity based on smooth inverse frequency and dependency parsing. Chongqing University of Posts and Telecommunications (2019). (In Chinese)
14. Ji, M., Wang, C., An, X., et al.: Method of sentence similarity calculation for intelligent customer service. Comput. Eng. Appl. **55**(13), 123–128 (2019). (In Chinese)
15. He, Y., Wang, Y.: Method of sentence similarity calculation for intelligent customer service. J. Yangtze Univ. (Self Sci. Edn.) (2019). (In Chinese)
16. Google: Neural architecture search for convolutional neural networks: CN201880022762.8[P], 19 November 2019. (In Chinese)

Fully Interval Integer Transhipment Problem - A Solution Approach

G. Sudha[1(✉)], G. Ramesh[1(✉)], D. Datta[2(✉)], and K. Ganesan[1(✉)]

[1] Department of Mathematics, SRM Institute of Science and Technology,
Kattankulathur, Tamil Nadu, India
{sudhag,ganesank}@srmist.edu.in

[2] Adjunct Faculty, Department of Mathematics, SRM Institute of Science and Technology,
(Former) Health Physics Division, Bhabha Atomic Research Centre, Trombay,
Mumbai 400085, India

Abstract. We present an improved interval Vogel's approximation algorithm for the initial feasible solution of a transhipment problem involving fully interval integer numbers in this paper. The interval variant of MODI algorithm is used to measure the optimality of the interval feasible solution. In a transportation problem, shipments are only permitted between source-sink pairs, while in a transhipment problem, shipments can be permitted both between sources and sinks. Furthermore, there may be points where units of a substance may be transferred from a source to a sink. To demonstrate the effectiveness of the proposed algorithm, a numerical example is given.

Keywords: Interval numbers · New interval arithmetic · Ranking · Interval initial feasible solution · Improved interval Vogel's approximation algorithm

1 Introduction

Goods are transported from specific sources to specific destinations in classic transportation problems (TP). However, in some situations, products can be delivered to a specific location either directly or through one or more intermediary nodes. Either of these nodes has the ability to function as an origin or a destination. Transhipment issues are models with these additional features. Thus, transhipment issues are a type of transportation problem in which merchandise are elated from a source to a destination via various intermediary nodes (sources/destinations), with the possibility of changing modes of transportation or consolidating or de-consolidating shipments. A transhipment problem's main goal is to find the lowest transportation cost while satisfying supply and demand at each node. These issues have found a lot of use in the e-commerce era, when the online shopping business is thriving thanks to its customers' convenience. To solve the transhipment problem, translate it to an equivalent TP first, and then use existing techniques to find an initial simple feasible solution. Furthermore, an optimal solution of equivalent TP can be obtained using the modified distribution (MODI) process. Orden

© IFIP International Federation for Information Processing 2021
Published by Springer Nature Switzerland AG 2021
Z. Shi et al. (Eds.): ICIS 2020, IFIP AICT 623, pp. 196–206, 2021.
https://doi.org/10.1007/978-3-030-74826-5_17

[17] introduced the concept of transhipment problem in the year 1956. When the decision parameters are known vaguely, the classical mathematical techniques are successfully applied to model and solve the transhipment problems. But actually there are problems that the charge coefficients and the supply and demand quantities may be uncertain due to many reasons. Hence we must deal with vague in order in formulating and solving real earth troubles. Hurt et al. [5] have developed an alternative formulations of the transhipment problem.

Ganesan et al. [3] discussed on arithmetic operations of interval numbers and some properties of interval matrices. Pandian et al. [11] studied a new method for finding an optimal solution of fully interval integer transportation problems. NagoorGani et al. [8] developed a transhipment problem in fuzzy environment. Rajendran et al. [12] discussed fully interval transhipment problems. Akilbasha et al. [1] formulated an innovative exact method for solving fully interval integer transportation problems.

Sophia Porchelvi et al. [14] studied a comparative study of optimum solution between interval transportation and interval transhipment problem. In this paper, we propose a new interval arithmetic and an improved interval Vogel's approximation algorithm to solve the problem of interval transhipment without translating into crisp form.

The structure of this article is given as follows: Sect. 2 deals with the basic definitions. Section 3 deals with the mathematical formulation of fully interval transhipment problem. Section 4 introduces improved interval Vogel's approximation algorithm under generalized interval arithmetic. Section 5 provides an example to illustrate the theory developed in this paper. Section 6 concludes this manuscript.

2 Preliminaries

The purpose of this segment is to provide some observations, ideas and results which are useful in our further consideration.

2.1 Interval Numbers

Let $\tilde{a} = [a_1, a_2] = \{x \in R: a_1 \leq x \leq a_2 \text{ and } a_1, a_2 \in R\}$ be an interval on the real line R. If $\tilde{a} = a_1 = a_2 = a$, then $\tilde{a} = [a, a] = a$ is a real number (or a degenerate interval). The terms interval and interval number are used interchangeably. The mid-point and width (or half-width) of an interval number $\tilde{a} = [a_1, a_2]$ are defined as $m(\tilde{a}) = \left(\frac{a_1 + a_2}{2}\right)$ and $w(\tilde{a}) = \left(\frac{a_2 - a_1}{2}\right)$. The interval number $\tilde{a}$ can also be expressed in terms of its midpoint and width as $\tilde{a} = [a_1, a_2] = \langle m(\tilde{a}), w(\tilde{a})\rangle$. We use IR to denote the set of all interval numbers definedon the real line R.

2.2 Ranking of Interval Numbers

Sengupta and Pal [13] suggested a easy and powerful index to compare any two intervals on IR through the satisfaction of decision-makers.

Definition 2.2.1. Let $\preceq$ be an extended order relation between the interval numbers $\tilde{a} = [a_1, a_2], \tilde{b} = [b_1, b_2]$ in IR, then for $m(\tilde{a}) < m(\tilde{b})$, we construct a premise $(\tilde{a} \preceq \tilde{b})$ which implies that $\tilde{a}$ is inferior to $\tilde{b}$ (or $\tilde{b}$ is superior to $\tilde{a}$).

An *acceptability function* $A_{\preceq} : \mathrm{IR} \times \mathrm{IR} \to [0, \infty)$ is defined as:

$$A_{\preceq}(\tilde{a}, \tilde{b}) = A(\tilde{a} \preceq \tilde{b}) = \frac{m(\tilde{b}) - m(\tilde{a})}{w(\tilde{b}) + w(\tilde{a})}, \text{ where } w(\tilde{b}) + w(\tilde{a}) \neq 0.$$

$A_{\prec}$ may be interpreted as the grade of acceptability of the first interval number to be inferior to the second interval number. For any two interval numbers $\tilde{a}$ and $\tilde{b}$ in IR either $A(\tilde{a} \preceq \tilde{b}) \geq 0$ (or) $A(\tilde{b} \succeq \tilde{a}) \geq 0$ (or) $A(\tilde{a} \preceq \tilde{b}) = 0$ (or) $A(\tilde{b} \succeq \tilde{a}) = 0$ (or) $A(\tilde{a} \preceq \tilde{b}) + A(\tilde{b} \preceq \tilde{a}) = 0$. If $A(\tilde{a} \preceq \tilde{b}) = 0$ and $A(\tilde{b} \preceq \tilde{a}) = 0$, then we say that the interval numbers $\tilde{a}$ and $\tilde{b}$ are equivalent (non-inferior to each other) and we denote it by $\tilde{a} \approx \tilde{b}$. Also if $A(\tilde{a} \preceq \tilde{b}) \geq 0$, then $\tilde{a} \preceq \tilde{b}$ and if $A(\tilde{b} \preceq \tilde{a}) \geq 0$, then $\tilde{b} \preceq \tilde{a}$.

2.3 A New Interval Arithmetic

Ming Ma et al. [7] suggested a new fuzzy arithmetic based on the index of locations and the index function of fuzziness. The location index number follows the usual arithmetic where as the fuzziness index functions are assumed to obey the lattice law which is the least upper bound and the greatest lower bound. That is for $a, b \in L$ we define $a \vee b = \max\{a, b\}$ and $a \wedge b = \min\{a, b\}$.

For any two intervals $\tilde{a} = [a_1, a_2], \tilde{b} = [b_1, b_2] \in \mathrm{IR}$ and for $* \in \{+, -, \cdot, \div\}$, the arithmetic operations on $\tilde{a}$ and $\tilde{b}$ are defined as:

$$\tilde{a} * \tilde{b} = [a_1, a_2] * [b_1, b_2] = \langle m(\tilde{a}), w(\tilde{a}) \rangle * \langle m(\tilde{b}), w(\tilde{b}) \rangle$$
$$= \langle m(\tilde{a}) * m(\tilde{b}), \max\{w(\tilde{a}), w(\tilde{b})\} \rangle.$$

In particular, here we should use following interval arithmetic operations like addition, subtraction, multiplication and division.

3 Main Results

3.1 Mathematical Formulation for Using Fully Interval Transhipment Problem

In the transshipment table $O_1, O_2, O_3, \ldots \ldots O_i \ldots \ldots O_m$ are sources form where goods are to be transported to destinations $D_1, D_2, D_3, \ldots \ldots D_j \ldots \ldots D_n$. Any of the sources can transport to any of the destinations. $\tilde{c}_{ij}$ is per unit transporting cost of goods from i^{th} source O_i to j^{th} destination D_j for all $i = 1, 2, \ldots m$ and $j = 1, 2, \ldots n$. $\tilde{x}_{ij}$ is the amount of goods transporting from i^{th} source O_i to j^{th} destination D_j. $\tilde{a}_i$ be the amount of goods available at the origins O_i and $\tilde{b}_j$ the demand at the destination D_j. The mathematical model of fully interval transportation problem is as follows (Table 1)

$$\text{Minimize } \tilde{Z} \approx \sum_{i=1}^{m} \sum_{j=1}^{n} \tilde{c}_{ij} \tilde{x}_{ij}$$

$$\text{subject to } \sum_{j=1}^{n} \tilde{x}_{ij} \approx \tilde{a}_i, \quad i = 1, 2, 3, \ldots, m$$

$$\sum_{i=1}^{m} \tilde{x}_{ij} \approx \tilde{b}_j, \quad j = 1, 2, 3, \ldots, n$$

$$\sum_{i=1}^{m} \tilde{a}_i \approx \sum_{j=1}^{n} \tilde{b}_j, \quad \text{where } i = 1, 2, 3, \ldots, m; \; j = 1, 2, 3, \ldots, n \text{ and}$$

$$\tilde{x}_{ij} \succeq \tilde{0} \quad \text{for all } i \text{ and } j$$

Table 1. Interval transhipment table

		1	2	...	i	...	m	m+1	m+2	...	m+j	...	m+n	
		O_1	O_2	...	O_i	...	O_m	D_1	D_2	...	D_j	...	D_n	supply
1	O_1	x_{11}	x_{12}	...	x_{1i}	...	x_{1m}	$x_{1,m+1}$	$x_{1,m+2}$		$x_{1,m+j}$	...	$x_{1,m+n}$	a_1
2	O_2	x_{21}	x_{22}	...	x_{2i}	...	x_{2m}	$x_{2,m+1}$	$x_{2,m+2}$		$x_{2,m+j}$	...	$x_{2,m+n}$	a_2
...	...	...	...	...	...	...	...	...	...	...	...	...	...	...
i	O_i	x_{i1}	x_{i2}	...	x_{ii}	...	x_{im}	$x_{i,m+1}$	$x_{i,m+2}$		$x_{i,m+j}$	...	$x_{i,m+n}$	a_i
...	...	...	...	...	...	...	...	...	...	...	...	...	...	...
m	O_m	x_{m1}	x_{m2}	...	x_{mi}	...	x_{mm}	$x_{m,m+1}$	$x_{m,m+2}$		$x_{m,m+j}$	...	$x_{m,m+n}$	a_m
m+1	D_1	$x_{m+1,1}$	$x_{m+1,2}$	...	$x_{m+1,i}$	...	$x_{m+1,m}$	$x_{m+1,m+1}$	$x_{m+1,m+2}$		$x_{m+1,m+j}$	...	$x_{m+1,m+n}$	...
m+2	D_2	$x_{m+2,1}$	$x_{m+2,2}$	...	$x_{m+2,i}$	...	$x_{m+2,m}$	$x_{m+2,m+1}$	$x_{m+2,m+2}$		$x_{m+2,m+j}$	...	$x_{m+2,m+n}$	...
...	...	...	...	...	...	...	...	...	...	...	...	...	...	...
m+j	D_j	$x_{m+j,1}$	$x_{m+j,2}$	...	$x_{m+j,i}^*$	...	$x_{m+j,m}$	$x_{m+j,m+1}$	$x_{m+j,m+2}$		$x_{m+j,m+j}$	...	$x_{m+j,m+n}$	...
...	...	...	...	...	...	...	...	...	...	...	...	...	...	...
m+n	D_n	$x_{m+n,1}$	$x_{m+n,2}$	...	$x_{m+n,i}$	...	$x_{m+n,m}$	$x_{m+n,m+1}$	$x_{m+n,m+2}$		$x_{m+n,m+j}$	...	$x_{m+n,m+n}$	...
Demand		...	...	...	...	...	...	b_1	b_2	...	b_j	...	b_n	

Since any source or destination may ship to any other source or destination in a transshipment problem, it would be convenient to number the origins and destinations sequentially, with the origins numbered 1 to m and the destinations numbered m + 1 to m + n (Fig. 1).

The unit cost of transport from receiver a point chosen as a shipper to the same location designated as a receiver is set to zero.

$$\tilde{x}_{i1} + \tilde{x}_{i2} + \ldots + \tilde{x}_{i,m+n} = \tilde{a}_i + (\tilde{x}_{1i} + \tilde{x}_{2i} + \ldots \tilde{x}_{m+n,i})$$

$$\tilde{x}_{i1} + \tilde{x}_{i2} + \ldots + \tilde{x}_{i,i-1} \ldots \tilde{x}_{i,i+1} + \ldots \tilde{x}_{i,m+n} = \tilde{a}_i + (\tilde{x}_{1i} + \tilde{x}_{2i} + \ldots \tilde{x}_{i-1,i} + \tilde{x}_{i+1,i} + \ldots + \tilde{x}_{m+n,i})$$

$$\sum_{\substack{j=1 \\ j \neq i}}^{n} \tilde{x}_{ij} \approx \tilde{a}_i + \sum_{\substack{j=1 \\ j \neq i}}^{m+n} \tilde{x}_{ji}, \quad i = 1, 2, 3, \ldots, m$$

$$\sum_{\substack{j=1 \\ j \neq i}}^{m+n} \tilde{x}_{ij} - \sum_{\substack{j=1 \\ j \neq i}}^{m+n} \tilde{x}_{ji} = \tilde{a}_i, \quad i = 1, 2, 3, \ldots, m$$

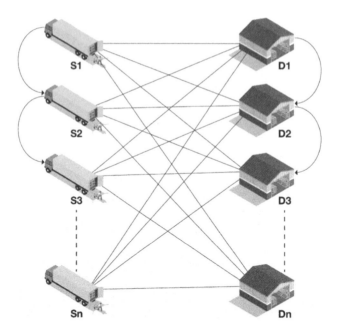

Fig. 1.

Similarly the total amount received at a destination D_j must be equal to its demand plus what it transships.

$$\tilde{x}_{1,m+j} + \tilde{x}_{2,m+j} + ... + \tilde{x}_{m+j-1,m+j} + \tilde{x}_{m+j+1,m+j} + ... + \tilde{x}_{m+n,m+j}$$

$$= \tilde{b}_{m+j} + (\tilde{x}_{m+j,1} + \tilde{x}_{m+j,2} + ... \tilde{x}_{m+j,m+j-1} + \tilde{x}_{m+j,m+j+1} + \ ... \ + \tilde{x}_{m+j,m+n})$$

$$\tilde{x}_{1,m+j} + \tilde{x}_{2,m+j} + ... + \tilde{x}_{m+j-1,m+j} + \tilde{x}_{m+j+1,m+j} + ... + \tilde{x}_{m+n,m+j}$$

$$= \tilde{b}_{m+j} + (\tilde{x}_{m+j,1} + \tilde{x}_{m+j,2} + ... \tilde{x}_{m+j,m} + \tilde{x}_{m+j,m+j} + \ ... \ + \tilde{x}_{m+j,m+n})$$

$$\sum_{\substack{i=1 \\ i \neq j}}^{m+n} \tilde{x}_{i,m+j} \approx \tilde{b}_{m+j} + \sum_{\substack{i=1 \\ i \neq j}}^{m+n} \tilde{x}_{m+j,i}, \ j = 1, 2, ...n$$

$$\sum_{\substack{i=1 \\ i \neq j}}^{m+n} \tilde{x}_{i,m+j} - \sum_{\substack{i=1 \\ i \neq j}}^{m+n} \tilde{x}_{m+j,i} = \tilde{b}_{m+j}, \ j = 1, 2, ...n$$

$$\sum_{\substack{i=1 \\ i \neq j}}^{m+n} \tilde{x}_{i,j} - \sum_{\substack{i=1 \\ i \neq j}}^{m+n} \tilde{x}_{j,i} = \tilde{b}_{j}, \ j = m+1, m+2, ...m+n.$$

$$\text{and } \tilde{x}_{ij} \succeq \tilde{0}, i = 1, 2, ...m+n, j \neq i.$$

Thus the transshipment difficulty may be written as

$$\text{Minimize } \tilde{Z} \approx \sum_{i=1}^{m} \sum_{\substack{j=1 \\ j \neq i}}^{m+n} \tilde{c}_{ij} \tilde{x}_{ij}$$

$$\text{subject to} \quad \sum_{\substack{j=1 \\ j \neq i}}^{m+n} \tilde{x}_{ij} + \sum_{\substack{j=1 \\ j \neq i}}^{m+n} \tilde{x}_{ji} \approx \tilde{a}_i, \quad i = 1, 2, 3, ..., m$$

$$\sum_{\substack{i=1 \\ i \neq j}}^{m+n} \tilde{x}_{ij} + \sum_{\substack{i=1 \\ i \neq j}}^{m+n} \tilde{x}_{ji} \approx \tilde{b}_j, \quad j = m+1, m+2, ..., m+n$$

$$\sum_{i=1}^{m} \tilde{a}_i \approx \sum_{j=m+1}^{m+n} \tilde{b}_j, \text{ and } \tilde{x}_{ij} \succeq \tilde{0}, i = 1, 2, ... m+n, j \neq i.$$

The beyond formulation is a linear indoctrination trouble, which is similar to but not identical to a transportation problem since xij's coefficients are -1. The problem, on the other hand, can be easily transformed into a normal transportation problem.

$$t_i = \sum_{\substack{j=1 \\ j \neq i}}^{m+n} \tilde{x}_{ij}, \quad i = 1, 2, 3, ..., m, \quad t_i + \tilde{x}_{ii} = \sum_{j=1}^{m+n} \tilde{x}_{ji}, \quad i = 1, 2, 3, ..., m$$

$$\text{and } t_j = \sum_{\substack{i=1 \\ i \neq j}}^{m+n} \tilde{x}_{ji}, \quad j = m+1, m+2, m+3, ..., m+n$$

$$t_i + \tilde{x}_{jj} = \sum_{i=1}^{m+n} \tilde{x}_{ji}, \quad j = m+1, m+2, m+3, ..., m+n$$

where t_i represents the total amount of transshipment through the ith origin and t_j represents the total amount shipped put from the jth destination as transshipment. Let $T > 0$ be sufficiently large number so that $ti \leq T$, for all i and $tj \leq T$ for all j. We now write $t_i + x_{ii} = T$ then the non negative slack variable $\tilde{x}_{ii}$ represents the difference between T and the actual amount of transshipment through the ith origin. Similarly, if we let $t_j + x_{jj} = T$ then the non negative slack variable $\tilde{x}_{jj}$ represents the difference between T and the actual amount of transshipment through the jth destination. It is self-evident that the total volume of products transshipped at any given time cannot exceed the total amount generated or obtained and hence $T = \sum_{i=1}^{m} \tilde{a}_i$.

The transshipment problem then reduces to

$$\text{Minimize } \tilde{Z} \approx \sum_{i=1}^{m+n} \sum_{\substack{j=1 \\ j \neq i}}^{m+n} \tilde{c}_{ij} \tilde{x}_{ij}$$

$$\text{subject to} \quad \sum_{j=1}^{m+n} \tilde{x}_{ij} = \tilde{a}_i + T, \quad i = 1, 2, 3, ..., m$$

$$\sum_{j=1}^{m+n} \tilde{x}_{ij} = T, \quad i = m+1, m+2, m+3, ..., m+n,$$

$$\sum_{i=1}^{m+n} \tilde{x}_{ij} = T, \quad j = 1, 2, ...m$$

$$\sum_{i=1}^{m+n} \tilde{x}_{ij} = \tilde{b}_j + T, \quad j = m+1, m+2, m+3, ..., m+n$$

$$\text{and } \tilde{x}_{ij} \succeq \tilde{0}, i = 1, 2, ...m+n, j = 1, 2, ...m+n.$$

$$\text{where } \tilde{C}_{ii} = 0, i = 1, 2, ...m+n.$$

A typical transportation problem with $(m + n)$ roots and $(m + n)$ destinations is described by the mathematical model above. The problem's solution includes $2m + 2n - 1$ basic variables. However, the remaining buffer stock is represented by $m + n$ of these variables appearing in the diagonal cells if they are omitted. Our interest has $m + n - 1$ basic variables.

4 Improved Interval Vogel's Approximation Algorithm

Row Opportunity Interval Cost Matrix
Each element of the same row is subtracted from the row's smallest interval expense.

Column Opportunity Interval Cost Matrix
The smallest cost of each column in the original interval tranhipment cost matrix is subtracted from each segment of the same column. (Note: the diagonal zero is ignored when determining the smallest interval cost). On the TOC matrix, the proposed algorithm is used. The following are the steps in detail:

Step 1: If either (full amount supply > full amount demand) or (total supply total demand), balance the given interval tranhipment problem.
Step 2: Convert the interval tranhipment problem in to interval transportation problem.
Step 3: state the all interval parameters supply, demand and unit transportation cost in terms of midpoint and half width form.
Step 4: Obtain the ROC, COC and then TOC matrix
Step 5: Identify the boxes in each row with the lowest and next-lowest interval transportation costs, and write the difference (penalty) along the table's side against the corresponding row.
Step 6: In each column, find the boxes with the lowest and next-lowest interval transportation costs, and write the difference (penalty) against the corresponding column.

Step 7: Determine the most severe penalty possible. If it's on the table's side, give the most space to the box with the lowest transportation interval cost in that row. If the box with the lowest interval cost of transportation in that column is below the table, give it the most money.

Step 8: Select the topmost row and the farthest left column if the penalties for two or more rows or columns are equal.

Step 9: The satisfied row or column does not need any further thought. If both the row and column are fulfilled at the same time, delete only one of them, and assign zero supply to the remaining row or column (or demand).

Step 10: Continue until all of the rows and columns are satisfied.

Step 11: Using the original balanced interval transhipment cost matrix, calculate the cumulative interval carrying cost for the possible allocations (Fig. 2).

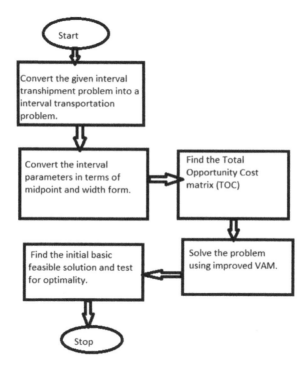

Fig. 2.

5 Numerical Example

To solve the following fully interval integer transhipment problem with 2 source and 2 destinations. The supply values of the source S1 and S2 are [4, 6], [5, 7] units respectively. The demand values of destination D1 and D2 are [3, 5], [6, 8] units respectively. Transportation cost per unit between various defined source and destination are given in the following table [12] (Tables 2 and 3).

Table 2. Interval transhipment problem

	S1	S2	D1	D2	Supply
S1	–	[1, 3]	[6, 9]	[2, 4]	[4, 6]
S2	[1, 3]	–	[8, 10]	[2, 4]	[5, 7]
D1	[0, 1]	[7, 9]	–	[3, 5]	–
D2	[1, 3]	[0, 2]	[1, 3]	–	–
Demand	–	–	[3, 5]	[6, 8]	

Table 3. Equivalent interval transportation problem of the interval transhipment problem

	S1	S2	D1	D2	Supply
S1	[0, 0]	[1, 3]	[6, 9]	[2, 4]	[13, 19]
S2	[1, 3]	[0, 0]	[8, 10]	[2, 4]	[14, 20]
D1	[0, 1]	[7, 9]	[0, 0]	[3, 5]	[9, 13]
D2	[1, 3]	[0, 2]	[1, 3]	[0, 0]	[9, 13]
Demand	[9, 13]	[9, 13]	[12, 18]	[15, 21]	[45, 65]

By applying the proposed algorithm, the initial basic feasible solution is (Table 4)

Table 4. Initial basic feasible solution

	S1	S2	D1	D2	Supply
S1	<0, 0> <11, 2>	<1, 1>	<11, 1.5>	<1, 1> <5, 3>	<16, 3>
S2	<1.5, 1>	<0, 0> <11, 2>	<14, 1>	<1, 1> <6, 3>	<17, 3>
D1	<0, 0.5>	<14.5, 1>	<0, 0> <11, 2>	<4.5, 1>	<11, 2>
D2	<2.5, 1>	<0, 1>	<1, 1> <4, 3>	<0, 0> <7, 3>	<11, 2>
Demand	<11, 2>	<11, 2>	<15, 3>	<18, 3>	<55, 3>

The route from source to desination is given as S1 to D2 = <5, 3>, S2 to D2 = <6, 3>, D2 to D1 = <4, 3>

Then the initial solution of the given interval transsshipment problem is Min z = <3, 1> <5, 3> + <3, 1> <6, 3> + <2, 1> <4, 3> = <41, 3> = [38, 44] (Figs. 3 and 4, Table 5).

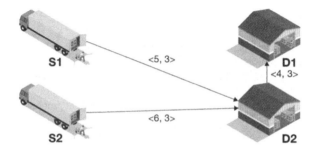

Fig. 3.

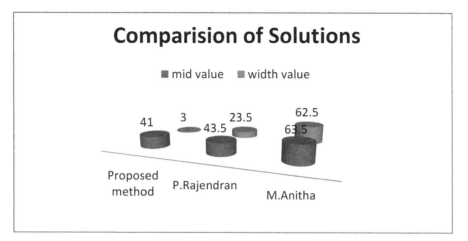

Fig. 4.

Table 5. Comparison of solutions

S.no	Solution by the proposed method	Solution by P.Rajendran and P.Pandian [12]	Solution by R.Sophiaporchelvi and M.Anitha [14]
1	[38, 44]	[20, 67]	[1, 126]

6 Conclusion

To discover the primary interval possible explanation to the interval shipment problem, we proposed an improved interval Vogel's approximation algorithm. The interval possible solution's optimality has been checked. A mathematical model is solved to demonstrate the suggested technique, and the outcome is compared to the results of other obtainable methods.

References

1. Akilbasha, A., Pandian, P., Natarajan, G.: An innovative exact method for solving fully interval integer transportation problems. Inf. Med. Unlock. **11**, 95–99 (2018)
2. Ganesan, K., Veeramani, P.: On Arithmetic Operations of Interval Numbers. Int. J. Uncert. Fuzziness Knowl. - Based Syst. **13**(6), 619–631 (2005)
3. Hurt, V., Tramel, T.: Alternative formulations of the transhipment problem. J. Farm Econ. **47**, 763–773 (1965)
4. Ma, M., Friedman, M., Kandel, A.: A new fuzzy arithmetic. Fuzzy Sets Syst. **108**, 83–90 (1999)
5. NagoorGani, A., Baskaran, R., Assarudeen, S.N.M.: Transhipment problem in fuzzy environment. Int. J. Math. Sci. Eng. Appl. **5**, 57–74 (2011)
6. Pandian, P., Natrajan, G.: A new method for finding an optimal solution of fully interval integer transportation problems, applied mathematical sciences. Applied **4**(37), 1819–1830 (2010)
7. Rajendran, P., Pandian, P.: Solving fully interval transhipment problems. Int. Math. Forum **41**, 2027–2035 (2012)
8. Sengupta, A., Pal, T.K.: Interval-valued transportation problem with multiple penalty factors. VU J. Phys. Sci. **9**, 71–81 (2003)
9. Porchelvi, S., Anitha, M.: Comparative study of optimum solution between interval transportation and interval transhipment problem. Int. J. Adv. Sci. Eng. **4**(4), 764–767 (2018)
10. Orden, A.: Transhipment problem. Manage. Sci. **3**, 276–285 (1956)

Vision Cognition

Novel Image Compression and Deblocking Approach Using BPN and Deep Neural Network Architecture

Rajsekhar Reddy Manyam[1], R. Sivagami[1], R. Krishankumar[1], V. Sangeetha[1], K. S. Ravichandran[1], and Samarjit Kar[2]([⊠])

[1] School of Computing, SASTRA University, Thanjavur 613401, TN, India
rajasekarreddy@it.sastra.edu, {sivagamiramadass,krishankumar, sangeetha}@sastra.ac.in, raviks@sastra.edu
[2] Department of Mathematics, NIT Durgapur, Durgapur, West Bengal, India
samarjit.kar@maths.nitdgp.ac.in

Abstract. Medical imaging is an important source of digital information to diagnose the illness of a patient. The digital information generated consists of different modalities that occupy more disk space, and the distribution of the data occupies more bandwidth. A digital image compression technique that can reduce an image's size without losing much of its important information is challenging. In this paper, a novel image compression technique based on BPN and Arithmetic coders is proposed. The high non-linearity and unpredictiveness of the interrelationship between the pixels present in the image to be compressed is handled by BPN. An efficient coding technique called Arithmetic coding is used to produce an image with a better compression ratio and lower redundancy. A deep CNN based image deblocker is used as a post-processing step to remove the artefacts present in the reconstructed image to improve the quality of the reconstructed image. The effectiveness of the proposed methodology is validated in terms of PSNR. The proposed method is able to achieve about a 3% improvement in PSNR compared with the existing methods.

Keywords: Image compression · MRI · Medical imaging · Arithmetic coders · Deblocker · CNN

1 Introduction

The advancement in sensor technology paved the way to generate more imaging data. A high-resolution medical image plays a key role in telemedicine techniques for diagnosing diseasesaccurately [1]. These raw data generated from the sensors occupy more disk space and high bandwidth when transferring data. As the need for transmission and digital storage of data has been increased abundantly, novel image compression techniques have gained more popularity among the research community.

Some of the traditional techniques for image compression are transform coding, predictive coding, and vector quantization. Transform based coding generates a set of

© IFIP International Federation for Information Processing 2021
Published by Springer Nature Switzerland AG 2021
Z. Shi et al. (Eds.): ICIS 2020, IFIP AICT 623, pp. 209–216, 2021.
https://doi.org/10.1007/978-3-030-74826-5_18

coefficients. The subsets of the coefficient allow good data representation and maintain good data compression. The transform-based compression highly depends on the choice of the transformation adopted. Discrete cosine transform is the one popular method highly used for image compression. Predictive coding techniques calculate the similarity between the neighbouring pixels and remove the redundancy within the image. Vector quantization is based on the codebook generated for compression. A hybrid approach to these techniques was proposed to achieve high data compression.

Many researchers have proposed different architectures for image compression with low information loss during compression and reconstruction for medical images. [2] proposed a three-stage approach based on convergence theory and an iterative process for reconstruction. However, the proposed method shows certain improvement in PSNR, the iterative method for reconstruction is computationally expensive. [3] proposed an improved version of ripplet transform-based coding. The singular value decomposition method is applied over the coefficients obtained from ripplet transforms to capture the essential feature information from coefficients. The entropy-based method is used to encode the low and high-frequency information of ripplet coefficients. This enhanced ripplet transform-based method showed significant improvement in performance than the JPEG compression and other transform-based compression methods. The choice of transform affects the compression system.

A recent review on image compression techniques [4] discusses all the existing and recent advancements to the standard methods and focuses on image compression for medical images. From the summary of the existing methods, it is clear that the standard compression techniques have their own pros like reduction in inter-pixel redundancy, high compression by block coding, and cons like edge degradation, highly sensitive to channel noise, blocking artefacts, computationally expensive and are highly sensitive to error. In this chapter a BPN and arithmetic coder based compression methods are proposed for image compression and to overcome the artifacts present in the compressed image, and deep learning-based Denoisng CNN is used as a post-processing step. The rest of the chapter deals with the methodology proposed and a detailed analysis of the outcome of the proposed work. The flow diagram for the proposed image compression methodology is given in Fig. 1.

2 Proposed Methodology

2.1 Image Preprocessing

The input image to be compressed is split into several non-overlapping sub-images. Here, the size of the sample input image is 256×256. A single image that is to be compressed is split into several non-overlapping 4×4 pixel length sub-images, as shown in Fig. 2. A total of $4 \times 4 \times 4096$ non-overlapping sub-images are obtained.

These sub-blocks are converted into a vector form. The vector form of each sub-block is further concatenated into a 16×4096 element vector. The intensity of the pixels varies from 0 to 255. A max normalization process is used as a preprocessing before feeding these element vectors as training samples to the neural network to ease the neural network's training process and reduce the complexity of mapping between the pixels. The input image to be compressed here is a greyscale MRI image. The maximum

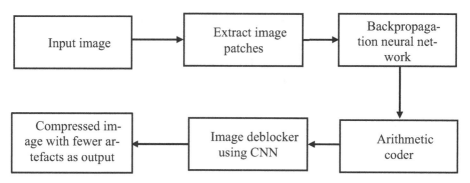

Fig. 1. Work flow for the proposed methodology

value in this greyscale is 255. A linear function is used for max normalization. Each pixel value is divided by 255, a maximum greyscale value. The normalized element vector ranges from 0 to 1. This normalized element vector is used as a training sample for the neural networks.

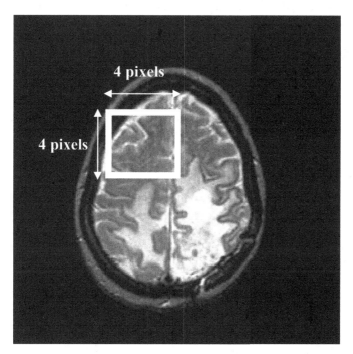

Fig. 2. Preprocessing of the input image with 4 × 4 sub-images

2.2 Backpropagation Neural Networks

The backpropagation algorithm is the most fundamental algorithm used to train a neural network through a chain rule effectively. The feed-forward neural network approximates the given problem with high accuracy. After each forward pass through a network, a backward pass is performed to adjust the model parameters such as weights and biases to reduce the error propagated. In simple terms, the forward and backward passes are the error-correction learning rule [5].

In the forward pass, the input vector is applied to the neurons in the input layer. The effect of this input vector is propagated through the hidden layer, and a set of output as a response is obtained from the output layer. The weights of all the layers are fixed in the forward pass. An error value is generated by subtracting the actual output and the response obtained during the forward pass from the output layer. If the error obtained from the output layer is above a tolerance value, a backward pass is carried out where the weights are adjusted based on the chain rule. This weight updation is performed iteratively until the actual response reaches the desired response.

A three-layered back propagation neural network consisting of an input layer, a hidden layer, and an output layer is used for image compression. The normalized pixel value of each sub-block image will be the input for the input nodes. For image compression, the compressed image's size will be equal to the input image's size with a reduced number of bits. Therefore, the number of neurons in the output layer is equal to the number of neurons in the input layer. The input itself acts as a target for training a neural network [6]. The input layer and the output layer of the neural network are fully connected to the hidden layer. Small random variables between -1 and $+1$ are initialized as weights for the neurons that connect the input, hidden, and output layers. A non-linear activation function like Sigmoid or tanh allows the network to learn the given data set's complex patterns. The training plot for BPN based compression is shown in Fig. 6.

2.3 Arithmetic Coding

Arithmetic coding is a lossless compression technique, and it is used in all types of data compression algorithms because of its advantages of flexibility and optimality. Arithmetic coding provides a better compression ratio, and redundancy is much reduced compared to Huffman Coding [7]. Unlike Huffman coding, a discrete number of bits for each code is not used in arithmetic coding [8]. Each pixel value is assigned with an interval value based on the sequence of event probabilities. Starting with lower limit 0 and upper limit 1, each interval is divided into several subintervals.

2.4 Image Deblocker

The image compression process causes the image to lose some information. So that the compressed image appears to is distorted, meaning that the image has some blocking artefacts. To reduce the effects of blocking artifacts, many algorithms have been proposed in the literature to deblock the artefact effect in the compressed image. Recent successful deep learning-based Denoising architecture has been used as a preprocessing step to reduce the artefact effects in the compressed image.

A built-in deep feed-forward Convolutional neural network, called DnCNN proposed by [9], is adopted for deblocking the artefacts in the compressed image. This DnCNN was originally designed to remove the noises present in the image. The same architecture can be trained in such a way to remove the artifacts and improve the resolution of the image. The network learns to estimate the residual information about the distorted image. The residual information is the difference between the original uncompressed image and the distorted image. The distortions here are the blocking artifacts.

The residual information is detected from the luminance of a color image. The luminance channel 'Y' is the brightness information present in the given image. The CNNs are trained using only the luminance channel 'Y' as the human perception is more sensitive to changes in brightness than to the changes in the color. The principle strategy adopted for estimating the residual information can be represented as follows.

$$Y_{Residual} = Y_{Compressed} - Y_{Original} \tag{1}$$

where $Y_{Original}$ is the luminance channel of the input uncompressed image. $Y_{Compressed}$ is the lossy compressed image. The CNN [10] network trained to learn and predict $Y_{Residual}$ from the training data. If the network is able to estimate the residual information effectively, an undistorted version of the uncompressed image can reconstructed by adding the predicted residual information with the compressed image.

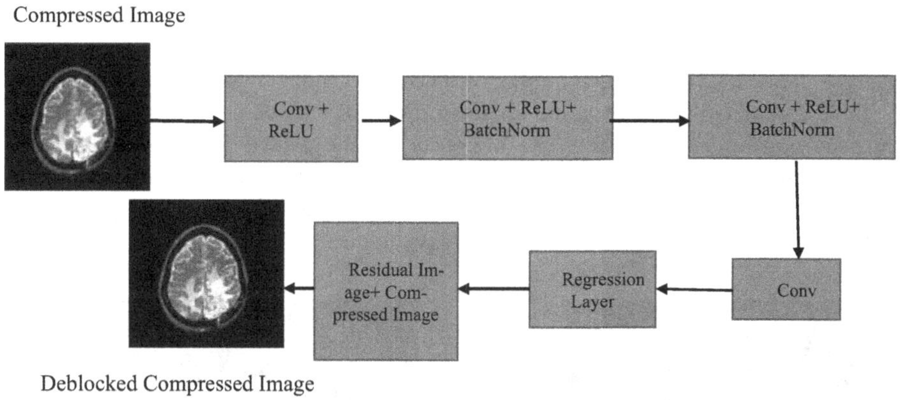

Compressed Image

Deblocked Compressed Image

Fig. 3. Architecture of the post processing using DeNoising CNN to remove the artefacts

The architecture for deblocking the artifacts present in the compressed image is given in Fig. 3. It has a series of convolution Layers consisting of convolution, ReLU - an activation Layer and batch normalization Layer. The DnCNN architecture adopted for this work has 19 Convolutional Layers. The last 20th layer has only a convolution 2D layer that maps the learned features to the resolution of the input image and final regression layer.

3   Performance Analysis and Discussion

The experimental part is done using MATLAB R2018b [12] on a computer with Intel i5 processor and CUDA enabled NIVIDIA GPU with compatibility 4.0. The BPN network has been trained with tangential sigmoid function with a maximum epoch of 1000 iterations with gradient method. The training plot of BPN based compression is shown in Fig. 6. The pretrained denoising convolutional neural networks adopted as post processing step has been trained with stochastic gradient descent optimization technique with initial learning rate of about 0.1 for a maximum of 30 epochs with a gradient threshold value of 0.005 to avoid exploding gradients and batch size has been initialized to 64.

For better visualization of the effect of compressed image without Denoising and with denoising is shown in Fig. 4 and Fig. 5. An image tool in MATLAB has been used for visualization. From the observation it is clear that, BPN and denoising based compressor has reconstructed the compressed image with reduced blocking artefacts. The quantitative analysis of proposed framework is done using peak signal to noise ratio (PSNR) as given below.

$$PSNR = 10 log_{10}\left(\frac{R^2}{MSE}\right) \tag{2}$$

$$MSE = \frac{\sum_{M,N}[I_1(m, n) - I_2(m, n)]^2}{M * N} \tag{3}$$

where R is the maximum fluctuation value in the image. If the input image has a double-precision floating-point data type, then R is 1. If it has an 8-bit unsigned integer data type, R is 255. M and N are number of rows and columns of the given input image.

Table 1 depicts the PSNR value obtained from BPN based compression and BPN with denoising as post processing. The inference from the result depicts that the BPN and DnCNN based compression achieved a 3% improvment in the total PSNR value compared with using only BPN based compression. An average of about 3.64% improvent in PSNR value was achieved by the proposed technique.

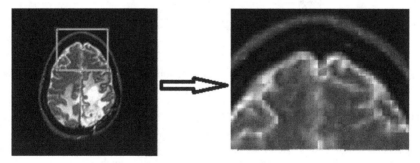

Fig. 4. Image visualization of the compressed image

Though the method proposed is able to achieve an average of 3% improvement in PSNR than the existing state of the art methods, training of neural networks is computationally expensive and data intensive. In future, further studies can be done on utilizing

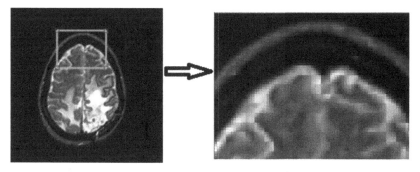

Fig. 5. Image visualization after applying denoising convolutional neural network

Table 1. Qualitative analysis of BPN based compression and BPN with Denoising CNN

Input image	PSNR using BPN (dB) [11]	PSNR using BPN and DnCNN
Image (1)	29.906	30.62
Image (2)	30.41	31.75
Image (3)	30.42	31.8
Image (4)	30.44	31.76
Image (5)	30.41	31.51
Image (6)	31.05	32.25
Image (7)	31.46	32.79
Image (8)	31.63	32.82

auto encoders an unsupervised technique for image compression and to more generalized architecture based only on deep learning dedicated for medical image compression.

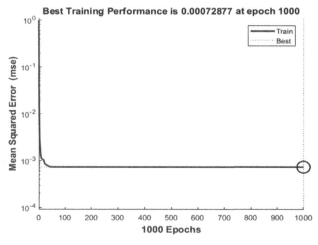

Fig. 6. Training plot of BPN in comparison with number of epochs and mean squared error

4 Conclusion

In this paper, a novel image compression technique based on BPN and Arithmetic coders is proposed. The high non-linearity and unpredictiveness of the interrelationship between the pixels of the image to be compressed are handled by BPN. An efficient coding technique called an arithmetic coding is used to produce an image with a better compression ratio and lower redundancy. Further, a deep CNN-based image deblocker is used as a post-processing step to remove the artefacts present in the reconstructed image. Post-processing is done to improve the quality of the reconstructed image. The effectiveness of the proposed methodology is validated by using PSNR. It is inferred from the results that PSNR has improved by three percent when compared with state-of-the-art methods.

Acknowledgment. The authors would like to thank the following funding under grant no: 09/1095(0033)18-EMR-I, 09/1095(0026)18-EMR-I, F./2015–17/RGNF-2015–17-PAM-83, SR/FST/ETI-349/2013.

References

1. Shapiro, J.M.: Embedded image coding using zerotrees of wavelet coefficients. IEEE Trans. Signal Process. **41**(12), 3445–3462 (1993)
2. Sunil, H., Hiremath, S.G.: A combined scheme of pixel and block level splitting for medical image compression and reconstruction. Alex. Eng. J. **57**, 767–772 (2018). https://doi.org/10.1016/j.aej.2017.03.001
3. Anitha, J., Sophia, P.E., Hoang, L., Hugo, V., De Albuquerque, C.: Performance enhanced ripplet transform based compression method for medical images. Measurement **144**, 203–213 (2019). https://doi.org/10.1016/j.measurement.2019.04.036
4. Hussain, A.J., Al-fayadh, A., Radi, N.: Neurocomputing Image compression techniques: a survey in lossless and lossy algorithms. Neurocomputing **300**, 44–69 (2018). https://doi.org/10.1016/j.neucom.2018.02.094
5. Savković-Stevanović, J.: Neural networks for process analysis and optimization: modeling and applications. Comput. Chem. Eng. **18**(11–12), 1149–1155 (1994)
6. Soliman, H.S., Omari, M.: A neural networks approach to image data compression. Appl. Soft Comput. **6**, 258–271 (2006). https://doi.org/10.1016/j.asoc.2004.12.006
7. St, H., Uhl, A.: Comparison of compression algorithms' impact on fingerprint and face recognition accuracy (2006)
8. Sibley, E.H., Willen, I.A.N.H., Neal, R.M., Cleary, J.G.: arithmetic coding for data compression. Commun. ACM **30**, 520–540 (1987)
9. Zhang, K., Zuo, W., Member, S., Chen, Y., Meng, D.: Beyond a Gaussian denoiser: residual learning of deep cnn for image denoising. IEEE Trans. Imageprocess. **26**, 3142–3155 (2017)
10. Sivagami, R., Srihari, J., Ravichandran, K.S.: Analysis of encoder-decoder based deep learning architectures for semantic segmentation in remote sensing images. In: Abraham, A., Cherukuri, A., Melin, P., Gandhi, N. (eds.) ISDA 2018. Advances in Intelligent Systems and Computing, vol. 941, pp. 332–341. Springer, Cham (2020). https://doi.org/10.1007/978-3-030-16660-1_33
11. Reddy, M.R., Deepika, M.A., Anusha, D., Iswariya, J., Ravichandran, K.S.: A new approach for image compression using efficient coding technique and BPN for medical images. In: Pandian, D., Fernando, X., Baig, Z., Shi, F. (eds.) ISMAC 2018, vol. 30, pp. 283–290. Springer, Cham (2019). https://doi.org/10.1007/978-3-030-00665-5_29
12. MATLAB R2018b, The MathWorks, Natick (2018)

Characterization of Orderly Behavior of Human Crowd in Videos Using Deep Learning

Shreetam Behera, Shaily Preetham Kurra, and Debi Prosad Dogra[✉]

School of Electrical Sciences, IIT Bhubaneswar, Bhubaneswar 752050, Odisha, India
{sb46,ksp10,dpdogra}@iitbbs.ac.in

Abstract. In the last few decades, understanding crowd behavior in videos has attracted researchers from various domains. Understanding human crowd motion can help to develop monitoring and management strategies to avoid anomalies such as stampedes or accidents. Human crowd movements can be classified as structured or unstructured. In this work, we have proposed a method using deep learning technique to characterize crowd behavior in terms of order parameter. The proposed method computes features comprised of motion histogram, entropy, and order parameter of the frames of a given crowd video. The features are fed to a Long Short Term Memory (LSTM) model for characterization. We have tested the proposed method on a dataset comprising of structured and unstructured crowd videos collected from publicly available datasets and our recorded video datasets. Accuracy as high as 91% has been recorded and the method has been compared with some of the recent machine learning algorithms. The proposed method can be used for real-time applications focusing on crowd monitoring and management.

Keywords: Crowd characterization · LSTM · Supervised learning · Crowd anomalies · Motion histogram · Crowd behaviour · Crowd monitoring

1 Introduction

Crowd motion provides important cues in understanding crowd behavior. Understanding the dynamics of the crowd motion can be used to understand different crowd anomalies such as stampede. Computer-vision algorithms have been popular among the researchers as it handles high volumes of data, reduces human intervention, and are often accurate and less error-prone. These algorithms can be used for crowd flow segmentation, behavioral analysis, and anomaly detection. Existing crowd analysis algorithms usually employ a combination of physics-based approaches and computer-vision approaches resulting in effective analysis of the crowd as mentioned in [10,12,30]. A Lagrangian particle dynamics-based framework has been proposed in [1] to segment crowd flows and to find flow

© IFIP International Federation for Information Processing 2021
Published by Springer Nature Switzerland AG 2021
Z. Shi et al. (Eds.): ICIS 2021, IFIP AICT 623, pp. 217–226, 2021.
https://doi.org/10.1007/978-3-030-74826-5_19

instabilities in the crowd. However, this technique is highly complex and computationally intensive. The approach mentioned in [11] can detect similar regions in the crowd. A force-field based approach has been used to understand the flow regions in a crowd in [2]. Social force models mentioned in [5,9,18] are also popular in defining crowd in terms of flow, interactions, and abnormalities. Wu et al. in [28] have described the crowd behavior in terms of bi-linear interactions of crowd flows by analyzing the curl and divergence of these flows. The crowd has been modeled as dynamical systems for behavioral analysis in [13,15,26,27] for behavioral analysis. Zhou et al. in [32] have analyzed crowd collective motion segments and analyzed these segments' collective behavior. In [4,25,33], Convolutional Neural Networks (CNN) have been used to perform crowd behavior analysis. The authors in [29] have developed a crowd anomaly algorithm based upon sparse representation-based scheme. There are a few similar approaches proposed in the areas of crowd monitoring systems and detecting abnormal behaviors [14,17,19,22]. Hao et al. in [7] have proposed a crowd anomaly detection scheme using spatio-temporal texture analysis. The texture analysis was performed based upon the grey-level co-occurrence matrix model. Ryan et al. in [20] have proposed an algorithm for detecting anomalous behaviors in crowds by analyzing the optical flow map as a texture. In [6], the authors have proposed a nearest-neighbor search approach for detecting anomalous behavior in crowd videos. Sabokrou in [21] has proposed a real-time anomaly detection method that performs scene localization in large crowds. The authors in [31] have proposed a model based on the social-force and entropy to analyze macrostates in the crowd. Their analysis shows that a disordered crowd motion has higher entropy as compared to orderly crowd motion. However, a few parameters are needed for a complete and accurate analysis of crowd behavior. In [3], fluctuations in energy and entropy curves have been used to characterize crowd in terms of speed and randomness. However, none of those mentioned above papers address the crowd characterization task in terms of ordered behavior, and none of them use a supervised learning algorithm for crowd characterization. This has motivated us to develop a method that can characterize crowd videos based on orderly behavior of the particles using supervised learning.

The rest of the paper is organized as follows. The proposed methodology is presented in Sect. 2. The results obtained from the proposed method is discussed in Sect. 3. The work has been concluded along with the future scope in Sect. 4.

2   Proposed Methodology

The human crowd can be characterized based on physical quantities such as energy, randomness, and orderliness. These physical quantities defined the typical behavior of the crowd. For example, in a marathon, as presented in Fig. 1a, the crowd can be seen moving in a single direction, which can be considered as structured crowd behavior with less random movements, characterized by lower entropy and higher order parameter. In another example, people moving inside a shopping mall as presented in Fig. 1e can be considered as highly random and is less structured in nature. This type of movement is considered as unstructured movement, and such movements are characterized by higher entropy and lower order parameter. These behaviors are illustrated in Figs. 2 and 3, respectively. Identifying these types of movements will certainly help to understand the cause of human movements leading to normal and abnormal activities. Now, the crowd characterization task can be mapped to a typical classification problem.

a　　　　　　b　　　　　c　　　　　d　　　　　e　　　　　f

Fig. 1. (*a–c*) Samples of structured crowd, and (*d–f*) unstructured crowd videos. (Best viewed in color) (Color figure online)

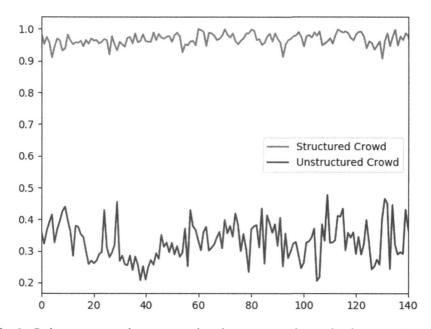

Fig. 2. Order parameter for structured and unstructured crowd, where y-axis represents ϕ or the order parameter and x-axis represents time of frame numbers. (Best viewed in color) (Color figure online)

2.1   Feature Extraction

Feature extraction is performed for each video segment consisting of 20 frames (Fig. 4). The feature vector consists of motion histogram, entropy, and order

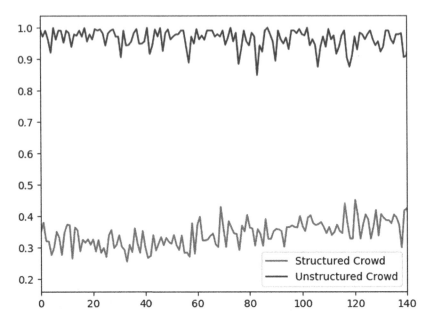

Fig. 3. Entropy for structured and unstructured crowd, where y-axis represents entropy and x-axis represents time of frame numbers. (Best viewed in color) (Color figure online)

parameter. For a sequence of 20 frames of a video, keypoints have been extracted using Shi-Thomasi corner detection [24]. Next, these keypoints (N) are fed to the optical flow process, as mentioned in [16] for the flow computation. The magnitudes and orientation of the keypoints (with v_x, v_y velocities across x and y axis, respectively) are calculated using (1) and (2), respectively.

$$m = \sqrt{(v_x)^2 + (v_y)^2} \tag{1}$$

$$\theta = \arctan\left((v_y)/(v_x)\right) \tag{2}$$

The obtained magnitude values are normalized within a range of $[0, 1]$. A 2D histogram is then constructed using the normalized magnitude and orientation values with m and n number of bins. Thus, $1/m$ and $360/n$ ranges are obtained for each keypoint. For example, if $m = 4$ and $n = 4$, the number of bins formed are 4 with the ranges of $[0, 0.25)$, $[0.25, 0.5)$, $[0.5, 0.75)$, and $[0.75, 1]$ and the orientation fall in the ranges of $[0, 90)$, $[90, 180)$, $[180, 270)$, and $[270, 360]$. Thus, we obtain a 4x4 matrix with 16 bins of the above-specified magnitude and orientation ranges. The count for each bin is increased when the number of keypoints with magnitude and orientation has been found to be within the aforementioned range. This is performed for all keypoints leading to a 2D histogram. Then the entropy and the order parameter are calculated using (3) and (4), respectively,

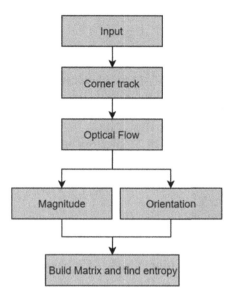

Fig. 4. Block diagram of the feature extraction technique.

$$S = \sum_{bin=1}^{m*n} -P_{bin}ln(P_{bin}) \tag{3}$$

where m and n are the sizes of the 2D Histogram such that $v_k = v_x\hat{i} + v_y\hat{j}$ is the velocity vector for each keypoint k. The final feature vector is represented as $[Hist_{m*n}, S, \phi]$. These features are then fed to the classifier for characterization (Fig. 5).

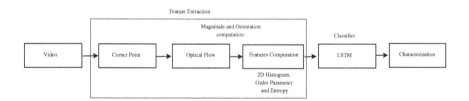

Fig. 5. Block diagram of the proposed characterization scheme.

$$\phi = \frac{1}{N} \sum_{k=1}^{N} |\frac{v_k(t)}{|v_k(t)|}| \tag{4}$$

2.2   Crowd Classification

The crowd characterization is considered as a binary classification problem, where a crowd's behavior is classified as structured or unstructured. In this work, Long Short Term Memory (LSTM)-based classifier [8] has been used for characterization as LSTM can deal with temporal data, and the features extracted for each sequence in this work are temporally related. The architecture of the LSTM characterization model is presented in Table 1. The proposed model consists of 2 layers. The first layer consists of 32 memory units of LSTM. The equations involved with a single memory unit are presented in (5)–(10).

$$ig_t = \sigma(W_{ig}x_t + Ub_{ig}hg_{t-1} + Vb_i cg_{t-1}) \tag{5}$$

$$fg_t = \sigma(W_{fg}x_t + Ub_{fg}h_{t-1} + Vb_f cg_{t-1}) \tag{6}$$

$$og_t = \sigma(W_{og}x_t + Ub_{og}hs_{t-1} + Vb_o cg_{t-1}) \tag{7}$$

$$\tilde{cg}_t = \tanh(W_{cg}x_t + Ub_{cg}hs_{t-1} \tag{8}$$

$$cg_t = fg_t^i \odot cg_{t-1} + ig_t \odot \tilde{cg}_t) \tag{9}$$

$$hs_t = og_t \odot \tanh(cg_t) \tag{10}$$

The second layer is a fully connected neural layer with a sigmoid activation function. Since, characterization is a binary classification problem, binary cross-entropy presented in (11) is considered as objective function for loss calculation.

$$L_{\text{entropy}} = -\sum_{i=1}^{N} [yTarget_i \ln(yPredict_i) + (1 - yTarget_i)\ln(1 - yPredict_i)]$$

$$\tag{11}$$

The model is trained with Adam's optimizer for a batch size of 16 and 100 epochs. In the above formulations, at each time step t, the output of the LSTM is controlled by input gate ig_t, a forget gate fg_t, an output gate og_t, a memory cell cg_t, and a hidden state hs_t. x_t is the input at the current time step, σ denotes the logistic sigmoid function, tanh, and $\odot$ stands for hyperbolic tangent function and element-wise multiplication. W, Ub, and Vb represent weight and bias matrices that are updated during the training stage, where N is the number of output scalar values in the model, $yTarget_i$ is the expected output, and $yPredict_i$ is the predicted output of the ith scalar value of the model.

Table 1. Architecture of the proposed LSTM-based model for characterization

Layers	Output shape	Number of parameters
LSTM	32	4352
Fully Connected Layer with sigmoid activation	1	33

3   Results

This section discusses about the results obtained from the proposed LSTM classification scheme.

3.1   Video Dataset

In this work, 251 videos have been selected from a publicly available video dataset [23] and our videos recorded during Puri Rath Yatra that happens every year in Odisha, India. There are 106 structured crowd videos consisting of linearly and non-linearly ordered motion flows, and 145 unstructured videos in this selection consist of crowd mixing and random pedestrian movements. The structured and unstructured videos have been labeled. The training and testing sets comprise of 80% and 20% videos of the dataset.

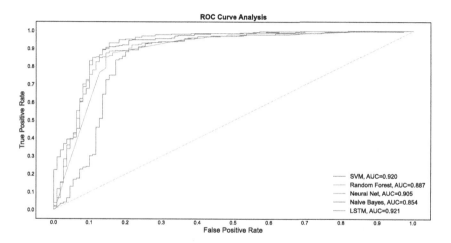

Fig. 6. ROC curves (receiver operating characteristic) for the classifiers used for crowd characterization. (Best viewed in color) (Color figure online)

3.2   Classification Results

The dataset is fed to the LSTM-based classifier that has been trained with 100 epochs and the outputs are evaluated in terms of the area under ROC curve (AUC). The proposed characterization scheme has also been compared with the existing popular machine learning algorithms like Support Vector Machine (SVM), Random Forest (RF), Multilayer Perceptron (MLP), Naive Bayes. The proposed algorithm outperforms all the above mentioned classification algorithms. Our proposed classification achieves accuracy and AUC values as high as 91% and 0.9218, respectively. The AUC and accuracy comparisons are illustrated in Figs. 6 and 7, respectively. The table depicting the values of each classifier is illustrated in Table 2.

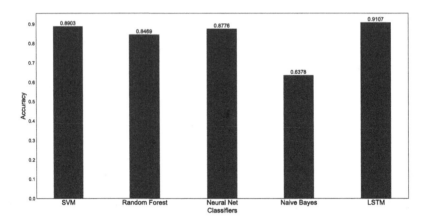

Fig. 7. Accuracy bar graph for the classifiers used for crowd characterization.

Table 2. Accuracy and area under ROC values for the classifiers used for crowd characterization.

Classifier	Accuracy	Area Under ROC
SVM	0.8903	0.920
Random Forest	0.8469	0.887
Neural Net	0.8776	0.905
Naive Bayes	0.6378	0.854
LSTM	0.9107	0.921

4   Conclusion and Future Scopes

Crowd characterization is a classification problem of identifying the crowd movements based on ordered behavior at a global level. The proposed method can successfully characterize the video sequences as structured and unstructured compared to the existing machine learning-based algorithms. The method can reduce the human load on crowd behavior analysis, segregating typical behaviors in the crowd more distinctively. Furthermore, the proposed method is robust and straightforward to implement. Thus, real-time applications are possible. In future, the method can be improved by integrating motion dynamics-based features, and deep learning-based features can be used for describing the crowd behavior. Furthermore, the proposed algorithm can further be extended to detect and localize the structured and unstructured behavior at local levels.

References

1. Ali, S., Shah, M.: A Lagrangian particle dynamics approach for crowd flow segmentation and stability analysis. In: IEEE Conference on Computer Vision and Pattern Recognition, pp. 1–6 (2007)

2. Ali, S., Shah, M.: Floor fields for tracking in high density crowd scenes. In: Forsyth, D., Torr, P., Zisserman, A. (eds.) ECCV 2008. LNCS, vol. 5303, pp. 1–14. Springer, Heidelberg (2008). https://doi.org/10.1007/978-3-540-88688-4_1

3. Behera, S., Dogra, D.P., Roy, P.P.: Characterization of dense crowd using Gibbs entropy. In: Chaudhuri, B.B., Kankanhalli, M.S., Raman, B. (eds.) Proceedings of 2nd International Conference on Computer Vision & Image Processing. AISC, vol. 704, pp. 289–300. Springer, Singapore (2018). https://doi.org/10.1007/978-981-10-7898-9_24

4. Cao, L., Zhang, X., Ren, W., Huang, K.: Large scale crowd analysis based on convolutional neural network. Pattern Recogn. **48**(10), 3016–3024 (2015)

5. Chaker, R., Al Aghbari, Z., Junejo, I.N.: Social network model for crowd anomaly detection and localization. Pattern Recogn. **61**, 266–281 (2017)

6. Colque, R.V.H.M., Caetano, C., de Andrade, M.T.L., Schwartz, W.R.: Histograms of optical flow orientation and magnitude and entropy to detect anomalous events in videos. IEEE Trans. Circuits Syst. Video Technol. **27**(3), 673–682 (2016)

7. Hao, Y., Xu, Z.J., Liu, Y., Wang, J., Fan, J.L.: Effective crowd anomaly detection through spatio-temporal texture analysis. Int. J. Autom. Comput. **16**(1), 27–39 (2019)

8. Hochreiter, S., Schmidhuber, J.: Long short-term memory. Neural Comput. **9**(8), 1735–1780 (1997)

9. Ji, Q.G., Chi, R., Lu, Z.M.: Anomaly detection and localisation in the crowd scenes using a block-based social force model. IET Image Proc. **12**(1), 133–137 (2017)

10. Junior, J.C.S.J., Musse, S.R., Jung, C.R.: Crowd analysis using computer vision techniques. IEEE Signal Process. Mag. **27**(5), 66–77 (2010)

11. Khan, S.D., Bandini, S., Basalamah, S., Vizzari, G.: Analyzing crowd behavior in naturalistic conditions: identifying sources and sinks and characterizing main flows. Neurocomputing **177**, 543–563 (2016)

12. Kok, V.J., Lim, M.K., Chan, C.S.: Crowd behavior analysis: a review where physics meets biology. Neurocomputing **177**, 342–362 (2016)

13. Kountouriotis, V., Thomopoulos, S.C., Papelis, Y.: An agent-based crowd behaviour model for real time crowd behaviour simulation. Pattern Recogn. Lett. **44**, 30–38 (2014)

14. Kratz, L., Nishino, K.: Anomaly detection in extremely crowded scenes using spatio-temporal motion pattern models. In: 2009 IEEE Conference on Computer Vision and Pattern Recognition, pp. 1446–1453. IEEE (2009)

15. Lim, M.K., Chan, C.S., Monekosso, D., Remagnino, P.: Detection of salient regions in crowded scenes. Electron. Lett. **50**(5), 363–365 (2014)

16. Lucas, B.D., Kanade, T., et al.: An iterative image registration technique with an application to stereo vision (1981)

17. Mahadevan, V., Li, W., Bhalodia, V., Vasconcelos, N.: Anomaly detection in crowded scenes. In: 2010 IEEE Computer Society Conference on Computer Vision and Pattern Recognition, pp. 1975–1981. IEEE (2010)

18. Mehran, R., Moore, B.E., Shah, M.: A streakline representation of flow in crowded scenes. In: Daniilidis, K., Maragos, P., Paragios, N. (eds.) ECCV 2010. LNCS, vol. 6313, pp. 439–452. Springer, Heidelberg (2010). https://doi.org/10.1007/978-3-642-15558-1_32

19. Mehran, R., Oyama, A., Shah, M.: Abnormal crowd behavior detection using social force model. In: IEEE Conference on Computer Vision and Pattern Recognition, CVPR 2009, pp. 935–942. IEEE (2009)

20. Ryan, D., Denman, S., Fookes, C., Sridharan, S.: Textures of optical flow for real-time anomaly detection in crowds. In: 2011 8th IEEE International Conference on Advanced Video and Signal-Based Surveillance (AVSS), pp. 230–235. IEEE (2011)
21. Sabokrou, M., Fathy, M., Hoseini, M., Klette, R.: Real-time anomaly detection and localization in crowded scenes. In: Proceedings of the IEEE Conference on Computer Vision and Pattern Recognition Workshops, pp. 56–62 (2015)
22. Saligrama, V., Chen, Z.: Video anomaly detection based on local statistical aggregates. In: 2012 IEEE Conference on Computer Vision and Pattern Recognition, pp. 2112–2119. IEEE (2012)
23. Shao, J., Change Loy, C., Wang, X.: Scene-independent group profiling in crowd. In: Proceedings of the IEEE Conference on Computer Vision and Pattern Recognition, pp. 2219–2226 (2014)
24. Shi, J., Tomasi, C., et al.: Good features to track. In: Computer Vision and Pattern Recognition, pp. 593–600 (1994)
25. Singh, K., Rajora, S., Vishwakarma, D.K., Tripathi, G., Kumar, S., Walia, G.S.: Crowd anomaly detection using aggregation of ensembles of fine-tuned convnets. Neurocomputing **371**, 188–198 (2020)
26. Solmaz, B., Moore, B.E., Shah, M.: Identifying behaviors in crowd scenes using stability analysis for dynamical systems. IEEE Trans. Pattern Anal. Mach. Intell. **34**(10), 2064–2070 (2012)
27. Wang, X., Yang, X., He, X., Teng, Q., Gao, M.: A high accuracy flow segmentation method in crowded scenes based on streakline. Optik-Int. J. Light Electron Opt. **125**(3), 924–929 (2014)
28. Wu, S., Su, H., Yang, H., Zheng, S., Fan, Y., Zhou, Q.: Bilinear dynamics for crowd video analysis. J. Vis. Commun. Image Represent. **48**, 461–470 (2017)
29. Yuan, Y., Wan, J., Wang, Q.: Congested scene classification via efficient unsupervised feature learning and density estimation. Pattern Recogn. **56**, 159–169 (2016)
30. Zhang, X., Yu, Q., Yu, H.: Physics inspired methods for crowd video surveillance and analysis: a survey. IEEE Access **6**, 66816–66830 (2018)
31. Zhao, Y., Yuan, M., Su, G., Chen, T.: Crowd macro state detection using entropy model. Physica A **431**, 84–93 (2015)
32. Zhou, B., Tang, X., Zhang, H., Wang, X.: Measuring crowd collectiveness. IEEE Trans. Pattern Anal. Mach. Intell. **36**(8), 1586–1599 (2014). https://doi.org/10.1109/TPAMI.2014.2300484
33. Zhou, S., Shen, W., Zeng, D., Fang, M., Wei, Y., Zhang, Z.: Spatial-temporal convolutional neural networks for anomaly detection and localization in crowded scenes. Signal Process. Image Commun. **47**, 358–368 (2016)

Perceptual Intelligence

Stability Analysis of Imprecise Prey-Predator Model

Anupam De$^{1(\boxtimes)}$ ⓘ, Debnarayan Khatua^2 ⓘ, Kalipada Maity^3 ⓘ,
Goutam Panigrahi^4 ⓘ, and Manoranjan Maiti^5 ⓘ

^{1} Department of Applied Sciences, Haldia Institute of Technology,
Haldia, West Bengal, India
http://hithaldia.in
^{2} Department of Basic Science and Humanities,
Global Institute of Science and Technology, Haldia, West Bengal, India
https://www.gisthaldia.org/
^{3} Department of Mathematics, Mugberia Gangadhar Mahavidyalaya,
Contai, West Bengal, India
http://www.mugberiagangadharmahavidyalaya.org
^{4} Department of Mathematics, National Institute of Technology Durgapur,
Durgapur, West Bengal, India
^{5} Department of Applied Mathematics with Oceanology and Computer
Programming, Vidyasagar University, Midnapore, West Bengal, India
http://www.vidyasagar.ac.in

Abstract. Since the last few decades, the prey-predator system delivers attractive mathematical models to analyse the dynamics of prey-predator interaction. Due to the lack of precise information about the natural parameters, a significant number of research works have been carried out to take care of the impreciseness of the natural parameters in the prey-predator models. Due to direct impact of the imprecise parameters on the variables, the variables also become imprecise. In this paper, we developed an imprecise prey-predator model considering both prey and predator population as imprecise variables. Also, we have assumed the parameters of the prey-predator system as imprecise. The imprecise prey-predator model is converted to an equivalent crisp model using "e" and "g" operator method. The condition for local stability for the deterministic system is obtained mathematically by analysing the eigenvalues of the characteristic equation. Furthermore, numerical simulations are presented in tabular and graphical form to validate the theoretical results.

Keywords: Prey-predator · Local stability · Imprecise environment · "e" and "g" operator method

1 Introduction

Mathematicians are provoked by the problems of understanding the biological phenomena. By quantitatively describing the biological problems, the mathematical researchers applied various mathematical tools to analyse and interpret

© IFIP International Federation for Information Processing 2021
Published by Springer Nature Switzerland AG 2021
Z. Shi et al. (Eds.): ICIS 2020, IFIP AICT 623, pp. 229–240, 2021.
https://doi.org/10.1007/978-3-030-74826-5_20

the results. Mathematical areas as calculus, differential equations, dynamical systems, stability theory, fuzzy set theory etc. are being applied in biology. Biological phenomena like prey-predator interaction, prey-predator fishery harvesting system, the prey-predator system with infection, etc. can be expressed by an autonomous or non-autonomous system of ordinary differential equations.

2   Related Work

Work in the area of theoretical biology was first introduced by Thomas Malthus in the late eighteenth century, which later is known as the Malthusian growth model. The Lotka [8] and Volterra [16] predator-prey equations are other famous examples. Till then, a significant development in the area of population dynamics has been made by the researchers. Kar [6] formulated and analysed a prey-predator harvesting problem incorporating a prey refuge, Chakraborty et al. [1], solved stage-structured prey-predator harvesting models. Qu and Wei [13] have presented bifurcation analysis in a stage-structure prey–predator growth model. Seo and DeAngelis [14] formulated a predator-prey model with a Holling response function of type I and many more. From the above works, it can be observed that the biological parameters whichever were used in the models are always fixed, but in reality they vary under dynamical ecological conditions. Not only that, due to lack of precise numerical information such as experimental part, data collection, measurement process, determining initial condition, some parameters become imprecise. Also, in deterministic dynamical system parameters need to be precisely defined. To have a rough estimation of the parameters, a huge amount information is needed to continue processed. Imprecise bio-mathematical models are more meaningful than the deterministic models. There is a long history of imprecise prey-predator model. To mention a few, Guo et al. [5] established fuzzy impulsive functional differential equation using Hullermeiers approach of a population model. Peixoto et al. [2] presented the fuzzy predator-prey model. Pal et al. [11], De et al. [3,4] investigated optimal harvesting of fishery-poultry system with interval biological parameters. Stability Analysis of Predator-Prey System with Fuzzy Impulsive Control done by Wang [17]. Tapaswini and Chakraverty [15] numerically solved of Fuzzy Arbitrary Order Predator-Prey Equations. Stability and bionomic analysis of fuzzy parameter based prey-predator harvesting model using UFM also done by Pal et al. [10]. Due to direct effect of the imprecise parameters on the variables, the variables also become imprecise in nature. But, in most of the research work it is found that only the parameters or the coefficients involved in the models are assumed to be imprecise. Khatua and Maity [7] have analysed the stability of fuzzy dynamical systems based on a quasi-level-wise system where all the variables along with the parameters are considered as imprecise variables and using "e" and "g" operator method the imprecise model converted to an equivalent crisp problem.

From previous research, though we found a significant amount of research in the area of the prey-predator system in imprecise environment, to the best of our knowledge, none of the articles have introduced the impreciseness in the variables along with the parameters. In this work, 1) We have developed an imprecise prey-predator model. We have considered both prey and predator population as imprecise variables. 2) Also, we have assumed the parameters, namely growth rate of prey, predation rate of the prey population, increase rate and decay rate of the predator population as imprecise parameters. 3) Following Khatua et al. [7], the imprecise prey-predator model is converted to equivalent crisp model using "e" and "g" operator method. The local stability analysis is done for the deterministic system mathematically. The Numerical result is presented in tabular and graphical form to validate the theoretical findings.

The rest of the paper is arranged in the following manner: In Sect. 3 Some mathematical preliminaries are mentioned. Section 4 is used for assumptions and notations. The prey-predator model is formulated in a crisp environment in Sect. 5. In Sect. 6, the model is transformed into an imprecise model, and after that, the imprecise model converted to equivalent crisp model using "e" and "g" operator method. Then the local stability of two different cases is analysed theoretically, numerically and presented graphically in this section. The results obtained in the numerical experiment are discussed in Sect. 7. Finally, the chapter is concluded in Sect. 8.

3   Mathematical Preliminaries

Mathematical preliminaries are recollected in this section.

3.1   Basic Concept of "e" and "g" Operators

Let $\mathbf{C}$ be a complex set i.e $\mathbf{C} = \{a + ib : a, b \in \Re\}$. Then "e" is a identity operator and "g" corresponds to a flip about the diagonal in the complex plane, i.e., $\forall a + ib \in \mathbf{C}$,

$$\begin{cases} e : a + ib \rightarrow a + ib, \\ g : a + ib \rightarrow b + ia. \end{cases} \tag{1}$$

3.2   Use of "e" and "g" Operators to Fuzzy Dynamical System

Let us consider the following non-homogeneous fuzzy dynamical system

$$\dot{\tilde{\mathbf{x}}}(t) = \mathbf{A}\tilde{\mathbf{x}}(t) + \mathbf{f}(t), \quad \tilde{\mathbf{x}}(0) = \tilde{\mathbf{x}}_0, \quad t \in [0, \infty) \tag{2}$$

where $\dot{\tilde{\mathbf{x}}}(t) = [\dot{\tilde{x}}_1(t) \cdots \dot{\tilde{x}}_n(t)]^T$ and $\mathbf{f}(t) = [f_1(t) \cdots f_n(t)]^T$.
Let $\overline{\mathbf{Y}}^\alpha(t) = [\bar{y}_1^\alpha(t) \cdots \bar{y}_n^\alpha(t)]^T$, $\underline{\mathbf{Y}}^\alpha(t) = [\underline{y}_1^\alpha(t) \cdots \underline{y}_n^\alpha(t)]^T$ be the solutions of quasi-level-wise system

$$\begin{cases} \dot{\underline{\mathbf{Y}}}^\alpha(t) + i\dot{\overline{\mathbf{Y}}}^\alpha(t) = B[\underline{\mathbf{Y}}^\alpha(t) + i\overline{\mathbf{Y}}^\alpha(t)] + (\underline{\mathbf{f}}(t) + i\overline{\mathbf{f}}(t)), \\ \underline{Y}^\alpha(0) = \underline{x}_0^\alpha, \quad \overline{Y}^\alpha(0) = \overline{x}_0^\alpha \end{cases} \tag{3}$$

where $\underline{f}(t) = \overline{f}(t) = f(t)$ and $\mathbf{B} = [b_{ij}]_{n \times n}$, $b_{ij} = \begin{cases} a_{ij}e & a_{ij} \geq 0 \\ a_{ij}g & a_{ij} < 0 \end{cases}$. Then

$\overline{x}_i^\alpha(t) = \max_{t \in (0,\infty)} \{\overline{y}_i^\alpha(t), \underline{y}_i^\alpha(t)\}$

$\underline{x}_i^\alpha(t) = \min_{t \in (0,\infty)} \{\overline{y}_i^\alpha(t), \underline{y}_i^\alpha(t)\}, i = 1, 2, \cdots, n$ are also the solutions of the fuzzy dynamical system (2).

Now if the dynamical system (3) is unstable, then moving the instability property to the level-wise system, we get to the same system as (3) i.e.,

$$\begin{cases} e[\underline{\dot{Y}}^\alpha(t) + i\overline{\dot{Y}}^\alpha(t)] = \bar{\mathbf{B}}[\underline{Y}^\alpha(t) + i\overline{Y}^\alpha(t)] + (\underline{f}(t) + i\overline{f}(t)), \\ \underline{Y}^\alpha(0) = \underline{x}_0^\alpha, \quad \overline{Y}^\alpha(0) = \overline{x}_0^\alpha \end{cases} \tag{4}$$

or

$$\begin{cases} g[\underline{\dot{Y}}^\alpha(t) + i\overline{\dot{Y}}^\alpha(t)] = \bar{\mathbf{B}}[\underline{Y}^\alpha(t) + i\overline{Y}^\alpha(t)] + (\underline{f}(t) + i\overline{f}(t)), \\ \underline{Y}^\alpha(0) = \underline{x}_0^\alpha, \quad \overline{Y}^\alpha(0) = \overline{x}_0^\alpha \end{cases} \tag{5}$$

where $\bar{\mathbf{B}} = [\bar{b}_{ij}]_{n \times n}$, $\bar{b}_{ij} = a_{ij}e$ or $a_{ij}g$.

4   Assumption and Notations

Here we have used the following notations: X, Y: the prey and predator population density at any time t respectively.

s: the natural growth rate of prey population.
K: the environmental carrying capacity at any time t.
δ: predation rate of prey population by prey population.
β: increase rate of predator population due to successful predation of prey.
γ: decay rate of predator population due to natural death.

We have assumed that the prey population do not have any decay due to natural death. The system is closed and there is no external harvesting of prey or predator.

5   Formulation of the Model in Crisp Environment

The present study analyses a prey-predator model. The prey population follows a logistic growth model with s as intrinsic growth rate and K to be environmental carrying capacity. The prey population decays due to predation by the predator at a rate δ and the predation function is XY. Again, the predator population increases by consuming the preys at a rate β and decays due to natural death at a rate γ. The mathematical formulation of the model is

$$\begin{cases} \frac{dX(t)}{dt} = sX(t)(1 - \frac{X(t)}{K}) - \delta X(t)Y(t) \\ \frac{dY(t)}{dt} = \beta X(t)Y(t) - \gamma Y(t) \end{cases} \tag{6}$$

This can be expressed as

$$\begin{cases} \frac{dX(t)}{dt} = sX(t) - \frac{sX^2(t)}{K} - \alpha X(t)Y(t) \\ \frac{dY(t)}{dt} = \beta X(t)Y(t) - \gamma Y(t) \end{cases} \tag{7}$$

6 Formulation and Stability Analysis of the Imprecise Model

Along with the prey X and the predator Y population, it has been considered that the predation rate δ, growth rate s, death rate γ and increase rate β to be imprecise. So the reformulated system become

$$
\begin{cases}
\frac{d\widehat{X}(t)}{dt} = \widehat{s}\widehat{X}(t) - \frac{\widehat{s}\widehat{X}^2(t)}{K} - \widehat{\delta}\widehat{X}(t)\widehat{Y}(t) \\
\frac{d\widehat{Y}(t)}{dt} = \widehat{\beta}\widehat{X}(t)\widehat{Y}(t) - \widehat{\gamma}\widehat{Y}(t)
\end{cases}
\tag{8}
$$

Now taking the imprecise variables and parameters to be interval numbers given by $\widehat{X} = [\underline{X}^\alpha, \overline{X}^\alpha]$, $\widehat{Y} = [\underline{Y}^\alpha, \overline{Y}^\alpha]$, $\widehat{\delta} = [\underline{\delta}^\alpha, \overline{\delta}^\alpha]$, $\widehat{\beta} = [\underline{\beta}^\alpha, \overline{\beta}^\alpha]$ and $\widehat{\gamma} = [\underline{\gamma}^\alpha, \overline{\gamma}^\alpha]$ and using Theorem-?? we have $\widehat{X} = [\underline{X}^\alpha, \overline{X}^\alpha]$ and $\widehat{Y} = [\underline{Y}^\alpha, \overline{Y}^\alpha]$ for the first form or $\widehat{X} = [\overline{X}^\alpha, \underline{X}^\alpha]$ and $\widehat{Y} = [\overline{Y}^\alpha, \underline{Y}^\alpha]$ for the second form. Now by using "e" and "g" operator method, and following [7] the system reduced to the following sub section:

6.1 Case-I

$$
\begin{cases}
e(\underline{\dot{P}}^\alpha(t) + i\overline{\dot{P}}^\alpha(t)) = e(\underline{s}^\alpha + i\overline{s}^\alpha)e(\underline{P}^\alpha(t) + i\overline{P}^\alpha(t)) \\
\qquad -g(\underline{s}^\alpha + i\overline{s}^\alpha)e(\underline{P}^\alpha(t) + i\overline{P}^\alpha(t))e(\underline{P}^\alpha(t) + i\overline{P}^\alpha(t))/K \\
\qquad -g(\underline{\delta}^\alpha + i\overline{\delta}^\alpha)e(\underline{P}^\alpha(t) + i\overline{P}^\alpha(t))e(\underline{N}^\alpha(t) + i\overline{N}^\alpha(t)) \\
e(\underline{\dot{N}}^\alpha(t) + i\overline{\dot{N}}^\alpha(t)) = e(\underline{\beta}^\alpha + i\overline{\beta}^\alpha)e(\underline{P}^\alpha(t) + i\overline{P}^\alpha(t))e(\underline{N}^\alpha(t) + i\overline{N}^\alpha(t)) \\
\qquad -g(\underline{\gamma}^\alpha + i\overline{\gamma}^\alpha)e(\underline{N}^\alpha(t) + i\overline{N}^\alpha(t))
\end{cases}
\tag{9}
$$

where
$$
\overline{X}^\alpha(t) = \max_{t\in[0,\infty)} \left\{ \overline{P}^\alpha(t), \underline{P}^\alpha(t) \right\}, \quad \underline{X}^\alpha(t) = \min_{t\in[0,\infty)} \left\{ \overline{P}^\alpha(t), \underline{P}^\alpha(t) \right\}, \quad \overline{Y}^\alpha(t) =
$$
$$
\max_{t\in[0,\infty)} \left\{ \overline{N}^\alpha(t), \underline{N}^\alpha(t) \right\}, \quad \underline{Y}^\alpha(t) = \min_{t\in[0,\infty)} \left\{ \overline{N}^\alpha(t), \underline{N}^\alpha(t) \right\}.
$$

6.2 Local Stability Analysis of Case-I

Form Eq. (9) it can be obtained that

$$
\begin{cases}
(\underline{\dot{P}}^\alpha(t) + i\overline{\dot{P}}^\alpha(t)) = (\underline{s}^\alpha + i\overline{s}^\alpha)(\underline{P}^\alpha(t) + i\overline{P}^\alpha(t)) \\
\qquad -(\overline{s}^\alpha + i\underline{s}^\alpha)(\underline{P}^\alpha(t) + i\overline{P}^\alpha(t))(\underline{P}^\alpha(t) + i\overline{P}^\alpha(t))/K \\
\qquad -(\overline{\delta}^\alpha + i\underline{\delta}^\alpha)(\underline{P}^\alpha(t) + i\overline{P}^\alpha(t))(\underline{N}^\alpha(t) + i\overline{N}^\alpha(t)) \\
(\underline{\dot{N}}^\alpha(t) + i\overline{\dot{N}}^\alpha(t)) = (\underline{\beta}^\alpha + i\overline{\beta}^\alpha)(\underline{P}^\alpha(t) + i\overline{P}^\alpha(t))(\underline{N}^\alpha(t) + i\overline{N}^\alpha(t)) \\
\qquad -(\overline{\gamma}^\alpha + i\underline{\gamma}^\alpha)(\underline{N}^\alpha(t) + i\overline{N}^\alpha(t))
\end{cases}
\tag{10}
$$

Therefore, the following system of ODE can be obtained

$$
\begin{cases}
\underline{\dot{P}}^\alpha(t) = \underline{s}^\alpha \underline{P}^\alpha(t) - \overline{s}^\alpha \underline{P}^{\alpha 2}(t)/K - \underline{\delta}^\alpha \underline{P}^\alpha(t)\underline{N}^\alpha(t) \\
\overline{\dot{P}}^\alpha(t) = \overline{s}^\alpha \overline{P}^\alpha(t) - \underline{s}^\alpha \overline{P}^{\alpha 2}(t)/K - \underline{\delta}^\alpha \overline{P}^\alpha(t)\overline{N}^\alpha(t) \\
\underline{\dot{N}}^\alpha(t) = \underline{\beta}^\alpha \underline{P}^\alpha(t)\underline{N}^\alpha(t) - \overline{\gamma}^\alpha \underline{N}^\alpha(t) \\
\overline{\dot{N}}^\alpha(t) = \overline{\beta}^\alpha \overline{P}^\alpha(t)\overline{N}^\alpha(t) - \underline{\gamma}^\alpha \overline{N}^\alpha(t)
\end{cases}
\tag{11}
$$

For simplicity let us consider the following:
$$\underline{s}^\alpha = s_1^\alpha,\ \overline{s}^\alpha = s_2^\alpha,\ \underline{\delta}^\alpha = \delta_1^\alpha,\ \overline{\delta}^\alpha = \delta_2^\alpha,\ \underline{\beta}^\alpha = \beta_1^\alpha,\ \overline{\beta}^\alpha = \beta_2^\alpha,\ \underline{\gamma}^\alpha = \gamma_1^\alpha,\ \overline{\gamma}^\alpha = \gamma_2^\alpha$$
$$\underline{P}^\alpha = P_1^\alpha,\ \overline{P}^\alpha = P_2^\alpha,\ \underline{N}^\alpha = N_1^\alpha,\ \overline{N}^\alpha = N_2^\alpha$$
The modified system of (11) is

$$
\begin{cases}
\dot{P_1}^\alpha = s_1^\alpha P_1^\alpha - \frac{s_2^\alpha P_1^{\alpha 2}}{K} - \delta_2^\alpha P_1^\alpha N_1^\alpha \\
\dot{P_2}^\alpha = s_2^\alpha P_2^\alpha - \frac{s_1^\alpha P_2^{\alpha 2}}{K} - \delta_1^\alpha P_2^\alpha N_2^\alpha \\
\dot{N_1}^\alpha = \beta_1^\alpha P_1^\alpha N_1^\alpha - \gamma_2^\alpha N_1^\alpha \\
\dot{N_2}^\alpha = \beta_2^\alpha P_2^\alpha N_2^\alpha - \gamma_1^\alpha N_2^\alpha
\end{cases}
\tag{12}
$$

The steady state solutions are given by $P_1^{\alpha *} = \frac{\gamma_2^\alpha}{\beta_1^\alpha}, P_2^{\alpha *} = \frac{\gamma_1^\alpha}{\beta_2^\alpha}, N_1^{\alpha *} = \frac{s_1^\alpha - s_2^\alpha \gamma_2^\alpha/\beta_1^\alpha K}{\delta_2^\alpha}, N_2^{\alpha *} = \frac{s_2^\alpha - s_1^\alpha \gamma_1^\alpha/\beta_2^\alpha K}{\delta_1^\alpha}.$
The Jacobian matrix for the system in steady state $(P_1^{\alpha *}, P_2^{\alpha *}, N_1^{\alpha *}, N_2^{\alpha *})$ is given by

$$
V = \begin{pmatrix}
-\frac{\gamma_2^\alpha s_2^\alpha}{\beta_1^\alpha K} & 0 & -\frac{\delta_2^\alpha \gamma_2^\alpha}{\beta_1^\alpha} & 0 \\
0 & -\frac{\gamma_1^\alpha s_1^\alpha}{\beta_2^\alpha K} & 0 & -\frac{\delta_1^\alpha \gamma_1^\alpha}{\beta_2^\alpha} \\
\frac{s_1^\alpha \beta_1^\alpha K - s_2^\alpha \gamma_2^\alpha}{K \delta_2^\alpha} & 0 & 0 & 0 \\
0 & \frac{s_2^\alpha \beta_2^\alpha K - s_1^\alpha \gamma_1^\alpha}{K \delta_1^\alpha} & 0 & 0
\end{pmatrix}
$$

and the corresponding eigenvalues are given by

$$
-\frac{s_1^\alpha \gamma_1^\alpha \pm \sqrt{\gamma_1^\alpha(-4K^2 s_2^\alpha \beta_2^\alpha + s_1^{\alpha 2}\gamma_1^\alpha + 4K s_1^\alpha \beta_2^\alpha \gamma_1^\alpha)}}{2K\beta_2^\alpha},
$$

$$
-\frac{s_2^\alpha \gamma_2^\alpha \pm \sqrt{\gamma_2^\alpha(-4K^2 s_1^\alpha \beta_1^\alpha + s_2^{\alpha 2}\gamma_2^\alpha + 4K s_2^\alpha \beta_1^\alpha \gamma_2^\alpha)}}{2K\beta_1^\alpha}.
$$

Clearly, the all the eigenvalues are negative if

$$
\sqrt{\gamma_1^\alpha(-4K^2 s_2^\alpha \beta_2^{2\alpha} + s_1^{\alpha 2}\gamma_1^\alpha + 4K s_1^\alpha \beta_2^\alpha \gamma_1^\alpha)} < s_1^\alpha \gamma_1^\alpha \text{and,}
\tag{13}
$$

$$
\sqrt{\gamma_2^\alpha(-4K^2 s_1^\alpha \beta_1^{\alpha 2} + s_2^{\alpha 2}\gamma_2^\alpha + 4K s_2^\alpha \beta_1^\alpha \gamma_2^\alpha)} < s_2^\alpha \gamma_2^\alpha.
\tag{14}
$$

The above theory is illustrated in the following numerical experiment:

Table 1. Input data

s_1^α	1.1	δ_1^α	0.01	β_1^α	.001	γ_1^α	0.085	K	150
s_2^α	1.15	δ_2^α	0.015	β_2^α	0.0015	γ_2^α	0.090		

Table 2. Output data

$P_1^{\alpha*}$	90	$N_1^{\alpha*}$	27.33
$P_2^{\alpha*}$	56.67	$N_2^{\alpha*}$	73.44

Table 3. Eigenvalues

-0.631575	$-0.207778 + 0.132856i$	$-0.207778 - 0.132856i$	-0.0584254

6.3 Numerical Experiment Case-I

The parameters are assumed as which shows the steady state solutions are stable.

6.4 Case-II

$$
\begin{cases}
g(\underline{\dot{P}}^\alpha(t) + i\overline{\dot{P}}^\alpha(t)) = g(\underline{s}^\alpha + i\overline{s}^\alpha)g(\underline{P}^\alpha(t) + i\overline{P}^\alpha(t)) \\
\qquad -g(\underline{s}^\alpha + i\overline{s}^\alpha)g(\underline{P}^\alpha(t) + i\overline{P}^\alpha(t))g(\underline{P}^\alpha(t) + i\overline{P}^\alpha(t))/K \\
\qquad -g(\underline{\delta}^\alpha + i\overline{\delta}^\alpha)g(\underline{P}^\alpha(t) + i\overline{P}^\alpha(t))g(\underline{N}^\alpha(t) + i\overline{N}^\alpha(t)) \qquad (15) \\
g(\underline{\dot{N}}^\alpha(t) + i\overline{\dot{N}}^\alpha(t)) = e(\underline{\beta}^\alpha + i\overline{\beta}^\alpha)g(\underline{P}^\alpha(t) + i\overline{P}^\alpha(t))g(\underline{N}^\alpha(t) + i\overline{N}^\alpha(t)) \\
\qquad -e(\underline{\gamma}^\alpha + i\overline{\gamma}^\alpha)g(\underline{N}^\alpha(t) + i\overline{N}^\alpha(t))
\end{cases}
$$

where $\overline{X}^\alpha(t) = \max\limits_{t \in [0,\infty)} \left\{ \overline{P}^\alpha(t), \underline{P}^\alpha(t) \right\}$, $\underline{X}^\alpha(t) = \min\limits_{t \in [0,\infty)} \left\{ \overline{P}^\alpha(t), \underline{P}^\alpha(t) \right\}$,
$\overline{Y}^\alpha(t) = \max\limits_{t \in [0,\infty)} \left\{ \overline{N}^\alpha(t), \underline{N}^\alpha(t) \right\}$, $\underline{Y}^\alpha(t) = \min\limits_{t \in [0,\infty)} \left\{ \overline{N}^\alpha(t), \underline{N}^\alpha(t) \right\}$.
 The stability of the above two systems given by (9) and (15) can be analysed as following:

6.5 Local Stability Analysis of Case-II

Form Eq. (15) it can be obtained that

$$
\begin{cases}
(\overline{\dot{P}}^\alpha(t) + i\underline{\dot{P}}^\alpha(t)) = (\overline{s}^\alpha + i\underline{s}^\alpha)(\overline{P}^\alpha(t) + i\underline{P}^\alpha(t)) \\
\qquad -(\overline{s}^\alpha + i\underline{s}^\alpha)(\overline{P}^\alpha(t) + i\underline{P}^\alpha(t))(\overline{P}^\alpha(t) + i\underline{P}^\alpha(t))/K \\
\qquad -(\overline{\delta}^\alpha + i\underline{\delta}^\alpha)(\overline{P}^\alpha(t) + i\underline{P}^\alpha(t))(\overline{N}^\alpha(t) + i\underline{N}^\alpha(t)) \qquad (16) \\
(\overline{\dot{N}}^\alpha(t) + i\underline{\dot{N}}^\alpha(t)) = (\overline{\beta}^\alpha + i\underline{\beta}^\alpha)(\overline{P}^\alpha(t) + i\underline{P}^\alpha(t))(\overline{N}^\alpha(t) + i\underline{N}^\alpha(t)) \\
\qquad -(\overline{\gamma}^\alpha + i\underline{\gamma}^\alpha)(\overline{N}^\alpha(t) + i\underline{N}^\alpha(t))
\end{cases}
$$

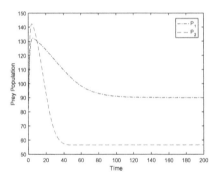

Fig. 1. Time series plot of prey population showing $\overline{X}^{\alpha}(t) = \max\limits_{t \in [0,\infty)} \{P_1(t), P_2(t)\}$, $\underline{X}^{\alpha}(t) = \min\limits_{t \in [0,\infty)} \{P_1(t), P_2(t)\}$

Fig. 2. Time series plot of predator population showing $\overline{Y}^{\alpha}(t) = \max\limits_{t \in [0,\infty)} \{N_1(t), N_2(t)\}$, $\underline{Y}^{\alpha}(t) = \min\limits_{t \in [0,\infty)} \{N_1(t), N_2(t)\}$

Therefore we have the following system of ODE

$$
\begin{cases}
\dot{\overline{P}}^{\alpha}(t) = \overline{s}^{\alpha}\overline{P}^{\alpha}(t) - \overline{s}^{\alpha}\overline{P}^{\alpha 2}(t)/K - \overline{\delta}^{\alpha}\overline{P}^{\alpha}(t)\overline{N}^{\alpha}(t) \\
\dot{\underline{P}}^{\alpha}(t) = \underline{s}^{\alpha}\underline{P}^{\alpha}(t) - \underline{s}^{\alpha}\underline{P}^{\alpha 2}(t)/K - \underline{\delta}^{\alpha}\underline{P}^{\alpha}(t)\underline{N}^{\alpha}(t) \\
\dot{\overline{N}}^{\alpha}(t) = \beta^{\alpha}\overline{P}^{\alpha}(t)\overline{N}^{\alpha}(t) - \underline{\gamma}^{\alpha}\overline{N}^{\alpha}(t) \\
\dot{\underline{N}}^{\alpha}(t) = \overline{\beta}^{\alpha}\underline{P}^{\alpha}(t)\underline{N}(t)^{\alpha} - \overline{\gamma}^{\alpha}\underline{N}^{\alpha}(t)
\end{cases}
\tag{17}
$$

For simplicity let us consider the following:
$\underline{s}^{\alpha} = s_1^{\alpha}$, $\overline{s}^{\alpha} = s_2^{\alpha}$, $\underline{\delta}^{\alpha} = \delta_1^{\alpha}$, $\overline{\delta}^{\alpha} = \delta_2^{\alpha}$, $\underline{\beta}^{\alpha} = \beta_1^{\alpha}$, $\overline{\beta}^{\alpha} = \beta_2^{\alpha}$, $\underline{\gamma}^{\alpha} = \gamma_1^{\alpha}$, $\overline{\gamma}^{\alpha} = \gamma_2^{\alpha}$
$\underline{P}^{\alpha} = P_1^{\alpha}$, $\overline{P}^{\alpha} = P_2^{\alpha}$, $\underline{N}^{\alpha} = N_1^{\alpha}$, $\overline{N}^{\alpha} = N_2^{\alpha}$ The modified system of (17) is

$$
\begin{cases}
\dot{P_1}^{\alpha} = s_1^{\alpha}P_1^{\alpha} - \frac{s_1^{\alpha}P_1^{\alpha 2}}{K} - \delta_1^{\alpha}P_1^{\alpha}N_1^{\alpha} \\
\dot{P_2}^{\alpha} = s_2^{\alpha}P_2^{\alpha} - \frac{s_2^{\alpha}P_2^{\alpha 2}}{K} - \delta_2^{\alpha}P_2^{\alpha}N_2^{\alpha} \\
\dot{N_1}^{\alpha} = \beta_2^{\alpha}P_1^{\alpha}N_1^{\alpha} - \gamma_2^{\alpha}N_1^{\alpha} \\
\dot{N_2}^{\alpha} = \beta_1^{\alpha}P_2^{\alpha}N_2^{\alpha} - \gamma_1^{\alpha}N_2^{\alpha}
\end{cases}
\tag{18}
$$

The steady state solutions are given by $P_1^{\alpha *} = \frac{\gamma_2^{\alpha}}{\beta_2^{\alpha}}, P_2^{\alpha *} = \frac{\gamma_1^{\alpha}}{\beta_1^{\alpha}}, N_1^{\alpha *} = \frac{s_1^{\alpha} - s_1^{\alpha}\gamma_2^{\alpha}/\beta_2^{\alpha}K}{\delta_1^{\alpha}}, N_2^{\alpha *} = \frac{s_2^{\alpha} - s_2^{\alpha}\gamma_1^{\alpha}/\beta_1^{\alpha}K}{\delta_2^{\alpha}}$.

The Jacobian matrix for the system in steady state $(P_1^{\alpha *}, P_2^{\alpha *}, N_1^{\alpha *}, N_2^{\alpha *})$ is given by

$$
V = \begin{pmatrix}
-\frac{\gamma_2^{\alpha}s_1^{\alpha}}{\beta_2^{\alpha}K} & 0 & -\frac{\delta_2^{\alpha}\gamma_1^{\alpha}}{\beta_1^{\alpha}} & 0 \\
0 & -\frac{\gamma_1^{\alpha}s_2^{\alpha}}{\beta_1^{\alpha}K} & 0 & -\frac{\delta_2^{\alpha}\gamma_1^{\alpha}}{\beta_1^{\alpha}} \\
\frac{s_1^{\alpha}\beta_2^{\alpha}K - s_1^{\alpha}\gamma_2^{\alpha}}{K\delta_1^{\alpha}} & 0 & 0 & 0 \\
0 & \frac{s_2^{\alpha}\beta_1^{\alpha}K - s_2^{\alpha}\gamma_1^{\alpha}}{K\delta_2^{\alpha}} & 0 & 0
\end{pmatrix}
$$

and the corresponding eigen values are given by

$$-\frac{s_2{}^\alpha\gamma_1{}^\alpha \pm \sqrt{s_2{}^\alpha\gamma_1{}^\alpha(-4K^2\beta_1^{\alpha2} + s_2{}^\alpha\gamma_1{}^\alpha + 4K\beta_1{}^\alpha\gamma_1{}^\alpha)}}{2K\beta_1{}^\alpha},$$

$$-\frac{s_1{}^\alpha\gamma_2{}^\alpha \pm \sqrt{s_1{}^\alpha\gamma_2{}^\alpha(-4K^2\beta_2^{2\alpha} + s_1{}^\alpha\gamma_2{}^\alpha + 4K\beta_2{}^\alpha\gamma_2{}^\alpha)}}{2K\beta_2{}^\alpha}.$$

Clearly, in this case also all the eigen values are negative if

$$\sqrt{s_2{}^\alpha\gamma_1{}^\alpha(-4K^2\beta_1^{\alpha2} + s_2{}^\alpha\gamma_1{}^\alpha + 4K\beta_1{}^\alpha\gamma_1{}^\alpha)} < s_2{}^\alpha\gamma_1{}^\alpha, \tag{19}$$

$$\sqrt{s_1{}^\alpha\gamma_2{}^\alpha(-4K^2\beta_2^{\alpha2} + s_1{}^\alpha\gamma_2{}^\alpha + 4K\beta_2{}^\alpha\gamma_2{}^\alpha)} < s_1{}^\alpha\gamma_2{}^\alpha. \tag{20}$$

Let us have the following numerical experiment to illustrate the above.

6.6   Numerical Experiment-2

The input parameters are assumed to be the same as in numerical experiment-1 in Sect. 6.3.

Table 4. Output data

$P_1^{\alpha*}$	60	$N_1^{\alpha*}$	66
$P_2^{\alpha*}$	85	$N_2^{\alpha*}$	23.22

Table 5. Eigenvalues

-0.578438	$-0.22 + 0.104881i$	$-0.22 - 0.104881i$	-0.0732288

which shows the steady state solutions are stable. The graphs for the above case are given by the following.

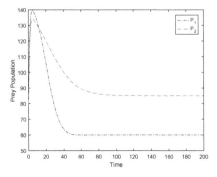
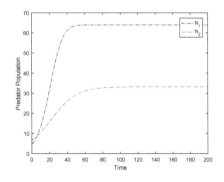

Fig. 3. Time series plot of prey population showing $\overline{X}^\alpha(t) = \max_{t\in[0,\infty)}\{P_1(t),P_2(t)\}$, $\underline{X}^\alpha(t) = \min_{t\in[0,\infty)}\{P_1(t),P_2(t)\}$

Fig. 4. Time series plot of predator population showing $\overline{Y}^\alpha(t) = \max_{t\in[0,\infty)}\{N_1(t),N_2(t)\}$, $\underline{Y}^\alpha(t) = \min_{t\in[0,\infty)}\{N_1(t),N_2(t)\}$

7   Discussion

In this work, we have analyzed a prey–predator model in imprecise environment. We have considered both prey and predator population as imprecise variables. Also, we have assumed the parameters namely growth rate of prey, predation rate of the prey population, increase rate and decay rate of the predator population as imprecise parameters. The imprecise model converted to two equivalent crisp models given by Eqs. (9) and (15). The non-zero steady state solutions are obtained for each of the cases. The local stability analysis is done with the help of eigen values of the jacobian matrices.

It can be observed form the system given by Eq. (9) is stable if the system satisfies the condition given by Eq. (13).

The numerical experiment-1 in Sect. 6.3 corresponds to the system Eq. (9). Table 1 shows the input data, Table 2 presents the output data and Table 3 gives the eigen values of the jacobian matrix at the steady state. From Table 3 we observe that the eigen values have negative real part, so the steady state solutions are stable.

Figures 1 and 2 represents time series plot for prey and predator populations respectively. It can be observed from both the figures that the graphical solutions are at a good agreement with the numerical values.

Similarly, form the system given by Eq. (15) is stable if the system satisfies the condition given by (19).

The numerical experiment-2 in Sect. 6.6 corresponds to the system (15). With the same the input data as on Numerical experiment-1, Table 4 presents the output data and Table 5 gives the eigen values of the jacobian matrix at the steady state. From Table 5 we observe that the eigen values have negative real part, so the steady state solutions are stable.

Figures 3 and 4 represents time series plot for prey and predator populations respectively. It can be observed from both the figures that the graphical solutions are at a good agrement with the numerical values.

In Fig. 1, we observe that the solution curves corresponding to the lower and upper values of the prey population for system (9), do not always remain as lower and upper values, rather the curves corresponding to lower value becomes upper value and the upper value becomes the lower value.

Again, from Fig. 4, it has been analyzed that the solution curves corresponding to the upper and lower values of the predator populations for system (15), do not always remain as upper and lower values, but the curves corresponding to lower value becomes upper value and the upper value becomes the lower value. Furthermore, it can be observed that in place of a single curve for each population, we are obtaining a lower and upper boundaries of the stable solutions. In case of interval approach to manipulate this type of problem we need to check the solution curves for different parametric values, but in this case we have obtained the boundaries at a single attempt. Though, we have checked the stability of only two cases, but we can obtain some more different cases with different combinations of "e" and "g" operators.

8   Conclusion

A prey-predator model is developed in the model in imprecise environment. Due to the environmental variation in different ecological conditions, the natural parameters vary. Some of the researchers developed imprecise models considering the natural parameters to be imprecise. These imprecise parameter are converted to interval numbers and depending upon the different parametric conditions the problems are solved. In the present work, the parameters like growth rate of prey, predation rate of the prey population, increase rate and decay rate of the predator population along with the prey and predator population are assumed to be imprecise. The imprecise model then converted to equivalent two different crisp model with the help of "e" and "g" operator method. Stability for both the crisp model are analyzed theoretically. With the help of numerical examples both the models are presented numerically and graphically. Numerical results and the graphical analysis for both the crisp model provides the upper and lower boundaries of the stable solutions for both the population rather than a single solution curve.

For different combinations of "e" and "g" we can obtain some more cases in crisp form. The model can be extended to a prey-predator harvesting model, prey-predator harvesting model with budget constraints etc.

References

1. Chakraborty, K., Das, S., Kar, T.K.: Optimal control of effort of a stage structured prey-predator fishery model with harvesting. Nonlinear Anal.: Real World Appl. **12**(6), 3452–3467 (2011)

2. da Silva Peixoto, M., de Barros, L., Carlos, R.B.: Predator-prey fuzzy model. Ecol. Model. **214**(1), 39–44 (2008)

3. De, A., Maity, K., Maiti, M.: Stability analysis of combined project of fish, broiler and ducks: dynamical system in imprecise environment. Int. J. Biomath. **8**(05), 1550067 (2015)

4. De, A., Maity, K., Maiti, M.: Fish and broiler optimal harvesting models in imprecise environment. Int. J. Biomath. **10**(08), 1750115 (2017)

5. Guo, M., Xue, X., Li, R.: Impulsive functional differential inclusions and fuzzy population models. Fuzzy Sets Syst. **138**(3), 601–615 (2003)

6. Kar, T.K.: Modelling and analysis of a harvested prey-predator system incorporating a prey refuge. J. Comput. Appl. Math. **185**(1), 19–33 (2006)

7. Khatua, D., Maity, K.: Stability of fuzzy dynamical systems based on quasi-level-wise system. J. Intell. Fuzzy Syst. **33**(6), 3515–3528 (2017)

8. Lotka, A.J.: Elements of Physical Biology. Williams and Wilkins Company, London (1925)

9. Ma, M., Friedman, M., Kandel, A.: A new fuzzy arithmetic. Fuzzy Sets Syst. **108**(1), 83–90 (1999)

10. Pal, D., Mahapatra, G.S., Samanta, G.P.: Stability and bionomic analysis of fuzzy parameter based prey-predator harvesting model using UFM. Nonlinear Dyn. **79**(3), 1939–1955 (2015). https://doi.org/10.1007/s11071-014-1784-4

11. Pal, D., Mahapatra, G.S., Samanta, G.P.: Optimal harvesting of prey-predator system with interval biological parameters: a bioeconomic model. Math. Biosci. **241**(2), 181–187 (2013)

12. Puri, M.L., Ralescu, D.: A differentials of fuzzy functions. J. Math. Anal. Appl. **91**(2), 552–558 (1983)

13. Qu, Y., Wei, J.: Bifurcation analysis in a time-delay model for prey-predator growth with stage-structure. Nonlinear Dyn. **49**(1–2), 285–294 (2007). https://doi.org/10.1007/s11071-006-9133-x

14. Seo, G., DeAngelis, D.L.: A predator-prey model with a holling type I functional response including a predator mutual interference. J. Nonlinear Sci. **21**(6), 811–833 (2011). https://doi.org/10.1007/s00332-011-9101-6

15. Tapaswini, S., Chakraverty, S.: Numerical solution of fuzzy arbitrary order predator-prey equations. Appl. Appl. Math. **8**(2), 647–672 (2013)

16. Volterra, P.: Lessons in the mathematical theory of the struggle for life. Seuil, Paris (1931)

17. Wang Y.: Stability analysis of predator-prey system with fuzzy impulsive control. J. Appl. Math. **2012** (2012)

Comparative Performance Study on Human Activity Recognition with Deep Neural Networks

Kavya Sree Gajjala[1], Ashika Kothamachu Ramesh[1], Kotaro Nakano[2], and Basabi Chakraborty[3]([✉])

[1] Graduate School of Software and Information Science, Iwate Prefectural University, Takizawa, Iwate, Japan
[2] Research and Regional Co-operative Division, Iwate Prefectural University, Takizawa, Iwate, Japan
[3] Faculty of Software and Information Science, Iwate Prefectural University, Takizawa, Iwate, Japan
basabi@iwate-pu.ac.jp

Abstract. Human activity recognition (HAR) plays an important role in every spheres of life as it assists in fitness tracking, health monitoring, elderly care, user authentication and management of smart homes. The assistive applications can be implemented on smartphones and wearable watches which are easily accessible and affordable. Now-a-days use of smart phones is ubiquitous, sensor data of diverse physical activities can be easily collected by in-built motion sensors. Many research works are proposed in this area using machine learning techniques including deep neural networks to develop smartphone based applications for human activity recognition. Our objective is to find an effective method from a variety of machine learning including deep learning models for low cost as well as high accuracy activity recognition. To fulfill the objective, a comparative performance study has been done in this work by simulation experiments on five publicly available bench mark data sets. The simulation results show that deep learning models, especially, 2D CNN and BI GRU can be promising candidates for developing smartphone based applications using motion sensor data.

Keywords: Human activity recognition · Smart phone sensors · Machine learning · Deep neural network

1 Introduction

Human actions such as walking, running, cooking, eating, lying down, sitting and so on are referred as human activities and monitoring them with emerging technologies for some useful purpose is the main objective of the research area of Human activity recognition(HAR). It is supportive in health monitoring [1], medical care, authentication, advance computing, sports and smart homes [2].

Z. Shi et al. (Eds.): ICIS 2020, IFIP AICT 623, pp. 241–251, 2021.
https://doi.org/10.1007/978-3-030-74826-5_21

HAR is one of the supporting technology for daily life care of aged people and the development of smart patient monitoring systems [3].

Activity recognition can be done through video based systems as well as other sensor based systems. Video based system [4] uses surveillance cameras to capture image or video for recognizing movements of people to identify their activities. But there are few shortcomings of video based systems such as intrusion to privacy and high cost. Sensor based systems use body sensors [5] or ambient sensors for the identification of person's movements. Now a days many sensors are embedded in smart phones [6], smart watches, spectacles, shoes and some non-movable objects like wall, furniture etc. Smartphones and watches have in-built accelerometer, gyroscope and magnetometer. Accelerometer helps to capture acceleration and velocity of the movement [7,8]. Gyroscope helps to capture orientation and angular velocity. Magnetometer is embedded with accelerometer and gyroscope. It identifies change in the magnetic field at certain position. From these sensors we can acquire tri variate time series data which can be used to recognize human activity with reasonable accuracy. Various research works have been done already in this field using several machine learning techniques including deep neural networks.

In this work, activity recognition from smartphone sensor data by machine learning including several deep neural networks models has been studied in order to find out an effective method suitable for developing a cost effective smart phone based health monitoring application. A comparative performance study of different techniques has been done by simulation experiments with multiple benchmark data sets. The following section describes some of the related works. Section 3 describes performance study of different models followed by the results and analysis in the Sect. 4. Last section comprises of summary and conclusion.

2   Related Work

Due to low cost, high degree of portability, and wide range of real world applications, wearable sensors based activity recognition with machine learning techniques became a popular research area. With the increasing popularity of deep neural networks, several researchers have proposed different deep neural network based techniques for human activity recognition. Here a brief summary of research works on traditional machine learning techniques and deep neural network-based techniques for human activity recognition problem has been presented.

2.1   Machine Learning Methods

Among traditional machine learning algorithms, Support Vector Machine (SVM) [9], K-Nearest Neighbor (KNN) [10], Decision tree(DT) [11], Naive Bayes(NB), Hidden Markov Model (HMM) [12], Random forest [13] classifiers have been used to recognize human activity in several works. Fan [11] used decision tree for classification of daily activities collected from wearable accelerometers. In [9], the authors presented a system architecture based on support vector machine (SVM) for HAR. HMM is used in [12] to classify the physical activities. In this approach the authors combine shape and optical flow features extracted from videos. Random forest is used in [13] for classifying daily activities and achieved accuracy more than 90%. Comparative study of HAR on machine learning algorithms can be found in [14].

2.2 Deep Neural Network Methods

A comprehensive study on HAR using deep neural networks can be found in [15]. Works on deep network based HAR can be categorized into three types. The first category, the most popular, uses Convolutional Neural Network (CNN). Ronao [16] used 1 dimensional CNN to classify the activity data recorded by smartphone sensors and compared their proposed model using SVM and DT. Authors in [17] used 2 dimensional CNN to classify six daily activities recorded from 12 volunteers. In [18], authors presented new architecture for CNN and handcrafted feature based methods to reduce the computation cost.

The second category uses Recurrent Neural Network (RNN) to capture time dependency of sensor data [19]. Among the RNN models, Long Short Term Memory (LSTM) network is the most popular one. In [20], authors applied LSTM to recognize daily activities and found LSTM is more accurate than machine learning methods. In [21], authors proposed BI LSTM for detecting activities. This method has two LSTM layers for extracting temporal features from both forward and backward direction. In [22], the authors used BI LSTM to classify 12 different activities recorded for 10 subjects. There is one more RNN model namely Gated Recurrent Unit (GRU). In [23], authors used GRU instead of LSTM and applied it to activity recognition. From the experiment they found GRU is efficient than CNN. In [24], authors presented new approach for using bidirection GRU model in human activity recognition and found BI GRU model gives promising and high quality recognition results.

The third category uses mixed model to identify human activity. This category represents the combination of CNN and RNN networks [25]. They observed that the combination attains higher accuracy because this model can utilize power of CNN in feature extraction and RNN in temporal dependencies among activities. Comparative study of HAR using hybrid models can be found in [26].

3 Comparative Performance Study

The performance study has been done by simulation experiments with five different bench mark data sets.

3.1 Datasets

The data sets used in this study are presented briefly in Table 1, the details can be found in the references noted in the Table.

Table 1. Details of datasets

Data sets	Sensors	Sampling frequency (HZ)	Activities	Subjects
WISDM [27]	Accelerometer	20	6	36
UCI HAR [28]	Accelerometer, Gyroscope	50	6	30
UCI HHAR [29]	Accelerometer, Gyroscope	50	6	9
Motion Sensor [30]	Accelerometer, Gyroscope	48	6	24
PAMAP2 [31]	Accelerometer, Gyroscope, Magnetometer	100	18	9

3.2 Methodology

Here a very brief introduction of the methods used in our study has been presented. K-Nearest Neighbour (KNN), Naive Bayes (NB), Decision tree (DT), eXtreme Gradient Boosting (XG Boost), Random Forest (RF) and Support Vector Machine (SVM) are used for our study. KNN is easily implementable, popular and computationally cheap instance based classifier. Naive Bayes is a probabilistic classification technique based on Bayes theorem. DT is another popular nonparametric supervised learning method used for classification while RF represents ensemble of DT. XG Boost is a newly developed, highly efficient and portable implementation of gradient boosted decision trees which provides high accuracy in pattern classification and regression problems. SVM is a discriminative classifier which aims to find a hyperplane with maximum margin so that error rate for classification is the least.

Among deep networks, we have used Convolutional Neural Networks (CNN) and two variants of Recurrent Neural Networks (RNN): Long Short Term Memory (LSTM) and Gated Recurrent Unit (GRU). CNN belongs to the class of multilayer feed forward network composed of convolution layer, max pooling layer, flatten layer and dense layer. These layers are stacked to form deep architecture for feature extraction from raw sensor data before classification. We have used one dimensional (1D CNN) and two dimensional (2D CNN) for our study. LSTM belongs to the class of recurrent neural network (RNN). This network captures temporal dependencies which can be used for prediction problems. Bi directional LSTM (BI LSTM) is an extension of LSTM which is comprised of 2 LSTM cells, and information flows both forward and backward direction. GRU is another variant of RNN similar to LSTM. But there are a few differences, such as LSTM has 4 gates and GRU has 2 gates. It is observed in certain tasks that GRU exhibits better performance than LSTM. Bi directional GRU (BI GRU) is a combination of two GRU's.

4   Simulation Experiments and Results

Simulation experiments with bench mark data sets have been done for selected traditional machine learning techniques and deep network models 1D CNN, 2D CNN, LSTM, BI LSTM, GRU and BI GRU mentioned in the previous section using Python 3.7 in Anaconda3. Performance of machine learning models in terms of classification accuracy and computational time is represented in Fig. 1 & Fig. 2 respectively. From Fig. 1 we can infer that **SVM** achieved the best performance with an average of 82% over all data set while Naive Bayes achieved accuracy of 71% which is the worst compared to the other models. We also observed that UCI HAR dataset achieved the highest accuracy for all the models. From Fig. 2, we can interpret that SVM takes longer time compared to the other models. Decision tree and XG Boost take lesser time to predict activity. By considering accuracy and time complexity we can conclude that **XG Boost** yields the best performance.

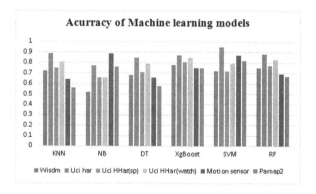

Fig. 1. Classification accuracy of traditional machine learning models

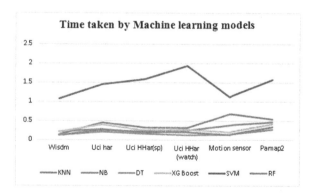

Fig. 2. Computation time for machine learning models

Deep learning models are implemented using tensor flow and Keras library for efficient performance. To stabilize and speed up training phase, batch normalization is used with a batch size of 80 data segments. The learning rate of the training is 10^{-3}. Sliding windows are used in this experiment to generate epochs with duration of 4s with 50% overlap. Each model is configured to run over 20–100 epochs using "sparse categorical loss entropy" as the loss function. After experimenting multiple times it is observed that every model have least loss value at epoch 50. Results of deep learning models are presented in Fig. 3 & Fig. 4. Figure 3 represents classification accuracy of deep learning models for multiple data sets. It is observed that the performance of all the models are better compared to traditional machine learning models. Among deep learning models **BI GRU, BI LSTM, 2D-CNN** achieved better results for all the data sets. 1D-CNN achieved the worst accuracy among DL models. Figure 4 represents computation time of different models. It is noticeable that BI LSTM consumes longer time compared to the other models. LSTM and BI GRU take almost the same time. GRU takes shorter time among RNN models whereas CNN models take less time compared to RNN models. Among all the models **2D CNN** takes the shortest time. UCI Har data set takes relatively higher time compared to other data sets.

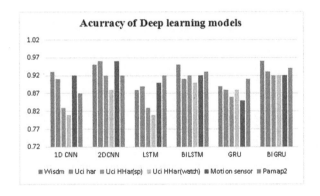

Fig. 3. Classification accuracy of deep learning models

Table 2 summarizes recognition accuracy of all the methods studied for all the data sets. For all the data sets, deep learning models produced better classification accuracy than traditional machine learning models. Tables 3, 4 and 5 represent activity wise accuracies of 2D CNN, BI LSTM, and BI GRU respectively. It seems that **BI-GRU** is the best model for discriminating different activities. All three models achieved high accuracy for sitting, standing and lying, as signal is constant throughout the activity. Upstairs and downstairs activities attained less accuracy compared to other activities, due to more similarity in them but BI GRU can distinguish these two activities better than other models. Walking attained satisfactory results in all the models. Riding bike achieved good results with all the models but there is slight inaccuracy as the signal values are similar to downstairs.

Figure 5 represents average accuracy of each activity for 2D CNN, BI LSTM and BI GRU. Three models achieved same accuracy for sitting, laying, jogging

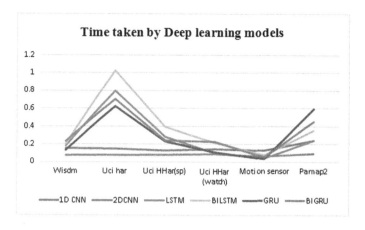

Fig. 4. Computation time of deep learning models

Table 2. Classification accuracy of ML and DL models with multiple datasets

Classifiers	Classification accuracy (%) for below Data sets					
	Wisdm	Uci har	Uci HHar(sp)	Uci HHar(watch)	Motion sensor	Pamap2
KNN	0.72	0.89	0.75	0.81	0.64	0.56
NB	0.52	0.77	0.66	0.66	0.89	0.76
DT	0.68	0.85	0.71	0.79	0.66	0.58
XG Boost	0.78	0.87	0.71	0.85	0.75	0.72
SVM	0.72	*0.95*	0.72	0.79	0.87	0.82
RF	0.75	0.88	0.77	0.83	0.69	0.67
1D CNN	0.93	0.91	0.83	0.81	0.85	0.87
2D CNN	*0.95*	**0.96**	**0.92**	0.88	0.89	0.92
LSTM	0.88	0.89	0.83	0.81	0.90	0.92
BILSTM	*0.95*	0.91	**0.92**	*0.90*	0.92	0.94
GRU	0.92	0.88	*0.86*	0.88	*0.93*	*0.95*
BIGRU	**0.96**	0.93	**0.92**	**0.92**	**0.94**	**0.96**

*sp-smartphone

Table 3. Activity wise accuracy of 2D CNN

Activities	Activity wise accuracy (%) for below Data sets					
	Wisdm	Uci har	Uci HHar(sp)	Uci HHar(watch)	Motion sensor	Pamap2
Walking	0.94	0.92	0.80	0.73	0.99	0.82
Sitting	1.00	0.96	1.00	0.93	1.00	1.00
Standing	1.00	0.98	1.00	0.93	1.00	1.00
Downstairs	1.00	0.88	0.66	0.59	0.90	0.90
Upstairs	0.83	0.96	0.80	0.80	0.94	0.84
Laying	NA	1.00	NA	NA	NA	1.00
Jogging	0.94	NA	NA	NA	0.97	0.92
Bike	NA	NA	0.97	0.90	NA	NA

*NA-Not Available

Table 4. Activity wise accuracy of BI LSTM

Activities	Activity wise accuracy (%) for below Data sets					
	Wisdm	Uci har	Uci HHar(sp)	Uci HHar(watch)	Motion sensor	Pamap2
Walking	0.95	0.90	0.80	0.93	0.94	0.95
Sitting	1.00	0.96	1.00	1.00	1.00	1.00
Standing	1.00	0.98	1.00	0.93	1.00	1.00
Downstairs	0.80	0.76	0.66	0.86	0.89	0.90
Upstairs	1.00	0.91	0.80	0.65	0.94	0.84
Laying	NA	1.00	NA	NA	NA	1.00
Jogging	0.95	NA	NA	NA	0.97	0.94
Bike	NA	NA	0.97	1.00	NA	NA

*NA-Not Available

Table 5. Activity wise accuracy of BI GRU

Activities	Activity wise accuracy (%) for below Data sets					
	Wisdm	Uci har	Uci HHar(sp)	Uci HHar(watch)	Motion sensor	Pamap2
Walking	0.94	0.90	0.83	0.80	0.89	0.94
Sitting	1.00	0.96	0.97	1.00	1.00	1.00
Standing	1.00	0.98	1.00	1.00	1.00	1.00
Downstairs	0.94	0.86	0.90	0.90	0.89	0.96
Upstairs	0.94	0.92	0.86	0.83	0.94	0.87
Laying	NA	1.00	NA	NA	NA	1.00
Jogging	0.94	NA	NA	NA	1.00	0.94
Bike	NA	NA	0.97	0.97	NA	NA

*NA-Not Available

and bike. For walking, BI LSTM and BI GRU has slight difference. For upstairs and downstairs, BI GRU gained superiority compared to the other models. From the results, it can be concluded that **BI GRU** model can discriminate well compared to the other models and this model can predict quickly and accurately.

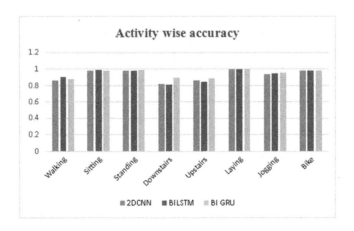

Fig. 5. Activity Wise accuracy for 2D CNN, BI LSTM, BI GRU

5   Conclusion

In this work we have studied performance of machine learning including deep learning tools for human activity recognition. This study has been done on multiple benchmark data sets and analysis of those results are presented. We observed that deep learning tools produce better classification accuracy for all the data sets. It is also observed that bidirectional LSTM and GRU perform better than uni-directional LSTM and GRU though the computational cost is slightly high. Activity wise accuracy for the models which achieved higher accuracy are represented. From that analysis, it is observed that BI GRU model provides the best performance and can be utilized for the implementation of HAR applications.

References

1. Majumder, S., Mondal, T., Jamal Deen, M.: Wearable sensors for remote health monitoring. Sensors **17**(1), 130–175 (2017)
2. Sarkar, J., Vinh, L.T., Lee, Y.-K., et al.: GPARS: a general purpose activity recognition system. Appl. Intell. **35**(2), 242–259 (2011)
3. Lentzas, A., Vrakas, D.: Non-intrusive human activity recognition and abnormal behavior detection on elderly people: a review. Artif. Intell. Rev. **53**, 1975–2021 (2020)
4. Subetha, T., Chitrakala, S.: A survey on human activity recognition from videos. In: International Conference on Information Communication and Embedded Systems (ICICES), Chennai, pp. 1–7 (2016)
5. Bulling, A., Blanke, U., Schiele, B.: A tutorial on human activity recognition using body worn inertial sensors. ACM Comput. Surv. **46**, 1–33 (2014)
6. Sousa Lima, W., Souto, E., El Khatib, K., Jalali, R., Gama, J.: Human activity recognition using inertial sensors in a smartphone: an overview. Sensors **19**(14), 3213 (2019)
7. Wang, J., Chen, R., Sun, X., She, M.F., Wu, Y.: Recognizing human daily activities from accelerometer signal. Procedia Eng. **15**, 1780–1786 (2011)
8. Bayat, A., Pomplun, M., Tran, D.A.: A study on human activity recognition using accelerometer data from smartphones. Procedia Comput. Sci. **34**, 450–457 (2014)
9. Anguita, D., Ghio, A., Oneto, L., Parra, X., Reyes-Ortiz, J.L.: Human activity recognition on smartphones using a multiclass hardware-friendly support vector machine. In: Bravo, J., Hervás, R., Rodríguez, M. (eds.) IWAAL 2012. LNCS, vol. 7657, pp. 216–223. Springer, Heidelberg (2012). https://doi.org/10.1007/978-3-642-35395-6_30
10. Paul, P., George, T.: An effective approach for human activity recognition on smartphone. In: IEEE International Conference on Engineering and Technology (ICETECH), Coimbatore, pp. 1–3 (2015)
11. Fan, L., Wang, Z., Wang, H.: Human activity recognition model based on decision tree. In: International Conference on Advanced Cloud and Big Data, Nanjing, pp. 64–68 (2013)

12. Kolekar, M.H., Dash, D.P.: Hidden Markov model based human activity recognition using shape and optical flow based features. In: Proceedings of the IEEE Region 10 Conference (TENCON), pp. 393–397 (2016)

13. Casale, P., Pujol, O., Radeva, P.: Human activity recognition from accelerometer data using a wearable device. In: Vitrià, J., Sanches, J.M., Hernández, M. (eds.) IbPRIA 2011. LNCS, vol. 6669, pp. 289–296. Springer, Heidelberg (2011). https://doi.org/10.1007/978-3-642-21257-4_36

14. Wu, W., Dasgupta, S., Ramirez, E.E., Peterson, C., Norman, G.J.: Classification accuracy's of physical activities using smartphone motion sensors. J. Med. Internet Res. **14**(5), e130 (2012)

15. Chen, K., Zhang, D., Yao, L., Guo, B., Yu, Z., Liu, Y.: Deep learning for sensor based human activity recognition: overview, challenges and opportunities (2020). https://arxiv.org/abs/2001.07416

16. Ronao, C.A., Cho, S.B.: Human activity recognition with smartphone sensors using deep learning neural networks. Expert Syst. Appl. **59**, 235–244 (2016)

17. Zebin, T., Scully, P.J., Ozanyan, K.B.: Human activity recognition with inertial sensors using a deep learning approach. In: Proceedings of the IEEE Sensors, pp. 1–3 (2016)

18. Gholamrezaii, M., Taghi Almodarresi, S.M.: Human activity recognition using 2D convolutional neural networks. In: 27th Iranian Conference on Electrical Engineering (ICEE), Yazd, Iran, pp. 1682–1686 (2019)

19. Hammerla, N.Y., Halloran, S., Ploetz, T.: Deep convolutional and recurrent models for human activity recognition using wearable's. J. Sci. Comput. **61**(2), 454–476 (2016)

20. Arifoglu, D., Bouchachia, A.: Activity recognition and abnormal behaviour detection with recurrent neural networks. Procedia Comput. Sci. **110**, 86–93 (2017)

21. Ishimaru, S., Hoshika, K., Kunze, K., et al.: Detecting reading activities by EOG glasses and deep neural networks. In: Proceedings of the ACM International Joint Conference on Pervasive and Ubiquitous Computing, pp. 704–711 (2017)

22. Aljarrah, A.A., Ali, A.H.: Human activity recognition using PCA and BiLSTM recurrent neural networks. In: 2nd International Conference on Engineering Technology and its Applications (IICETA), Al-Najef, Iraq, pp. 156–160 (2019)

23. Yao, S., Hu, S., Zhao, Y., Zhang, A., Abdelzaher, T.: DeepSense: a unified deep learning framework for time-series mobile sensing data processing. In: Proceedings of the 26th International Conference on World Wide Web, pp. 351–360 (2017)

24. Alsarhan, T., Alawneh, L., Al-Zinati, M., Al-Ayyoub, M.: Bidirectional gated recurrent units for human activity recognition using accelerometer data. In: Proceedings of IEEE Sensors, pp. 1–4 (2019)

25. Xia, K., Huang, J., Wang, H.: LSTM-CNN architecture for human activity recognition. IEEE Access **8**, 56855–56866 (2020)

26. Abbaspour, S., Fotouhi, F., Sedaghatbaf, A., Fotouhi, H., et al.: A comparative analysis of hybrid deep learning models for human activity recognition. Sensors (Basel) **20**(19), 5707 (2020)

27. Kwapisz, J.R., Weiss, G.M., Moore, S.A.: Activity recognition using cell phone accelerometers. ACM SIGKDD Explor. Newsl. **12**(2), 74–82 (2011)

28. Anguita, D., Ghio, A., Oneto, L., Parra, X., Reyes Ortiz, J.L.: A public domain dataset for human activity recognition using smartphones. In: Proceedings of ESANN (2013)

29. Stisen, A., et al.: Smart devices are different: assessing and mitigating mobile sensing heterogeneities for activity recognition. In: Proceedings of the 13th ACM Conference on Embedded Networked Sensor Systems, pp. 127–140. ACM (2015)
30. Malekzadeh, M., Clegg, R.G., Cavallaro, A., Haddadi, H.: Protecting sensory data against sensitive inferences. In: Proceedings of the Workshop on Privacy by Design in Distributed Systems (W-P2DS18) (2018)
31. Reiss, A., Stricker, D.: Introducing a new benchmarked dataset for activity monitoring. In: Proceedings of the 16th International Symposium on Wearable Computers, pp. 108–109 (2012)

Intelligent Robot

Beam and Ball Plant System Controlling Using Intuitionistic Fuzzy Control

Onur Silahtar[1]([✉]), Özkan Atan[1], Fatih Kutlu[2], and Oscar Castillo[3]

[1] Department of Electrical and Electronics Engineering, Engineering Faculty,
Van Yuzuncu Yil University, 65080 Van, Turkey
onursilahtar@yyu.edu.tr
[2] Department of Mathematics, Faculty of Science, Van Yuzuncu Yil University,
65080 Van, Turkey
[3] Department of Computing Science, Tijuana Institute of Technology, Tijuana, Mexico

Abstract. In this study, simplified "beam and ball (BNB) system" is controlled using "intuitionistic fuzzy control (IFC)" method. It is aimed to keep the ball on it in balance by applying DC voltage in different magnitudes to the DC motor of the system called "ball and beam plant", which has a beam on which a DC motor is attached to the middle point and a ball moving without friction. In order to better observe the effect of this new generation controller applied to the system, parameters such as the torque of the motor, the mass of the beam and the ball, internal and external disturbance, friction etc. were ignored and the system is simplified. The position and velocity of the ball on the beam is taken as input for the controller, while the controller output is chosen as a DC voltage. After I-Fuzzification, I-Inference and I-Defuzzification process are performed in controller block, performance and efficiency of the system are discussed in terms of steady state error, setting time, maximum overshoot, chattering.

Keywords: Beam and Ball (BNB) System · Controller design · Intuitionistic fuzzy control (IFC)

1 Introduction

The under actuated mechanical systems represent a challenge for the control. An active field of research exists, due to the applications of these systems such as aircrafts, spacecraft, flexible and legged robots [1]. BNB system is one of the most used subsystems among these nonlinear systems. As shown in Fig. 1, a simple BNB system consists of a beam, ball, and an activator connected to the balance point of beam [2]. Here, by means of the DC motor, the beam is rotated clockwise and counterclockwise to keep the ball in the middle of the beam. Thus, according to the position and speed information of the ball, voltage control signal is applied to the DC motor and the control of the system is completed in this way. For this, various control methods have been applied for the same or more developed BNB systems.

© IFIP International Federation for Information Processing 2021
Published by Springer Nature Switzerland AG 2021
Z. Shi et al. (Eds.): ICIS 2020, IFIP AICT 623, pp. 255–262, 2021.
https://doi.org/10.1007/978-3-030-74826-5_22

Olfati-Saber and Megretski developed a new version of back stepping procedure for systems that do not have a triangular structure [1]. Yu and Ortiz proposed an asymptotically stabilizing PD controller that ensures that, for a well-defined set of initial conditions, the ball remains on any point of the bar [3]. Keshmiri et al. are designed Linear Quadratic Regulator (LQR) considering two Degrees-of-Freedom and coupling Dynamics [4]. Almutairi and Zribi are designed a static and a dynamic sliding-mode controllers using a simplified model of the system [5]. Modelling and control of ball and beam system using PID controller is developed by Maalini et al. [6]. Particle swarm optimization algorithm is presented as a robust and highly useful optimization technique to tune the gains of the PID controllers in the two feedback loops of the classic Ball and Beam Control System by Rana et al. [7].

In recent years, fuzzy logic control method in Beam ball system control has also been examined by many researchers and various studies have been performed. For example Amjad et al. presented the design and analysis of a FLC for the outer loop and a PD controller for the inner loop of a ball and beam system. Additionally, a classical PID controller based on ITAE equations for the beam position has been also designed to compare the performance of both types of controllers [8]. Based on the T-S fuzzy modeling, the dynamic model of the ball and beam system has been formulated as a strict feedback form with modeling errors by Chang et.al. Then, an adaptive dynamic surface control (DSC) has been utilized to achieve the goal of ball positioning subject to parameter uncertainties. The robust stability of the closed-loop system has been preserved by using the Lyapunov theorem. In addition to simulation results, the proposed T-S fuzzy model-based adaptive dynamic surface controller has been applied to a real ball and beam system for practical evaluations [9]. Oh et.al, introduced a design methodology for an optimized fuzzy cascade controller for ball and beam system by exploiting the use of hierarchical fair competition-based genetic algorithm (HFCGA) [10]. Castillo et al. described the design of a fuzzy logic controller for the ball and beam system using a modified Ant Colony Optimization (ACO) method for optimizing the type of membership functions, the parameters of the membership functions and the fuzzy rules [11]. Combined with the conventional dynamic surface control, an adaptive fuzzy scheme is proposed for the equilibrium balance of the ball for beam and ball system by Chang et al. [12].

In the above mentioned studies, it can be seen that many modern control methods, including fuzzy control, are presented to control the beam and ball system. However, in this study, an intuitionistic fuzzy controller was designed and simulated for a sampled beam and ball system for the first time in the literature.

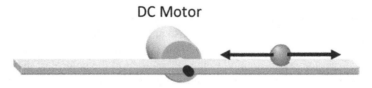

Fig.1. Sampled beam and ball system

2   Preliminaries

After Zadeh's [13] impressive description of fuzzy set theory, many studies have been conducted by many researchers. Then this theory was developed over the years and several Fuzzy Logic Controllers (FLCs) were designed for various linear or nonlinear systems. With the definition of the intuitionistic fuzzy logic concept, which is a more generalized version of the fuzzy logic concept, by Atanassov [14], IFLCs that are more applicable and functional to many systems in real life began to be developed. Because it has become easier to design a controller for systems that involve uncertainties such as disturbance.

An intuitionistic fuzzy set A on a non-empty set X named as universe of discourse given by a set of ordered triples $A = \{(x, \mu_A(x), \eta_A(x)) : x \in X\}$ where $\mu_A(x) : X \to I$, and $\eta_A(x) : X \to I$ are functions defined such that $0 \leq \mu(x) + \eta(x) \leq 1$ for all $x \in X$, $\mu(x)$ and $\eta(x)$ represent the degree of membership and degree of non-membership of x in A respectively. For each $x \in X$, the intuitionistic fuzzy index of x in A can be defined as follows: $\pi_A(x) = 1 - \mu_A(x) - \eta_A(x)$. π_A is called the degree of hesitation or indeterminacy [15].

3   Ball and Beam Mathematical Model

As shown Fig. 2, the right and left directions in the sense of movement of the ball are respectively $(-)$ and $(+)$ and the middle of the bar, that is the point where the DC motor is connected, is taken as balance or the stability point. It is also assumed that the dc motor will rotate clockwise when induced with $(+)$ voltage and counterclockwise when induced with $(-)$ voltage ideally and without friction.

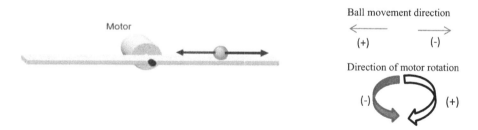

Fig. 2. Directions for BNB systems

Here, the distance of ball to the stability (balance) point and the velocity of ball are assumed as the error (e) and the derivative of error ($\dot{e}$), respectively. Thanks to the voltage to be obtained according to the e and $\dot{e}$ controller inputs, it is aimed that the motor rotates the beam and the ball remain stable in the stability point. The simplified BNB system is depicted in Fig. 2 and modeled (for simulation purposes) by [16]:

$$\ddot{e} = 9.81 \sin kv$$

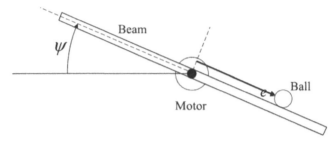

Fig. 3. Beam and ball system

In Fig. 3, e is the position of the ball along the beam (with $e = 0$ defined as the center of the beam), and ψ is the beam angle commanded by the motor (with $\psi = 0$ defined as horizontal). The input to the ball and beam system is the voltage v supplied to the motor. The beam angle ψ is proportional to v (i.e. $\psi = kv$) [2].

4 Controller Design

I-Fuzzification, I-İnference ve I-Defuzzification processes are required respectively for an IFC (Intuitionistic Fuzzy Controller) design.

As seen in Fig. 4, while e and $\dot{e}$ crisp values are sent to the controller input, crisp v value is taken from the controller output.

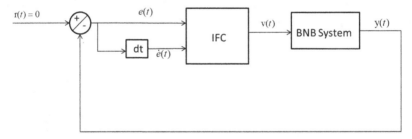

Fig. 4. BNB system with controller

a. **I-Fuzzification:**

First of all, IFC crisp inputs e and $\dot{e}$ need to fuzzification. For this, using the triangular membership functions in Fig. 5, membership degrees for both inputs are obtained. (with NL: Negative Large, NS: Negative Small, Z:Zero, PS: Positive Small, PL: Positive Large). Also, using the singleton membership functions as in Fig. 6, the output (v) is fuzzification [2].

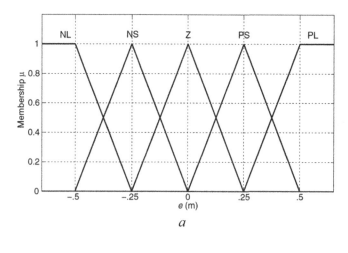

a

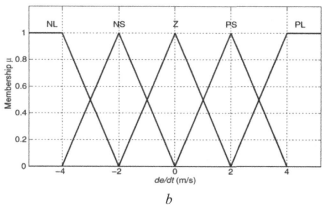

b

Fig. 5. a. Triangular fuzzy sets for ball position b. Triangular Fuzzy Sets for ball velocity.

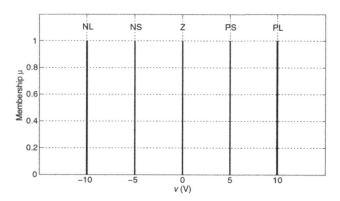

Fig. 6. Singleton fuzzy sets for v

After the membership degrees of inputs are obtained, the degrees of non-membership degrees of inputs are found by using a sugeno type generator [17]. Thus, IFPs (İntuitionistic fuzzy pairs) consisting of membership and non-membership degrees $\langle \mu_A, v_A \rangle$ are obtained.

b. **I-Inference:**

After obtaining the intuitionistic fuzzy input and output sets, the rule base is created. Since there are 5 intuitionistic fuzzy sets for each input, a total of 5 x 5 = 25 rule bases are obtained as shown Table 1 [2].

Table 1. Rule base

v	$\dot{e}$				
		NL	NS	Z	PS	PL
e	NL	NL	NL	NL	NS	Z
	NL	NL	NL	NS	Z	PS
	Z	NL	NS	Z	PS	PL
	PS	NS	Z	PS	PL	PL
	PL	Z	PS	PL	PL	PL

In IFC, although the same rule base is defined in the inference part of the system as in the FC design, the inference process is performed by using IFPs, not just degrees of membership. In this study intersection and implication methods are chosen as "min T-norm" and "Mamdani implication" respectively [18].

c. **I-Defuzzification:**

Finally the IFC design is completed by converting the new intuitionistic fuzzy pairs obtained from inference process to the v crisp output using the "a novel center of area (COA)" type intuitionistic defuzzification method [19].

5 Numerical Solution

The controllers designed above are simulated using Matlab/Simulink. In the simulation, 4th Runga-Kutta method is used with the step size of 0.001s. Also initial conditions are chosen as $[e, \dot{e}] = [0.3 \, \text{m}, -2.5 \, \text{m/s}]$. Additionally $k = 0.2$ and sugeno constant $\lambda = 1$ are chosen. Simulation results are obtained as in Figs. 7, 8 and 9.

Considering the simulation results, it is observed that the system reached stability in approximately 1.86 s, the steady state error is approximately 1.068×10^{-5} m and the maximum overshoot is 0.18 m. Additionally, no chattering is observed.

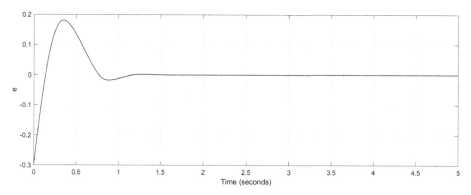

Fig. 7. Ball position (e)

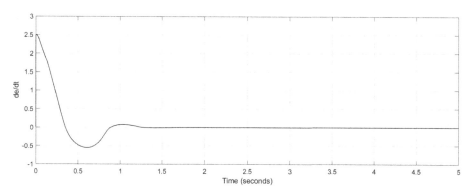

Fig. 8. Ball velocity (de/dt)

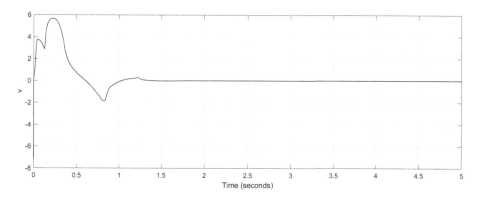

Fig. 9. Applied voltage (control output) (v)

6 Conclusion

In this paper a novel IFC has been designed for a simplified BNB system for the first time in literature. The performance of this proposed control method discussed in terms of steady state error, setting time, maximum overshoot, chattering. It has been seen that the steady state error and maximum overshoot were low and no chattering occurred. In future studies, it is planned to apply the designed IFC to more complex BNB systems. Also, using Intuitionistic fuzzy logic, it is aimed to design more complex controllers such as IFSMC and IFSAMC and to apply them to complex BNB systems.

References

1. Aoustin, Y., Formal'skii, A.M.: Beam-and-ball system under limited control: Stabilization with large basin of attraction. In: 2009 American Control Conference. IEEE (2009)
2. Lilly, J.H.: Fuzzy Control and Identification. Wiley, Hoboken (2011)
3. Yu, W., Floriberto, O.: Stability analysis of PD regulation for ball and beam system. In: Proceedings of 2005 IEEE Conference on Control Applications. CCA 2005. IEEE (2005)
4. Keshmiri, M., et al.: Modeling and control of ball and beam system using model based and non-model based control approaches. Int. J. Smart Sens. Intell. Syst. **5**(1) (2012)
5. Almutairi, N.B., Mohamed, Z.: On the sliding mode control of a ball on a beam system. Nonlinear Dyn. **59**(1–2), 221 (2010)
6. Maalini, P.V.M., Prabhakar, G., Selvaperumal, S.: Modelling and control of ball and beam system using PID controller. In: 2016 International Conference on Advanced Communication Control and Computing Technologies (ICACCCT). IEEE (2016)
7. Rana, M.A., Zubair, U., Zeeshan, S.: Automatic control of ball and beam system using particle swarm optimization. In: 2011 IEEE 12th International Symposium on Computational Intelligence and Informatics (CINTI). IEEE (2011)
8. Amjad, M., et al.: Fuzzy logic control of ball and beam system. In: 2010 2nd International Conference on Education Technology and Computer, vol. 3. IEEE (2010)
9. Chang, Y.-H., Wei-Shou, C., Chia-Wen, C.: TS fuzzy model-based adaptive dynamic surface control for ball and beam system. IEEE Trans. Ind. Electron. **60**(6), 2251–2263 (2012)
10. Oh, S.-K., Jang, H.-J., Pedrycz, W.: The design of a fuzzy cascade controller for ball and beam system: a study in optimization with the use of parallel genetic algorithms. Eng. Appl. Artif. Intell. **22**(2), 261–271 (2009)
11. Castillo, O., et al.: New approach using ant colony optimization with ant set partition for fuzzy control design applied to the ball and beam system. Inf. Sci. **294**, 203–215 (2015)
12. Chang, Y.-H., et al.: Adaptive fuzzy dynamic surface control for ball and beam system. Int. J. Fuzzy Syst. **13**(1), 1–7 (2011)
13. Zadeh, L.A.: Fuzzy sets. Inf. Control **8**(3), 338–353 (1965)
14. Atanassov, K.T.: Intuitionistic fuzzy set. Fuzzy Sets Syst. **20**, 87–96 (1986)
15. Atanassov, K.T.: On intuitionistic fuzzy sets theory. Springer, Berlin (2014). https://doi.org/10.1007/978-3-642-29127-2
16. Close, C.M., Frederick, D.K.: Modeling and Analysis of Dynamic Systems. Houghton Mifflin, Boston (1978)
17. Chaira, T.: Fuzzy Set and Its Extension. Wiley, Hoboken (2019)
18. Lin, L., Xia, Z.-Q.: Intuitionistic fuzzy implication operators: expressions and properties. J. Appl. Math. Comput. **22**(3), 325–338 (2006)
19. Kutlu, F., Atan, Ö., Silahtar, O.: Intuitionistic fuzzy adaptive sliding mode control of nonlinear systems. Soft. Comput. **24**(1), 53–64 (2019). https://doi.org/10.1007/s00500-019-04286-8

Application of Pinching Method to Quantify Sensitivity of Reactivity Coefficients on Power Defect

Subrata Bera[✉]

Safety Research Institute, Atomic Energy Regulatory Board, Mumbai 400094, India
sbera@aerb.gov.in

Abstract. Reactor power affects the temperature of fuel and coolant of a nuclear power plant. The change in temperature of fuel and coolant modify the reactivity of a nuclear reactor. The change in reactivity due to the change of reactor power is known as power defect. The power defect depends on the various parameters such as reactivity coefficients due to thermal hydraulics feedback, the response of fuel and coolant temperature due to variation of reactor power, etc. The reactivity coefficients are significantly varied due to different operating conditions, changes in fuel characteristics and fuel burn-up during fuel residence time inside the reactor. This wide variation of reactivity coefficient are in general technically specified in the form of lower and upper bound for safety analysis purpose. A thought experiment has been carried out considering those reactivity coefficients contain stochastic variability and ignorance (i.e., lack of knowledge). The uncertainty involved in reactivity coefficients are captured by defining them with probability box (p-box). After propagating the p-box of reactivity coefficients through the theoretical model of power defect, the p-box of power defect has been generated. In the pinching method, one of two reactivity coefficients will be fixed at their average value and observation on the change of area of p-box of power defect has been made for sensitivity analysis. Based on the reduction of area of p-box of power defect, the sensitivity of these two reactivity coefficients has been analyzed. The parametric studies of variation of sensitivity for five different power drops (i.e., 10%, 25%, 50%, 75% and 100%) have been studied and quantified in this paper. It is found that the reactivity coefficient due to coolant temperature is more sensitive than reactivity coefficient due to the fuel temperature on power defect. It is also found that the sensitivity does not depend on amount of power drop.

Keywords: Power defect · Probability-box · Pinching method · Nuclear power plant · Reactivity · Thermal hydraulic feedback

1 Introduction

The basic principle of power manoeuvrings of a nuclear power plant (NPP) is reactivity management. The reactivity is the measure of the rate of nuclear

© IFIP International Federation for Information Processing 2021
Published by Springer Nature Switzerland AG 2021
Z. Shi et al. (Eds.): ICIS 2020, IFIP AICT 623, pp. 263–272, 2021.
https://doi.org/10.1007/978-3-030-74826-5_23

fission reactions occurring inside a reactor. Positive reactivity signifies that reactor is in supercritical sate. Reactivity equals to zero means reactor in critical state i.e., self sustaining chain reaction of nuclear fission. Negative reactivity implies that the reactor is in subcritical state, i.e., shut-down condition. The power raises with increase of reactivity and decrease of reactivity lowers the power. Apart from the physical reactivity devices such as control rods, neutron poison addition/removal system, the variation of reactivity is possible through thermal hydraulic feedback mechanisms such as temperature of fuel and coolant. In the pressurized light water reactor system, the reactivity coefficient of fuel temperature (RCFT) and reactivity coefficient for coolant temperature (RCCT) are desired to be negative by design to enhance the inherent safety feature [1]. The characteristics of negative reactivity coefficients enables decrement of reactivity during heating up and increment of reactivity during cooling down. If there is a need to shut down the reactor from its hot full power state, then there will be a change of reactivity that may reduce the effective control rod worth due to the cooling effect of fuel and coolant. Hence, power defect [2] should be considered during evaluation of reactivity margin available in shut-down condition. The reactor is operated from very low power during the start-up operation to its full power operation. Again, with continuous extraction of power leads to the change in fuel depletion and build-up of neutron absorbing fission products. The reactivity coefficients are significantly varied due to different operating conditions, changes in fuel characteristics and fuel burn-up during fuel residence time inside the reactor. This wide variation of reactivity coefficient are in general technically specified in the form of lower and upper bound for safety analysis purpose. A thought experiment has been carried out considering those reactivity coefficients contain random variability and ignorance (i.e., lack of knowledge). The variability of a parameter is analysed by probabilistic method whereas ignorance by possibilistic approach such as interval arithmetic [3], fuzzy set theory [4], evidence theory [5], etc. However, combined effect can be studied by second order Monte-carlo method [6], fuzzy-stochastic response surface [7] formulation and modelling with probability box (aka p-box) [8]. The hybrid uncertainty involved in reactivity coefficients are captured by defining them with p-box, which is bounded by lower and upper cumulative distribution functions (CDFs). After propagating the p-box of reactivity coefficients through the theoretical model of power defect, the p-box of power defect has been generated. In the pinching method [9], one of two reactivity coefficients will be fixed at their average distribution. The change in area of the p-box of an input variable affects the area of the output variable. The fractional change of area of p-box of the output variable from its unpinched condition is the indicator for the sensitivity of the pinched input variable [8]. The calculation methodology for power defect, notation of probability box, random sampling methodology from a normal distribution, methodology for sensitivity analysis, parametric studies and results obtained are discussed in the subsequent sections.

2   Calculation Methodology

2.1   Calculation of Power Defect

The variation of reactivity depends of various parameters such as fuel temperature, coolant temperature, coolant density, fuel burn-up, thermal hydraulics feedbacks, etc. and hence the power defect also depends on various properties such as thermal hydraulic feedbacks, control characteristics, etc. For simplicity, here the power defect is estimated for variation of two parameters such as fuel temperature and coolant temperature. Hence, let us consider that the reactivity is a function of fuel temperature (T_F) and Coolant temperature (T_C) as represented in Eq. (1).

$$\rho = \rho\left(T_F, T_C\right) \tag{1}$$

To account for the changes in reactivity due to changes in reactor power which in turn depends on changes in fuel temperature and coolant temperature, the change of reactivity can be represented in Eq. (2).

$$d\rho = \left(\frac{\delta\rho}{\delta T_F}\frac{\delta T_F}{\delta P} + \frac{\delta\rho}{\delta T_C}\frac{\delta T_C}{\delta P}\right)dP \tag{2}$$

where, P is reactor power; ρ is reactivity; $\frac{\delta\rho}{\delta T_F}, \frac{\delta\rho}{\delta T_C}$ are the RCFT and RCCT respectively; and $\frac{\delta T_F}{\delta P}, \frac{\delta T_C}{\delta P}$ are response of fuel temperature (PRFT) and coolant temperature (PRCT) due to the change of reactor power respectively. In this study, RCFT and RCCT are considered as hybrid uncertain variables defined by p-box. Random sampling-based algorithm is used to generate the p-box of power defect from the reactivity coefficients.

2.2   Notation for Probability Box

Probability bounding analysis can be carried out by defining probability box or simply 'p-box'. P-box is a structure for probability distribution of a hybrid uncertain variable X i.e., $F_X := \left[\underline{F_X}(x), \overline{F_X}(x)\right]$. $\left[\underline{F_X}(x), \overline{F_X}(x)\right]$ denote the set of monotonically increasing functions $F_X(x)$ of real variable X in the interval $[0, 1]$ such that $\underline{F_X}(x) \leq F_X(x) \leq \overline{F_X}(x)$ [10]. The functions $\overline{F_X}$ and $\underline{F_X}$ are interpreted as bounds on CDFs. The region specified by the pair of distributions $\left[\underline{F_X}(x), \overline{F_X}(x)\right]$ is known as the "probability box" or "p-box". This means that if $\left[\underline{F}(x), \overline{F}(x)\right]$ is a p-box for a random variable X whose distribution $F_X(x)$ is unknown except that it is within the p-box, then $\underline{F_X}(x)$ is a lower bound on $F_X(x)$, which is the (imprecisely known) probability that the random variable X is smaller than x. Likewise, $\overline{F_X}(x)$ is an upper bound on the same probability. From a lower probability measure P for a random variable X, one can compute the upper and lower bounds on the distribution functions using Eqs. (3) and (4).

$$\overline{F}_X(x) = 1 - \underline{P}(X > x) \tag{3}$$

$$\underline{F}_X\left(x\right) = \underline{P}\left(X \le x\right) \tag{4}$$

When the bounds of the p-box coincide for every $x \in X$, i.e., $\underline{F}_X\left(x\right) = \overline{F_X\left(x\right)}$, the corresponding p-box degenerates into a single CDF, as it is usual in standard probability theory.

The values for RCFT and RCCT are generally reported as range of values in the safety analysis. The interval for RCFT and RCCT considered in this study are $[-3.2, -2.1]$ and $[-63.0, -4.0]$ in unit of $pcm/°C$ respectively to demonstrate the application of pinching methodology in carrying out the sensitivity analysis. In the p-box formulation, lower and upper bound of each reactivity coefficients are considered as random variables and following a normal distribution. The normal distribution is defined with bounding value as mean and 10% of bounding value as standard deviation for this sensitivity analysis. The graphical representation of the p-box for RCFT and RCCT are shown in the Fig. 1.

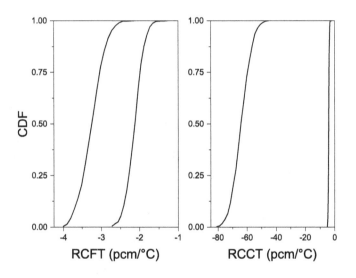

Fig. 1. Representation of RCFT and RCCT in p-box form

2.3   Random Sampling from Normal Distribution

Random sampling techniques include conversion of probability density function (PDF) to CDF, mapping with that of uniform random number distribution in the interval $[0, 1]$. Requirement of mapping on uniform distribution is based on the fact that most of the computer based random number generators generate uniform random number $(i.e., \xi)$ in closed interval $[0, 1]$. In the random sampling method ξ will be used to draw a sample from other distributions utilising the property of CDF of $0 \le F\left(x\right) \le 1.0$. After mapping of CDF of uniform distribution with desired distribution, inverse CDF corresponding to the uniform random number ($i.e., \xi$) gives the random number belonging to the desired PDF. Graphically, these steps are explained for continuous distribution function in the Fig. 2.

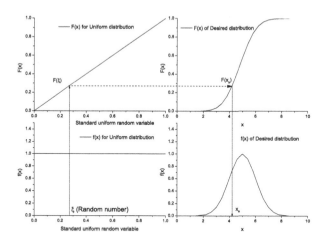

Fig. 2. General sampling techniques for continuous PDF

In this paper, variability of the reactivity coefficients in power defect assessment is considered as normal distribution. This unbounded symmetric uni-modal distribution is mathematically defined as a two parameters distribution function. These two parameters are location (i.e., mean) and scale (i.e., standard deviation). A normal distribution with mean (μ) and standard deviation (σ) is mathematically represented as $N(\mu, \sigma)$. The probability distribution of the distribution $N(\mu, \sigma)$ is given in Eq. (5).

$$p(x) = \frac{1}{\sqrt{2\pi}\sigma} e^{-\frac{1}{2}\left(\frac{x-\mu}{\sigma}\right)^2} \tag{5}$$

There are many approaches to draw a random sample for the normal distribution such as inverse transform sampling, Ziggurat algorithm [11,12], BOX-Muller algorithm [13], Wallace method [14]. Generation of a sample from the given normal distribution involves two steps process. First step is to generate random sample form a standard normal distribution $N(0, 1)$. Second step is to transform the generated random sample to the desired normal distribution using Eq. (6).

$$x = x_s \sigma + \mu \tag{6}$$

The random number i.e. x_s for distribution $N(0, 1)$ is generated using Box-Muller algorithm, which requires two uniformly distributed random numbers in range $[0, 1]$. The algorithm is shown in Eq. (7) and Eq. (8).

$$x_{s,i} = \sqrt{-2ln\xi_i}cos(2\pi\xi_{i+1}) \tag{7}$$

$$x_{s,i+1} = \sqrt{-2ln\xi_i}sin\left(2\pi\xi_{i+1}\right) \tag{8}$$

This algorithm is very simple to implement and also fast.

2.4   Methodology for Sensitivity Analysis

Interval form of the RCFT and RCCT are converted to p-box by assuming normal distribution with bound value as mean and 10% bound value as standard deviation for sensitivity analysis. Monte-Carlo sampling [15] of size 1000 from the both lower and upper bounding distributions of p-box has been drawn. Each pair of samples (i.e., lower and upper) are propagated through the theoretical model power defect to obtain the p-box for power defect. The methodology for p-box analysis is graphically presented in the Fig. 3.

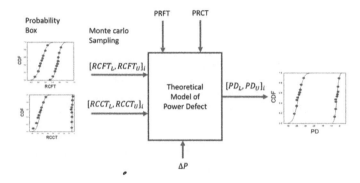

Fig. 3. Methodology for probability box analysis for a theoretical model

The area between two cumulative distribution functions (CDFs) is the measure of uncertainty. If one of the p-box of input variables is pinched, i.e., fixed to its average distribution, then the area of the p-box of the output variation will reduce. The amount of reduction of area can be give relative measure of sensitivity. Let's assume the area of the p-box of an output variable without pinching of input variable as a base case. And reduced area with pinching of input variable is a pinched case. Then sensitivity of the input variable can be measured by using Eq. (9).

$$sensitivity\,(\%) = 100 \times \left(1 - \frac{Area\,[Pinched]}{Area\,[Base]}\right) \tag{9}$$

3   Results and Discussions

3.1   Temperature Response with Respect to the Power Variation

System thermal hydraulic code RELAP5 [16] has been used to model reactor systems with primary heat transport system, reactor core, pressurizer, secondary heat transport system, etc. and estimate the fuel temperature and coolant temperature response with variation of reactor power through steady state simulation for each power level.

At the ten discrete power levels i.e., 10%, 20%, 30%, 40%, 50%, 60%, 70%, 80%, 90% and 100% of full power, the discrete values of fuel temperature and coolant temperature are obtained and fitted with line equation with R^2 value is 0.999. The formulae for the fitted curve of fuel temperature and coolant temperature response are given in Eqs. (10) and (11) respectively [2].

$$T_F = 270.8756 + 5.21779P \tag{10}$$

$$T_C = 276.10307 + 0.30274P \tag{11}$$

Equations (10) and (11) are used to obtain, slope of the response of fuel temperature (PRFT) and coolant temperature (PRCT) with respect to the power variation respectively, to be used in Eq. (2) for power defect estimation. The values of PRFT and PRCT are 5.21779 and 0.30274 respectively.

3.2   Estimation of Probability Box for Power Defect

P-box for power defect i.e., the change of reactivity with change in reactor power, has been generated from probability box definition of reactivity coefficients. P-box for power defect has been generated for five different power drops (i.e., 10%, 25%, 50%, 75% and 100%). The p-box for RCFT and RCCT have been shown in the left and middle figure respectively on the top layer of Fig. 4.

The right figure of the top layer depicts the p-box of power defect corresponding to power drop of 75% full power. The input variables, RCFT and RCCT are not pinched on the top layer figures. Therefore, p-box obtained for the power defect is the base case. The area of the p-box of power defect for the base case is found to be 17.92.

The variation of the p-box area of power defect for the base case with five discrete power drops i.e., 10%, 25%, 50%, 75% and 100% of full power has been analyzed and given in the Fig. 5.

It is found that the variation of p-box area is linearly proportional to the power drop. It is also found that the increment p-box area of amount 0.239 per unit percent of power drop.

In the middle layer, the p-box of RCFT is pinched to single normal distribution with mean and standard deviation are −2.65 and 0.265 respectively. The pinched distribution of RCFT has been shown in left figure of middle layer of Fig. 4. The p-box of RCCT is kept unchanged. The p-box of power defect with RCFT in pinched condition has been generated and shown in right figure in the middle layer in Fig. 4. It is found that area of the p-box of power defect is reduced by 24% of base case.

In the bottom layer, the p-box of RCCT is pinched to single normal distribution with mean and standard deviation are −33.5 and 3.35 respectively. The pinched distribution of RCCT has been shown in middle figure of bottom layer of Fig. 4. The p-box of RCFT is kept unchanged. The p-box of power defect

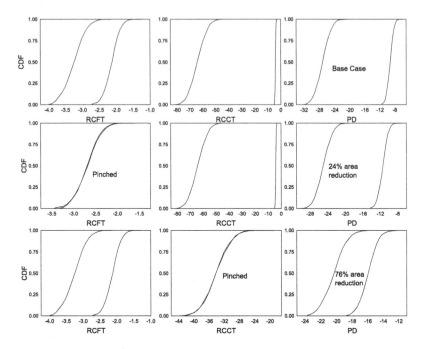

Fig. 4. Reduction of uncertainty through pinching

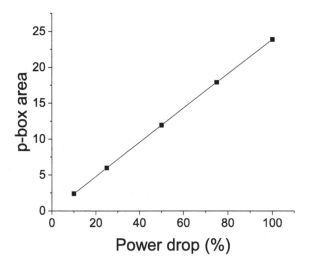

Fig. 5. Variation of p-box area of power defect with power drop

with RCCT in pinched condition has been generated and shown in right figure in the bottom layer in Fig. 4. It is found that area of the p-box of power defect is reduced by 76% of base case.

3.3   Sensitivity Analysis

Sensitivity in terms of the reduction of p-box area due to the pinching of input variables such as RCFT and RCCT has been estimated using Eq. (9). Sensitivity analysis for five discrete power drops, i.e., 10%, 25%, 50%, 75% and 100% of full power has been carried out. The result obtained from this analysis is shown in the Table 1. It is found that the sensitivity does not change with the power drop.

Table 1. Variation of sensitivity with power drop

Power drop %	Sensitivity %	
	RCFT Pinched	RCCT Pinched
10	25.40	75.55
25	25.40	75.55
50	25.40	75.55
75	25.40	75.55
100	25.40	75.55

4   Conclusions

The methodology for estimation of sensitivity of the theoretical model for power defect has been developed and demonstrated using p-box and pinching technique. In this study, interval bound data of input variables are transformed into p-box formulation. P-box of input variables propagated through the theoretical model of power defect using Monte Carlo based sampling techniques. Finally, the p-box of the output variable, i.e., power defect, has been generated for sensitivity analysis. Pinching method, where single distribution with average properties is used on the desired reactivity coefficient, is applied to study the variation of p-box area of power defect of the nuclear reactor. It is found that the area of p-box of power defect is linearly increasing with power drop. However, the sensitivity of input variable does not change with power drop. It is found that the RCCT is more sensitive than RCFT on power defect.

References

1. Duderstadt, J.J., Hamilton, L.J.: Nuclear Reactor Analysis. Wiley, New York (1976)

2. Bera, S., Lakshmanan, S.P., Datta, D., Paul, U.K,. Gaikwad, A.J.: Estimation of epistemic uncertainty in power defect due to imprecise definition of reactivity coefficients. In: DAE-BRNS Theme meeting on Advances in Reactor Physics (ARP-2017), DAE Convention Centre, Anushaktinagar, Mumbai-400094, 6–9 December (2017)

3. Chalco-Cano, Y., Lodwick, W.A., Bede, B.: Single level constraint interval arithmetic. Fuzzy Sets Syst. **257**, 146–168 (2014)

4. Simic, D., Kovacevic, I., Svircevic, V., Simic, S.: 50 years of fuzzy set theory and models for supplier assessment and selection: a literature review. J. Appl. Log. **24**, 85–96 (2017)

5. Hui, L., Shangguan, W.-B., Dejie, Yu.: An imprecise probability approach for squeal instability analysis based on evidence theory. J. Sound Vib. **387**, 96–113 (2017)

6. Ferson, S., Ginzburg, L.R.: Different methods are needed to propagate ignorance and variability. Reliab. Eng. Saf. Syst. **54**, 133–144 (1996)

7. Bera, S., Datta, D., Gaikwad, A.J.: Uncertainty analysis of contaminant transportation through ground water using fuzzy-stochastic response surface. In: Chakraborty, M.K., Skowron, A., Maiti, M., Kar, S. (eds.) Facets of Uncertainties and Applications. Springer Proceedings in Mathematics & Statistics, vol. 125, pp. 125–134. Springer, New Delhi (2015). https://doi.org/10.1007/978-81-322-2301-6_10

8. Ferson, S., Hajagos, J.G.: Arithmetic with uncertain numbers: rigorous and (often) best possible answers. Reliab. Eng. Saf. Syst. **85**, 135–152 (2004)

9. Ferson, S., Troy Tucker, W.: Sensitivity analysis using probability bounding. Reliab. Eng. Syst. Saf. **91**, 1435–1442 (2006)

10. Tang, H., Yi, D., Dai, H.-L.: Rolling element bearing diagnosis based on probability box theory. Appl. Math. Model. (2019). https://doi.org/10.1016/j.apm.2019.10.068

11. Doornik, J.A.: An Improved Ziggurat Method to Generate Normal Random Samples. University of Oxford, Oxford (2005)

12. Leong, P., Zhang, G., Lee, D.-U., Luk, W., Villasenor, J.: A comment on the implementation of the Ziggurat method. J. Stat. Softw. Art. **12**(7), 1–4 (2005)

13. Okten, G., Goncu, A.: Generating low-discrepancy sequences from the normal distribution: Box-Muller or inverse transform? Math. Comput. Model. **53**(5), 1268–1281 (2011)

14. Riesinger, C., Neckel, T., Rupp, F.: Non-standard pseudo random number generators revisited for GPUs. Future Gener. Comput. Syst. **82**, 482–492 (2018)

15. Boafo, E., Gabbar, H.A.: Stochastic uncertainty quantification for safety verification applications in nuclear power plants. Ann. Nucl. Energy **113**, 399–408 (2018)

16. RELAP5/MOD3.2 Code manual, Idaho National Engineering Laboratory, Idaho Falls, Idaho 83415, June 1995

Computation with Democracy: An Intelligent System

Kamalika Bhattacharjee[1] and Sukanta Das[2(✉)]

[1] Department of Computer Science and Engineering, National Institute of Technology, Tiruchirappalli, Tiruchirappalli 620015, Tamil Nadu, India
kamalika@nitt.edu
[2] Department of Information Technology, Indian Institute of Engineering Science and Technology, Shibpur, Howrah 711103, West Bengal, India
sukanta@it.iiests.ac.in

Abstract. The notion of artificial intelligence is primarily centered around a *functional approach* where it is tested whether the machine *acts* intelligently. This paper reports a work-in-progress where we advocate for the idea of the machine itself *being* inherently intelligent by incorporating two features in the machine – feature of the life and feature of a democratic system. The first property of the life is *self-replication*, where the machine can reproduce itself indefinitely. The second is *self-organization* which gives the machine ability to reorganize and heal itself. *Fault-tolerance* is another important property of an intelligent system that helps it to achieve reliability and context-sensitive nature. Only a *many-body system* where the elements, like a democratic system, work collectively by interacting among themselves can achieve these properties. Such a machine can be *computationally universal*. Hence, we claim that, a *democratic* model of computation can be a *better* choice for an intelligent system.

Keywords: Functional approach · Self-replication · Self-organization · Fault-tolerance · Computational universality · Democracy · Cellular Automata (CAs)

1 Introduction

The history of artificial intelligence (AI) is at least as old as the history of computing. One of the significant measures of testing whether a machine is *intelligent* is Turing Test (TT), a behavioral test introduced by Alan Turing in 1950 [27]. In TT, a machine is judged based on its response to some set of questions and passing TT indicates that the machine *shows* intelligent behavior. According to Turing, such a machine should be declared as a thinking machine. Nevertheless, this test is based on a *functional approach* where the machine is declared intelligent if it is successful in performing some functions – in other

© IFIP International Federation for Information Processing 2021
Published by Springer Nature Switzerland AG 2021
Z. Shi et al. (Eds.): ICIS 2020, IFIP AICT 623, pp. 273–282, 2021.
https://doi.org/10.1007/978-3-030-74826-5_24

words, *acts* intelligently. But it does not matter to this approach if or not the system is inherently intelligent like a living organism. One can see the famous *Chinese Room Argument* [25] which differentiates between what *is* intelligent and what *acts* intelligently. In fact, the concept of *strong AI* is developed to address this issue [14]. In this work, we want to advocate for the idea of machines which *are* intelligent by going beyond this functional approach.

The formal notion of computing is attributed to Alan Turing and Alonzo Church and is now primarily associated with Turing Machines (TM) [28]. Turing Machine is an abstract model of computation where a finite controller works on an infinite input/output tape based on some predefined rules and computes the output. The modern computer architecture is derived from this model of computation and it is inherently the heart of all machines and artifacts designed to show any kind of intelligent behavior. Obviously, this model is an *autocratic* model of computation. Here, by an autocratic model, we indicate a system where every component is controlled by one or few special components having the supreme authority; so, those components do not have any role in the decision executed by the system and have no independence. Conversely, alternative models of computation have also been proposed, the popular examples of which are Artificial Neural Network (ANN) [21] and Cellular Automata (CAs) [4]. These models may be called as *democratic* as they are composed of many components which enjoy some independence and can interact with each other to make their own decisions. However, these individual decisions can impact the final decision of the system as the system acts on the collective decision or consensus taken by the whole system. In the pioneering work of Langton on *Artificial Life* [16,17], he has categorically used this class of models in general, and cellular automata in particular. Can such a democratic system be more *intelligent* than an autocratic system? This work targets to give an affirmative answer to this question.

We claim that, any *living* element in nature is more *intelligent* than a non-living element. Therefore, the paramount criterion of an intelligent system is that it has to emulate a living element. However, the preliminary sign of life is it can *replicate* itself. That is, it can create copies of itself which inherit all its characteristics. If we look into any living element, starting with even a cell at the microscopic level, it can replicate itself. Because of this self-replication, any species *survives* in the evolution – its characteristics are not lost. So, the first property of our proposed system is *self-replication*.

The other two important signs of life are *self-organization* and *self-resilience/fault-tolerance*. This indicates the ability to work correctly even if some part of the system is damaged, repair or heals oneself automatically and upgrade oneself according to the demand of the environment. Due to this fault-tolerance and self-organization capabilities, the *fittest* species have survived in evolution. These features also indicate the re-evaluation and continuous learning process which provide an intelligent species the ability to make *correct* decisions and not to repeat the same mistake. Therefore, the next properties required by our system to be intelligent are self-organization and fault-tolerance.

However, a system that satisfies the above mentioned necessary properties of being intelligent, is inherently a *many-body system* where all entities act together and all decisions are taken collectively. This many-body system can be an ideal choice to model the intelligence in human society as well as the physiological and natural systems. To find an analogy, we can look at the neuron structure of the human brain which is a perfect example of a many-body system working on a network. Each of these entities of a many-body system can do some amount of work on their own, and can be *non-homogeneous* in nature, that is, they may have diversity in behavior, property and structure, but as a whole the system runs on the result of their collective effort. A decision is taken when all elements of the system come to a *consensus*. In case of society also, democracy talks about unity and consensus allowing individuals to maintain their diversity and independence. One may find several instances in our lives where to come to a conclusion about any crucial task collective decision is taken. Think for example of a medical board, or a review committee. Here, the collective decision is more reliable and dependable which as a whole makes the system more intelligent. These examples indicate that a many-body system is more reliable than any autocratic system where instead of a single entity, every element has worked together for consensus. So, even if some of the elements are at fault, the system can show resilience to that fault and work correctly. That is, the system is also more available. Similarly, if each of the elements individually decides according to an interaction with its neighboring elements, whereas, the system works only on the collective decision, then this system becomes more sensitive to the context and environment. So, such a many-body system with self-organization and self-replication properties can be an ideal choice for designing intelligent systems. In the next sections, we briefly survey these three necessary properties of an intelligent system – self-reproduction, self-organization and fault-tolerance, which are, in some way, overlapping and related to each other.

2   Self-replication

The journey of self-replicating machines was initiated by John von Neumann and Stanislaw Ulam [30]. Von Neumann was looking for the logical organization sufficient for a non-trivial self-reproduction. His idea was to abstract the logical form from the natural self-reproduction problem: if self-reproduction can be performed by machine whose behavior can be described as an algorithm, then there exists a Turing machine which can exhibit the same algorithm. So, in turn, a Turing machine exists which could demonstrate its own replication. This idea gave birth to a new mathematical model of computation, named *cellular automata*. A cellular automaton (CA) is a discrete dynamical system comprising of an orderly network of cells, where each cell is a finite state automaton. The next state of the cells are decided by the states of their neighboring cells following a local update *rule*. The snapshot of states of all cells at any time instant is called a *configuration*. Hence, the behavior of a CA is represented by its *global transition function* by which the CA hops from one configuration to another. The beauty of a CA is, it works on very simple rules on a well-define lattice and can be characterized easily, but the simple local interaction and computation of cells result in a huge complex behavior when the cells act together.

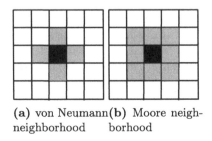

(a) von Neumann neighborhood **(b)** Moore neighborhood

Fig. 1. Neighborhood dependencies for 2-dimensional CAs. The black cell is the cell under consideration, and its neighbors are the shaded cells and the black cell itself

John von Neumann's CA is an infinite 2-dimensional square grid, where each square box, called cell, can be in any of the possible 29 states. The next state of each cell depends on the state of itself and its four orthogonal neighbors (see Fig. 1a). This CA can not only model biological self-reproduction, but is also computationally universal. In fact, his machine, an *universal constructor* can construct any machine described on its input tape along with a copy of the input tape attached to the constructed machine. So, self-reproduction happens when the description of the universal constructor itself is given as input, which will keep on reproducing a copy of itself along with its own description indefinitely [16]. It was a path-breaking research showing a existence of a model which is not only powerful enough to reproduce itself (and thus maintain that level of complexity) but also initiate a process of indefinite growth in complexity. Since then, it has been an area of research with a target to reduce complexity and size of the self-reproducing configuration keeping the system computationally universal, see [3, 4, 13, 23] for some related survey. In fact, the research field of *Artificial Life*, a term coined by Christopher Langton [17], was introduced as a descendant of von Neumann's seminal work and is closely related with the field of AI.

Cellular automata have also been utilized to design the famous *Game of Life*, a CA introduced by John Conway and popularized by Martin Gardener [9], where replication and transformation of moving entities can be observed just like in real life (see Fig. 2). In Conway's Game of Life, the local rule is $f : \{0,1\}^9 \rightarrow \{0,1\}$, where state 0 represents a dead cell and state 1 represents an alive cell and the neighborhoods dependency is like Fig. 1b (which was proposed by [22]). The rules of the game are very simple [19]:

- Birth: a cell that is dead at time t, will be alive at time $t + 1$, if exactly 3 of its eight neighbors were alive at time t.
- Death: a cell can die by:
    - Overcrowding: if a cell is alive at time t and 4 or more of its neighbors are also alive at time t, the cell will be dead at time $t + 1$.
    - Exposure: If a live cell at time t has only 1 live neighbor or no live neighbors, it will be dead at time $t + 1$.
- Survival: a cell survives from time t to time $t + 1$ if and only if 2 or 3 of its neighbors are alive at time t.

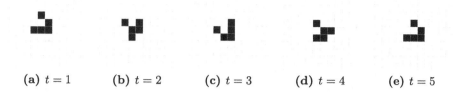

(a) $t = 1$ **(b)** $t = 2$ **(c)** $t = 3$ **(d)** $t = 4$ **(e)** $t = 5$

Fig. 2. Movement of *Glider* at time t in Game-of-Life.

However, this simple cellular automaton exhibits a number of powerful phenomena, such as (1) self-reproduction and self-organization, (2) growth of bacterial colony, (3) computational universality, etc. Figure 2 shows an example behavior of Game-of-Life CA where the initial pattern at t = 0 is a glider. It is similar to movement of a living body to another location. Figure 3 shows an example of self-replicating behavior by Game-of-Life. Here, the initial pattern is replicated to generate two images after 12 iterations and four images after 36 iterations. For some online resources on the beautiful life-like patterns formed by it, one can see [1]. Garden-of-Life proves that although individually the cells can have very limited computational ability, but when all the cells work collectively, they are as powerful as the universal Turing machine and can create copies of itself. In this way, the cells with limited computational ability can appear to us *intelligent* when they act together. This shows the power of collective intelligence which can keep on running without human intervention. This work, in fact, motivated a range of researchers around the globe to think of a Life-like intelligent system for various purposes. Like other CAs, the signature property of this Game-of-Life is that it is a decentralized system, as it has no centralized control mechanisms and works collectively to achieve a common goal. These systems have similarities with the notion of democracy. It is accepted in the history of human civilization that a democratic society is more advanced than others. We believe that a computing system can perform in a better way if the laws of democracy are followed by the system.

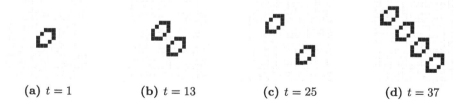

(a) $t = 1$ **(b)** $t = 13$ **(c)** $t = 25$ **(d)** $t = 37$

Fig. 3. Self-replication in Game-of-Life.

3   Self-organization and Artificial Life

Cellular Automata (CAs) have been explored since the 1950s to model biological phenomena such as self-reproduction, sociological phenomena such as growth of a colony, etc. Later, the idea of Artificial Life has been introduced in the domain of Artificial Intelligence (AI) by further exploring the cellular automata. It has been dreamt by researchers like Christopher Langton [17] that artificial and intelligent life would be created around the principle of CAs. Langton argued that there is a possibility that life could emerge from the interaction of inanimate artificial molecules [16,17]. This molecular logic of life is a dynamic distributed logic. To model this logic, a machine needs to satisfy the following properties. First, it must support massive parallelism - they must have huge computing elements in order to provide for the simultaneous interactions of many operators. Second, the computing elements need only to be locally connected. Third, the machine must support the motion of operators through the field of processing elements. Cellular automata provide the logical universes within which these artificial molecules can be embedded. This initiated the field of Artificial Life (ALife) which has since emerged through the interaction of Biology and Computer Science, but also with important contributions from Physics, Mathematics, Cognitive Science and Philosophy.

Artificial life is the study of artificial systems that exhibit behavior characteristic of natural living systems. This includes computer simulations, biological and chemical experiments, and purely theoretical endeavors. The ultimate goal is to extract the logical form of living systems [17]. It was introduced as a direct lineal descendant and inheritor of von Neumann's seminal research programme, begun in the late 1940s, combining "automata theory" with problems of biological self-organization, self-reproduction, and the evolution of complexity [5,7,31]. The question of "what is life?" and its possible answers are the fundamental query in artificial life which expects to underpin the scope, problems, methods, and results in terms of its investigations [3,23]. Discovering how to make such self-reproducing patterns more robust so that they can evolve to increasingly more complex states is probably the central problem in the study of artificial life. Artificial life has also inherited some roots from AI, particularly that differentiates between "weak" and "strong" AIs, a notion associated with the philosopher John Searle [25]. An analogous distinction can be drawn within artificial life itself, between merely simulating and actually realizing "real" life [18]. Many researchers from diverse backgrounds share the ALife approach and apply it in their own discipline. They seek to understand, through synthetic experiments, the organizational principles underlying the dynamics of living organisms. Then, these principles are used for synthesizing models or artificial systems with life-like properties [13]. For example, artificial life research is carried out in the computational world working with real chemical systems [20], systems biology [32], synthetic biology [24], epistemological modelling [6]. Thus, integration of self-reproducing programs with self-organization, self-maintenance and individuation (autopoiesis), and the demonstration of sustained evolutionary growth of complexity in a purely virtual world remains the key "grand challenge" problem in the field of artificial life [10].

Several other interesting results on the learning, self-organization, and associative memory problems have also been reported in the literature [2,17,31]. One of the important problems here is to organize a reliable system utilizing unreliable elements - as in the case of the brain, thousands of neurons die out every day and an individual neuron is not reliable, but the behavior of the brain is relatively stable and reliable. Therefore, the study of this field targets to develop new powerful methods in the field of information processing that can correct or detect any errors [12]. Cellular automata can be regarded as one of the simple mathematical models for the neural networks in the brain. Fault-tolerant cellular automata (FCA) belong to this larger category of reliable computing devices built from unreliable components, in which, the error probability of the individual components is not required to decrease as the size of the device increases. In such a model it is essential that the faults are assumed to be transient: they change the local state but not the local transition function. Such a machine has to use massive parallelism, as well as, a self-correction mechanism is to be built into each part of it. In cellular automata, it has to be a property of the transition function of the cells. Due to the homogeneity of cellular automata, since large groups of errors can destroy large parts of any kind of structure, "self-stabilization" techniques are introduced in conjunction with traditional error-correction [8].

4   Fault Tolerance and Distributed System

Another class of many-body systems is *distributed systems*, defined as a collection of autonomous independent processing components which interact with each other via an interconnecting communication link to achieve a common goal [26]. A distributed system has to be fault-tolerant – it should be able to continue functioning in the presence of faults. Fault tolerance is related to dependability that includes availability, reliability, safety and maintainability. These features are related; a highly maintainable system shows a high degree of availability, which in turn means that the system will be reliable and fault-tolerant. Ethernet is the first widespread implementation of distributed system, whereas, *ARPANET* is one of the first large-scale distributed system. Success of ARPANET eventually leads to the *Internet* and *World Wide Web (www)*, the most successful example of the power and potential of distributed systems.

Research has been done to design distributed systems which can detect and repair its faults automatically without effecting its availability. Some of the proposed methods are incorporating redundancy [29], allowing *graceful degradation* [11] etc. However, to work even if some parts of the system break down, the system has to come to a consensus. Moreover, there are several types of faults that the system has to deal with, like – *crash failure, omission failure, value failure, timing failure, state transition failure* and *arbitrary failure* [26]. Among them, arbitrary failure, also known as *Byzantine failure* is considered as the most general and most difficult class of failures. In case of Byzantine failure, some of the components may behave arbitrarily, but there is imperfect information on whether a component has failed. The name comes from an allegory of the Byzantine Generals Problem [15] where some of the generals are unreliable (traitors) but to conclude the system has to decide on a consensus. In a Byzantine fault, a component such as a server can inconsistently appear both failed and functioning to failure-detection systems, presenting different symptoms to different observers. So, it is difficult for the other components to declare it failed and shut it out of the network, because they need to first reach a consensus regarding which component has failed in the first place. Therefore, the system has to work knowing some if its components are unreliable, but not knowing which they are. Byzantine fault tolerance (BFT) is the dependability of a distributed system to such conditions. In was shown that, even if one third of the components of a system behave arbitrarily, still the distributed system can work effectively and tolerate the fault [15]. That is, although the system has unreliable components, but if two-third of the components are reliable, the system can intelligently take correct decision *masking* its faults. For other kinds of faults, it is far easier for a distributed system to detect and deal with those faults. This shows the effectiveness of many-body system in handling failures and reliability of such a system in decision making capability. This is the sign of intelligence in a many-body system.

5   To Move Forward

This discussion indicates that a many-body system where the components work together for achieving a common target based on interaction and communication may be more coherent as an intelligent system. Such a system needs to have self-reproducibility, self-organization and fault-tolerance – its universal computation power may follow as a consequence. This system can be resilient to any kind of faults, more available and have self-healing capability. It can also be context-sensitive to the changes in its environment and its decisions can be more reliable as they are taken collectively. Figure 4 shows an abstract diagram for the model. Here, a many-body system is shown which takes any statement as input and gives the decision as '*Yes*' if the particles of the system reaches to a consensus as yes and '*No*' otherwise. The *yes* state is shown in *white* and *no* in *black*. Our claim is, such a *democratic* model of computation can be a better choice for an intelligent system than an autocratic one. However, what we need here is a metric to measure the intelligence of a machine. We propose to include following parameter in the metric: abilities of self-reproduction, self-organization, fault-tolerance and capabilities of following democratic values.

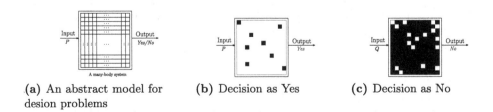

(a) An abstract model for desion problems   (b) Decision as Yes   (c) Decision as No

Fig. 4. Working of a many-body system

In a CA, both time and space are discrete and the cells are updated in parallel. Similarly, in case of our model also, the elements are individual entities whose movement and behavior is apparently independent of each other. There are also other features of many-body systems having self-organization and self-reproduction which suggest that this can be modeled efficiently by CA. So we want to exploit the inherent locality, massive parallelism and computational ability of cellular automata to design a many-body system where the components work together for achieving a common target based on interaction and communication and may be more coherent as an intelligent system. Such a system may not be exactly a CA, but will inherit all features of self-organization and self-replication offered by a CA. The future work involves development of a class of many-body systems that can imitate the basic features of living elements and that uses the laws of democracy to make the system intelligent in a classical sense. These intelligent systems can revolutionize the existing era of machine intelligence and help the scientists, researchers as well as industry to reach a new level of AI going beyond the traditional capability of intelligent acts performed by machines. Additionally, the incorporation of the essence of democratic values in the computing model can place the definition of intelligence to a different level.

Acknowledgment. The authors are grateful to Prof. Mihir K. Chakraborty for his valuable comments, discussion and inputs on this work.

References

1. Conwaylife.com. https://www.conwaylife.com/
2. Amari, S.-I.: Learning patterns and pattern sequences by self-organizing nets of threshold elements. IEEE Trans. Comput. **100**(11), 1197–1206 (1972)
3. Banzhaf, W., McMullin, B.: Artificial life. In: Rozenberg, G., Bäck, T., Kok, J.N. (eds.) Handbook of Natural Computing. Springer, Heidelberg (2012). https://doi.org/10.1007/978-3-540-92910-9_53
4. Bhattacharjee, K., Naskar, N., Roy, S., Das, S.: A survey of cellular automata: types, dynamics, non-uniformity and applications. Nat. Comput. **19**(2), 433–461 (2020)
5. Yovits, M.C., Cameron, S., (eds.): Self-Organizing Systems – Proceedings of an Interdisciplinary Conference, on Computer Science and Technology and their Application, 5–6 May 1959. Pergamon, New York (1960)
6. Cariani, P.: Some epistemological implications of devices which construct their own sensors and effectors. In: Towards a Practice of Autonomous Systems, pp. 484–493 (1992)
7. Chopard, B., Droz, M.: Cellular Automata Modelling of Physical Systems. Cambridge University Press, Cambridge (1998)
8. Gács, P.: Reliable cellular automata with self-organization. J. Stat. Phys. **103**(1–2), 45–267 (2001). https://doi.org/10.1023/A:1004823720305
9. Gardner, M.: On cellular automata self-reproduction, the Garden of Eden and the Game of 'Life'. Sci. Am. **224**(2), 112–118 (1971)
10. Gershenson, C., Trianni, V., Werfel, J., Sayama, H.: Self-organization and artificial life. Artif. Life **26**(3), 391–408 (2020)

11. González, O., Shrikumar, H., Stankovic, J.A., Ramamritham, K.: Adaptive fault tolerance and graceful degradation under dynamic hard real-time scheduling. In: Proceedings Real-Time Systems Symposium, pp. 79–89. IEEE (1997)
12. Harao, M., Noguchi, S.: Fault tolerant cellular automata. J. Comput. Syst. Sci. **11**(2), 171–185 (1975)
13. Heudin, J.C.: Artificial life and the sciences of complexity: history and future. In: Feltz, B., Crommelinck, M., Goujon, P. (eds.) Self-organization and Emergence in Life Sciences, pp. 227–247. Springer, Dordrecht (2006). https://doi.org/10.1007/1-4020-3917-4_14
14. Hunt, K.J.: Introduction to Artificial Intelligence, E. Charniak and D. Mcdermott. Addison-Wesley, Reading (1985). ISBN 0-201-11946-3. Int. J. Adapt. Control Signal Process. 2(2), 148–149 (1988)
15. Shostak, R., Lamport, L., Pease, M.: The Byzantine generals problem. ACM Trans. Program. Lang. Syst. **4**(3), 382–401 (1982)
16. Langton, C.G.: Self-reproduction in cellular automata. Physica D **10**(1–2), 135–144 (1984)
17. Langton, C.G.: Studying artificial life with cellular automata. Physica D **22**(1–3), 120–149 (1986)
18. Levy, S.: Artificial life: A report from the frontier where computers meet biology (1992)
19. Li, J., Demaine, E., Gymrek, M.: Es.268 The Mathematics in Toys and Games, Spring. Massachusetts Institute of Technology, MIT Opencourseware (2010). http://ocw.mit.edu. Accessed 2010
20. Luisi, P.L., Varela, F.J.: Self-replicating micelles-a chemical version of a minimal autopoietic system. Orig. Life Evol. Biosph. **19**(6), 633–643 (1989). https://doi.org/10.1007/BF01808123
21. McCulloch, W.S., Pitts, W.: A logical calculus of the ideas immanent in nervous activity. Bull. Math. Biophys. **5**(4), 115–133 (1943)
22. Moore, E.F.: Machine models of self-reproduction. In: Proceedings of Symposia in Applied Mathematics, vol. 14, pp. 17–33 (1962)
23. Reggia, J.A., Chou, H-H., Lohn, J.D.: Cellular automata models of self-replicating systems. In: Advances in Computers, vol. 47, pp. 141–183. Elsevier (1998)
24. Regis, E.: What is Life?: Investigating the Nature of Life in the Age of Synthetic Biology. Oxford University Press, Oxford (2009)
25. Searle, J.R.: Minds, brains, and programs. Behav. Brain Sci. **3**, 417–457 (1980)
26. Tanenbaum, A.S., Van Steen, M.: Distributed Systems: Principles and Paradigms. Prentice-Hall, Upper Saddle River (2007)
27. Turing, A.M.: Computing machinery and intelligence. Mind **59**(236), 433–460 (1950)
28. Turing, A.M.: On computable numbers, with an application to the Entscheidungs problem. In: Proceedings of the London Mathematical Society, vol. 2, issue 1, pp. 230–265 (1937)
29. von Neumann, J.: Probabilistic logics and the synthesis of reliable organisms from unreliable components. Autom. Stud. **34**, 43–98 (1956)
30. von Neumann, J.: Theory of Self-Reproducing Automata. University of Illinois, Urbana & London (1966). Edited and Completed by Arthur W. Burks
31. Von Neumann, J.: The General and Logical Theory of Automata **1951**, 1–41 (1951)
32. Wolkenhauer, O.: Why systems biology is (not) called systems biology. BIOforum Europe **4**, 2–3 (2007)

Medical Artificial Intelligence

Economy and Unemployment Due to COVID19: Secondary Research

Moitri Chakraborty[1], Madhumita Ghosh[1], Akash Maity[1], Dipanwita Dutta[1], Sayan Chatterjee[2(✉)], and Mainak Biswas[1]

[1] Department of Electrical Engineering, Techno International New Town, Rajarhat, Kolkata 700156, India
[2] Cognizant Technology Solutions, Kolkata, India

Abstract. The coronavirus pandemic has hit a hard blow on the world economy and employment rates. Countries like India, with a high population, have faced major economic degradation and high unemployment rates. Most of the countries are expected to face a major economic recession as most internal and external economic activities have ceased to operate due to the worldwide lockdown and quarantine measures being taken. This might affect the socioeconomic relationships between countries. It has also affected the economically challenged sector of the world largely. In India, about 41 lakh people lost their jobs, including several migrant workers. Several G7 countries have ensured subsidies as the jobless rates vary from 30million in the US to 1.76 million in Japan.

Keywords: GDP · Year-on-year · Subsidies · Recessions · G7 countries · World Bank

1 Introduction

The COVID19 pandemic has created a truly global crisis for the first time since the Second World War. Life, as we knew it in 2019, has changed forever. As countries all around the world begin to tentatively emerge out of lockdown, they are being confronted with a reality that had seemed impossible only a few months ago. This apparent "new normal" is reflecting the true extent of the vast divide between the rich and the poor and, indeed, sees it extended even further [1]. Millions of people are affected, died, lost their jobs due to COVID19 [2]. The virus has brought even superpowers like the USA and the UK to their collective knees and devastated the world economy. The effects of this devastation have been felt most by those most vulnerable. Unemployment levels have soared as companies cut losses in order to save their own existence. This led to a loss of spending power of the average consumer and further restricted the flow of currency in the global market sending stock markets spiraling downwards. As economists begin to cautiously hope with various vaccines slowly becoming available, this paper will analyze the damage that has been done by the pandemic to the global economy and job market [3].

Published by Springer Nature Switzerland AG 2021
Z. Shi et al. (Eds.): ICIS 2020, IFIP AICT 623, pp. 285–290, 2021.
https://doi.org/10.1007/978-3-030-74826-5_25

2 Data Analysis

India's economy has faced a significant contraction amidst the pandemic. In the first quarter of 2020, the Indian GDP grew by 3.1% yet recording the slowest GDP growth since quarterly data was made available in 2004 [4]. As the Country followed a nationwide lockdown to control the spread of coronavirus from the end of March 2020. Figure 1 represents the GDP growth in India in the financial year 2020-21 in all the quarters. The country experienced the first wave of recession in the second quarter of the same year, registering a contraction of year-on-year Extending the lockdown time and again has resulted in the postpone of economic activities throughout the country, recording the greatest plunge in the country's economy till date. The nation recorded a 7.5% year on year shrink in the third quarter of 2020, lesser than what the economists predicted would be 8.8%. The pandemic however seems to be far from controlled, and the GDP is still seen falling in the fourth quarter of the year. However, the government has announced a 10$ billion stimulus package to balance the entire economic scenario of the country.

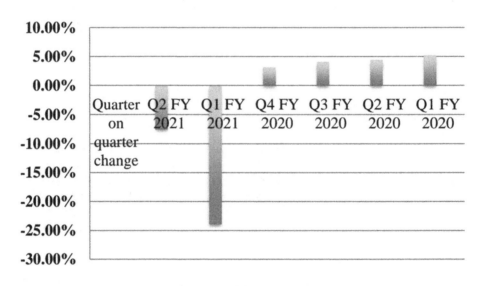

Fig. 1. GDP in India.

3   Effect on World Economy

As the spread of the virus continues to haunt all over the world, the word economy has witnessed massive contraction. It is said to be one of the largest blow to the world economy till date with a baseline forecast of 52% contraction in the global GDP. The lockdown imposed all over the world has brought most of the economic activities to a halt. The Corona Virus outbreak is predicted to cause major recession waves in most countries in the world is the greatest recession since World War I, as reported by the World Bank in June 1020.

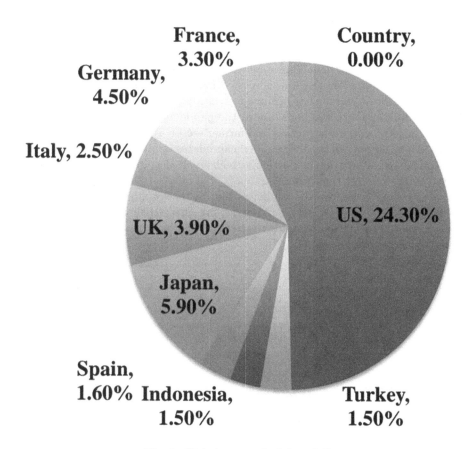

Fig. 2. Global economical downfall.

4   Employment Rates in India

Majority of the Indian population who work belong to the low-waged informal sector, having the lowest income jobs. The catastrophe seems to be awaiting a huge shock to all the global industries and advanced economies most countries depending on global trades, tourism. Exports and external financing will bear a large scar post-pandemic. The per capita GDP all over the world is estimated to fall by 6.2%. Figure 2 shows the economic downfall in each of the countries [5]. Several migrant workers all over India migrate from one place to other all over India to work as day laborers, daily wagers, etc. No limitation on commutation inside the country makes it easier for these people to move around in search of work. Figure 3 represents the rate of unemployment due to COVID19, from January to September 2020 [6]. Figure 4 represents the unemployment rate in India from 2010–2020 [7]. The rate of unemployment hiked all the way from 7% in mid-March to the beginning of May. The other major sector is the government sector including government offices. Railways, banks, and army forces being the major recruiters. The next formal sector of white collared jobs in the IT sectors and other private sectors. This sector saw the major no. of deployments. Employees from all these sectors faced work crises; layoffs, unpaid leaves, and a major chunk of them were deployed. Unemployed was a problem in the country since 2017–2018, reaching almost 6.1%. A sharp decline in the number of jobs in the formal sector has occurred. According to a report in August 2020, about 4.1 million youth lost jobs due to the pandemic. The majority of the job losses belonged to the construction and farmer sector. Applications for jobs have increased as a number of highly qualified candidates apply for recruitment. Most of them lack experience in the particular tech stack, despite being highly skilled. Several employees lost their jobs due to a lack of ability to cope up with the new normal of online work from home. According to the Mumbai-based think tank, the urban areas being the major red zones are affected the most. Urban areas [8] have recorded 29.22% unemployment rates, whereas rural areas have 26.69% unemployment rates.

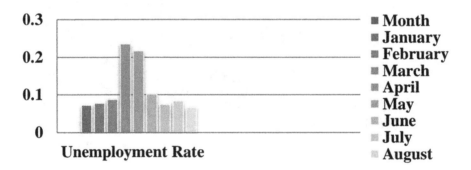

Fig. 3. The unemployment rate in India during COVID19.

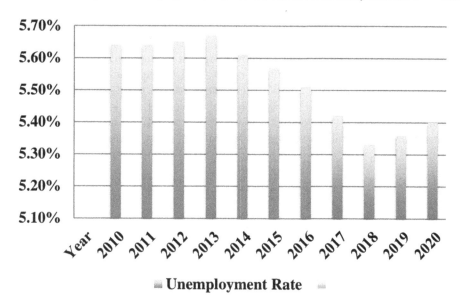

Fig. 4. Unemployment rate in India from 2010 to 2020.

4.1 Employment Rates in India

The pandemic wreaked havoc on job crises all-over the world, especially on the economically vulnerable sector of women and youths. The crisis might subside, but the economic and unemployment consequences might take a long time to refurbish back to normal. Most of the Asian countries that have a huge population are affected by the Corona Virus. Countries namely, India, Iraq. Iran, Turkey, Indonesia. Several European Governments released the unemployment crisis with wage subsidies. In April 2020 3.8 million Americans claimed unemployment that being the fourth consecutive fall in the number of claims. US private sectors shed 20million jobs in April. In the first quarter of 2020, the rate of unemployment was 4% and predicted to be increased to 15% by the third quarter. In Canada current reports from the countries, the official bureau says the unemployment rate in April was 13%, up from 5.2% points in March. More than 7.2 million people have applied for emergency unemployment assistance, in this recent on-going crisis. Unlike the UK and other G7 countries, Japan's unemployment rate is rising slowly. In March, it was 2.5%, with 1.76 million unemployed – an increase of 20,000 from March 2020 [9]. In the UK, its main statistics office shows employment at a record high and unemployment at around 4%. However, KPMG forecasts this will rise to just under 9% during the lockdown period. Report said that more than 2 million wanted to avail that privilege [10]. A quarter of the UK's employed workforces have registered for the government's job retention scheme, which pays 80% of an employee's wages. In France, a scheme named chômagepartiel (partial employment or short-time working) is implemented by the state to support more than 10 million workers in France's private sector. The unemployment rate in Germany has risen far less rapidly than in countries such as the US. This is because of a government scheme to subsidize the wages of struggling employers and employees called the Kurzarbeit or short-time work program. By the end of April, more than 10 million people were benefitted from this. Unemployment graph rose in the month of April 2020 and also jobless is 5.8% [11]. In March, Italy recorded the lowest employment rate in 9 years, the unemployment rate dropped to 8.4%. The government has promised around 75 billion of support for families and companies [12].

5 Conclusion

The Coronavirus outbreak undermined economic activities all over the world leading to major unemployment problems. The pandemic might gone away within a few months as the news of the vaccine coming out is all over the town, but the consequent outcomes might persist post-pandemic and might take years to recover. It is expected the GDP as well as the employment rate not only in India but also worldwide will rise within a short time.

References

1. Shipsy. https://www.shipsy.in/. Accessed 11 Nov 2020
2. Sohrabi, C., Alsafi, Z., O'Neill, N., Khan, M., Kerwan, A., Al-Jabir, A.: World health organization declares global emergency: a review of the 2019 novel coronavirus (COVID-19). Int. J. Surg. **76**, 71–76 (2020)
3. Fana, M., Torrejón Pérez, S., Fernández-Macías, E.: Employment impact of COVID-19 crisis: from short term effects to long terms prospects. J. Ind. Bus. Econ. **47**(3), 391–410 (2020). https://doi.org/10.1007/s40812-020-00168-5
4. Jamir, I.: Forecasting Potential Impact of COVID-19 Outbreak on India's GDP Using ARIMA Model
5. World Bank. Central African Republic Economic Update, October 2020: The Central African Republic in Times of COVID-19-Diversifying the Economy to Build Resilience and Foster Growth (2020)
6. Joshi, A., Bhaskar, P., Gupta, P.K.: Indian economy amid COVID-19 lockdown: a prespective. J. Pure Appl. Microbiol. **14**, 957–961 (2020)
7. Statista. https://www.statista.com/. Accessed 11 Nov 2020
8. Verma, C., Shakthisree, S.: Socio-economic impact of COVID-19 on India and some remedial measures. J. Econ. Policy Anal. **1**(1), 132–140 (2020)
9. World Economic Forum. https://www.weforum.org/. Accessed 11 Nov 2020
10. The Economic Times. https://economictimes.indiatimes.com/. Accessed 11 Nov 2020
11. BW BUSINESSWORLD. https://www.businessworld.in/. Accessed 11 Nov 2020
12. BBC. https://www.bbc.com/. Accessed 11 Nov 2020

Optimal Control of Dengue-Chikungunya Co-infection: A Mathematical Study

Anupam De[1]([envelope]) [ORCID], Kalipada Maity[2] [ORCID], Goutam Panigrahi[3] [ORCID], and Manoranjan Maiti[4] [ORCID]

[1] Department of Applied Sciences, Haldia Institute of Technology, Haldia, West Bengal, India
[2] Department of Mathematics, Mugberia Gangadhar Mahavidyalaya, Bhupatinagar, West Bengal, India
[3] Department of Mathematics, National Institute of Technology Durgapur, Durgapur, West Bengal, India
[4] Department of Applied Mathematics with Oceanology and Computer Programming, Vidyasagar University, Midnapore, West Bengal, India
http://hithaldia.in, http://www.mugberiagangadharmahavidyalaya.org, http://www.vidyasagar.ac.in

Abstract. In the present work, we developed a mathematical model for dengue-Chikungunya co-infection to analyze the disease transmission dynamics and interrelationship. We considered the essence of time dependent optimal control with the help of Pontryagin's maximum principle. The result of diverse combinations of control measures is analyzed and shown graphically.

Keywords: Dengue-Chikungunya · Co-infection · Optimal control

1 Introduction

Dengue and chikungunya infections are arboviral diseases which are among the leading public health problem to the entire world [27]. The dengue virus (DENV) infections have been found mainly in Asia, the Pacific, South and Central America and the Caribbean [21]. According to Londhey et al. [16] "Dengue, chikungunya causes a major matter of concern on the health care system. Both the diseases are transmitted through the same female mosquitos *Aedes aegypti* and *Aedes albopictus*. Since both dengue and chikungunya viruses are transmitted through a common vector, they often co-circulate in mosquito and are transmitted to human beings as co-infections following the mosquito bite".

Patients with dengue and chikungunya virus infection generally present with a self-limited febrile disease [20]. Due to the overlap of the symptoms, it is not always easy to distinguish the two infections clinically. The climatic condition during the rainy season and subsequent months favor mosquito breeding places and thereby amplify the number of mosquitoes as well as a sudden increase

© IFIP International Federation for Information Processing 2021
Published by Springer Nature Switzerland AG 2021
Z. Shi et al. (Eds.): ICIS 2020, IFIP AICT 623, pp. 291–302, 2021.
https://doi.org/10.1007/978-3-030-74826-5_26

of dengue and chikungunya. In recent past, cases of dengue-chikungunya have been confirmed in Africa, Southeast Asia, the Pacific Islands, the Caribbean and Latin America [16]. An effective vaccine for both the diseases are not so frequently available.

The first case of dengue-chikungunya co-infection has been reported in Thailand in 1962 where four co-infected cases among 150 patients was detected with either infected by dengue or by chikungunya; In 1963, three co-infected cases out of 144 infected patients and in 1964, 12 co-infected cases out of 334 infected patients was reported (cf. [10]). In the Southern part of India, the first case of co-infection was reported in 1964 (cf. [5]), where, among the patient with a dengue-like illness, 2% was found to have co-infection. In Myanmar by an active investigation, 36 out of 539 dengue and/or the chikungunya positive patients have been diagnosed to be co-infected in 1970; 8 out of 129 in 1971 and 11 out of 244 in 1972 (c.f. [14]). In the following 30 years, no reported new cases of dengue chikungunya were found, inspite of continuous CHIKV and DENV endemicity in Africa and Asia (cf. [10]).

In 2006, chikungunya-dengue co-infection was found in Malaysia (cf. Nayar et al. [22]), Sri Lanka (cf. Kularatne et al. [15]), Madagaskar (cf. Ratsitorahina et al. [26]) and India (cf. Chahar et al. [7]). Numerous studies have been reported on the co-circulation and co-infection of chikungunya–dengue in Africa (cf. Baba et al. [2] and Caron et al. [6]), Western Pacific region (cf. Chang et al. [8]), South-East Asia (cf. Afreen et al. [1]) and others during 2006–2012.

In the above discussions, it is evident that the spread and severity of the disease have been analyzed by the researchers. But very few articles have been formulated with a mathematical model of dengue chikungunya co-infection. A preliminary mathematical model has been formulated by Isea and Lonngren [11] to analyze the dynamic transmission of dengue chikungunya and zika. Malaria-Schistosomiasis co-infection model has been formulated by Bakare and Nwozo [3]. Mathematical models for HIV/TB Co-infection have been analyzed by Boralin and Omatola [4] and Optimal control theory was applied by Mallela et al. [19]. A co-infection model on Malaria-Cholera model was analyzed by Okosun and Makinde [24].

In this paper, A mathematical model for co-infection of dengue and chikungunya is formulated. Further, the problem is improvised to an optimal control problem and solved using Pontryagin's Maximum Principle [25].

The rest of the paper is arranged as follows: In Sect. 2, the co-infection model is formulated. A time-dependent optimal control problem is formulated with bednet as a new time-dependent control measure is obtained at Sect. 3 and presented the result graphically with the help of numerical experiment in Sect. 4. The results are discussed in Sect. 5 and sensitivity analysis are discussed in detail in Sect. 6. The paper is concluded in Sect. 7.

2   Formulation of Co-infection Model

The co-infection model sub-divides the total human host population N_h into sub-classes of susceptible (S_h), only dengue infected (I_{hd}), only chikungunya infected I_{hc}, dengue-chikungunya co-infected (I_{hdc}), only dengue recovered R_{hd}, only chikungunya recovered (R_{hc}) and dengue-chikungunya co-infection recovered (R_{hdc}) human population.

Also the total mosquito vector population N_m sub-divided into susceptible (S_m), dengue infected (I_{md}), chikungunya infected (I_{mc}) and co-infected I_{mdc} mosquito population.

The major mosquito vector for the transmission of dengue and chikungunya is the female Aedes aegypti species and they can carry and transmit both the disease simultaneously. Initial susceptible population recruitment rate is $\mu_h N_h$. The susceptible human population is infected by the dengue infected, chikungunya infected and dengue-chikungunya co-infected mosquito population through biting at a rate B per day. So a susceptible individual become infected at a rate $B\alpha_1 \frac{S_h I_{md}}{N_h}$ by dengue infected, $B\alpha_2 \frac{S_h I_{mc}}{N_h}$ by chikungunya infected and $B\alpha_3 \frac{S_h I_{mdc}}{N_h}$ by dengue-chikungunya co-infected mosquito. A dengue infected human individual becomes co-infected individual at a rate $B\alpha_2 \frac{I_{hd} I_{mc}}{N_h}$ by the bite of infected chikungunya infected mosquito and at a rate $B\alpha_3 \frac{I_{hd} I_{mdc}}{N_h}$ by the bite of dengue-chikungunya co-infected mosquito. Similarly, a chikungunya infected human individual becomes co-infected at a rate $B\alpha_1 \frac{I_{hc} I_{md}}{N_h}$ and $B\alpha_3 \frac{I_{hc} I_{mdc}}{N_h}$ by a dengue infected and dengue-chikungunya co-infected mosquito respectively. The I_{hd}, I_{hc} population recovered naturally at a rate $\eta_1 I_{hd}$ and $\eta_2 I_{hc}$ and by treatment at a rate $\tau_1 u_1$ and $\tau_2 u_2$ respectively. The co-infected human population I_{hdc} recover totally at a rate $(\eta_{dc}(1-c_1))I_{hdc}$, from dengue only at a rate $(\eta_{dc}c_1 d_1)I_{hdc}$ and from chikungunya only at a rate $(\eta_{dc}c_1(1-d_1))I_{hdc}$ naturally and totally at a rate $(\tau_3 u_3(1-c_2))I_{hdc}$, from dengue only at a rate $(\tau_3 u_3 c_2 d_2)I_{hdc}$ and from chikungunya only at a rate $(\tau_3 u_3 c_2(1-d_2))I_{hdc}$ by treatment. $\gamma_1 I_{hd}, \gamma_2 I_{hc}$ and $\gamma_3 I_{hdc}$ are the disease induced deaths for dengue chikungunya and co-infected classes. The recovered population R_{hd}, R_{hc} and R_{hdc} again becomes susceptible at a rate $\rho_1 R_1, \rho_2 R_2$ and $\rho_3 R_{hdc}$. μ_h is the natural death rate of each of the subclasses of human population.

The diseases transmitted to the susceptible mosquito population when they come in the contact of dengue infected, chikungunya infected and dengue-chikungunya co-infected human population through biting. So a susceptible mosquito become infected at a rate $B\delta_1 \frac{S_m I_{hd}}{N_m}$ by dengue infected, $B\delta_2 \frac{S_m I_{hc}}{N_m}$ by chikungunya infected and $B\delta_3 \frac{S_m I_{hdc}}{N_m}$ by dengue-chikungunya co-infected human. Again dengue infected mosquitoes become co-infected at a rate $B\delta_2 \frac{I_{md} I_{dc}}{N_m}$ by contact of infected chikungunya infected human and at a rate $B\delta_3 \frac{I_{md} I_{hdc}}{N_m}$ by the contact of dengue-chikungunya co-infected human. Similarly, A chikungunya infected mosquito becomes co-infected at a rate $B\delta_1 \frac{I_{mc} I_{hd}}{N_m}$ and $B\delta_3 \frac{I_{mc} I_{hdc}}{N_m}$ by a dengue infected and dengue-chikungunya co-infected human respectively.

Mosquitoes die at a rate μ_m naturally and $\tau_4 u_4$ by the use of pesticides. Further, it is assumed that there is no disease induced death in mosquito population. The mathematical formulation is:

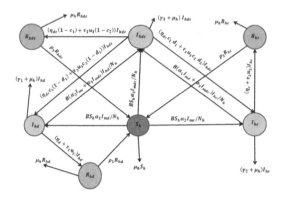

Fig. 1. Pictorial representation of Human Infection

$$
\begin{cases}
\dot{S}_h = \mu_h N_h - \frac{BS_h}{N_h}(\alpha_1 I_{md} + \alpha_2 I_{mc} + \alpha_3 I_{mdc}) - \mu_h S_h + \rho_1 R_{hd} + \rho_2 R_{hc} + \rho_3 R_{hdc} \\
\dot{I}_{hd} = B\alpha_1 \frac{S_h I_{md}}{N_h} - (\eta_d + \tau_1 u_1 + \gamma_1 + \mu_h)I_{hd} - B(\alpha_2 I_{mc} + \alpha_3 I_{mdc})\frac{I_{hd}}{N_h} \\
\qquad + (\eta_{dc}c_1(1 - d_1) + \tau_3 u_3 c_2(1 - d_2))I_{hdc} \\
\dot{I}_{hc} = B\alpha_2 \frac{S_h I_{mc}}{N_h} - (\eta_c + \tau_2 u_2 + \gamma_2 + \mu_h)I_{hc} - B(\alpha_1 I_{md} + \alpha_3 I_{mdc})\frac{I_{hc}}{N_h} \\
\qquad + (\eta_{dc}c_1 d_1 + \tau_3 u_3 c_2 d_2)I_{hdc} \\
\dot{I}_{hdc} = B\alpha_3(S_h + I_{hd} + I_{hc})\frac{I_{mdc}}{N_h} + \frac{B}{N_h}(\alpha_1 I_{md}I_{hc} + \alpha_2 I_{mc}I_{hd}) \\
\qquad - (\eta_{dc} + \tau_3 u_3 + \gamma_3 + \mu_h)I_{hdc} \\
\dot{R}_{hd} = (\eta_d + \tau_1 u_1)I_{hd} - (\rho_1 + \mu_h)R_{hd} \\
\dot{R}_{hc} = (\eta_c + \tau_2 u_2)I_{hc} - (\rho_2 + \mu_h)R_{hc} \\
\dot{R}_{hdc} = (\eta_{dc}(1 - c_1) + \tau_3 u_3(1 - c_2))I_{hdc} - (\rho_3 + \mu_h)R_{hdc} \\
\dot{S}_m = \mu_m N_m - \frac{BS_m}{N_m}(\delta_1 I_{hd} + \delta_2 I_{hc} + \delta_3 I_{hdc}) - (\mu_m + \tau_4 u_4)S_m \\
\dot{I}_{md} = B\delta_1 \frac{I_{hd}S_m}{N_m} - (\mu_m + \tau_4 u_4)I_{md} - B(\delta_2 I_{hc} + \delta_3 I_{hdc})\frac{I_{md}}{N_m} \\
\dot{I}_{mc} = B\delta_2 \frac{I_{hc}S_m}{N_m} - (\mu_m + \tau_4 u_4)I_{mc} - B(\delta_1 I_{hd} + \delta_3 I_{hdc})\frac{I_{mc}}{N_m} \\
\dot{I}_{mdc} = B\delta_3(S_m + I_{md} + I_{mc})\frac{I_{hdc}}{N_m} + \frac{B}{N_m}(\delta_1 I_{hd}I_{mc} + \delta_2 I_{hc}I_{md}) - (\tau_4 u_4 + \mu_m)I_{mdc}
\end{cases}
\tag{1}
$$

with $S_h(0) \geq 0$, $I_{hd}(0) \geq 0$, $I_{hc}(0) \geq 0$, $I_{hdc}(0) \geq 0$, $S_m(0) \geq 0$, $I_{md}(0) \geq 0$, $I_{mc}(0) \geq 0$, $I_{mdc}(0) \geq 0$. Pictorially the above model is presented in Fig. 1 and 2.

3   Application of Optimal Control

In this section, optimal control for the dengue-chikungunya co-infection problem is calculated. In place of fixed control, time-dependent optimal control is used. Since the best way to prevent chikungunya is with avoidance of mosquito bites (cf. [18]), in this section, one additional control measure $u_5(t)$ (namely the use of bed-net or mosquito repellent) is used as a new control variable and the system is reduced to the following:

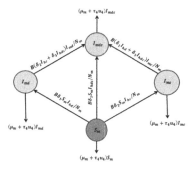

Fig. 2. Pictorial representation of Mosquito Infection

$$
\begin{cases}
\dot{S}_h = \mu_h N_h - (1 - u_5)\frac{BS_h}{N_h}(\alpha_1 I_{md} + \alpha_2 I_{mc} + \alpha_3 I_{mdc}) - \mu_h S_h + \rho_1 R_{hd} + \rho_2 R_{hc} + \rho_3 R_{hdc} \\
\dot{I}_{hd} = (1 - u_5)B\alpha_1 \frac{S_h I_{md}}{N_h} - (\eta_d + \tau_1 u_1 + \gamma_1 + \mu_h)I_{hd} - (1 - u_5)B(\alpha_2 I_{mc} + \alpha_3 I_{mdc})\frac{I_{hd}}{N_h} \\
\qquad + (\eta_{dc}c_1(1 - d_1) + \tau_3 u_3 c_2(1 - d_2))I_{hdc} \\
\dot{I}_{hc} = (1 - u_5)B\alpha_2 \frac{S_h I_{mc}}{N_h} - (\eta_c + \tau_2 u_2 + \gamma_2 + \mu_h)I_{hc} - (1 - u_5)B(\alpha_1 I_{md} + \alpha_3 I_{mdc})\frac{I_{hc}}{N_h} \\
\qquad + (\eta_{dc}c_1 d_1 + \tau_3 u_3 c_2 d_2)I_{hdc} \\
\dot{I}_{hdc} = (1 - u_5)B\alpha_3(S_h + I_{hd} + I_{hc})\frac{I_{mdc}}{N_h} + (1 - u_5)\frac{B}{N_h}(\alpha_1 I_{md}I_{hc} + \alpha_2 I_{mc}I_{hd}) \\
\qquad - (\eta_{dc} + \tau_3 u_3 + \gamma_3 + \mu_h)I_{hdc} \\
\dot{R}_{hd} = (\eta_d + \tau_1 u_1)I_{hd} - (\rho_1 + \mu_h)R_{hd} \\
\dot{R}_{hc} = (\eta_c + \tau_2 u_2)I_{hc} - (\rho_2 + \mu_h)R_{hc} \\
\dot{R}_{hdc} = (\eta_{dc}(1 - c_1) + \tau_3 u_3(1 - c_2))I_{hdc} - (\rho_3 + \mu_h)R_{hdc} \\
\dot{S}_m = \mu_m N_m - (1 - u_5)\frac{BS_m}{N_m}(\delta_1 I_{hd} + \delta_2 I_{hc} + \delta_3 I_{hdc}) - (\mu_m + \tau_4 u_4)S_m \\
\dot{I}_{md} = (1 - u_5)B\delta_1 \frac{I_{hd}S_m}{N_m} - (\mu_m + \tau_4 u_4)I_{md} - (1 - u_5)B(\delta_2 I_{hc} + \delta_3 I_{hdc})\frac{I_{md}}{N_m} \\
\dot{I}_{mc} = (1 - u_5)B\delta_2 \frac{I_{hc}S_m}{N_m} - (\mu_m + \tau_4 u_4)I_{mc} - (1 - u_5)B(\delta_1 I_{hd} + \delta_3 I_{hdc})\frac{I_{mc}}{N_m} \\
\dot{I}_{mdc} = (1 - u_5)B\delta_3(S_m + I_{md} + I_{mc})\frac{I_{hdc}}{N_m} + (1 - u_5)\frac{B}{N_m}(\delta_1 I_{hd}I_{mc} + \delta_2 I_{hc}I_{md}) \\
\qquad - (\tau_4 u_4 + \mu_m)I_{mdc}
\end{cases}
\tag{2}
$$

Also, the cost and side effects of applying the control measures are analyzed.

Here, the objective is not only to control the epidemics i.e., minimize the total number of infected population, but also to minimize the cost of applying the control and the side effect of applying the controls. In order to minimize both cost and side effect, one needs to minimize the squares of the control variables.

Following De et al. [9], the objective function of the present optimal control problem is given by

$$
J = \min_{u_1,u_2,u_3,u_4,u_5} \int_0^T (A_1 I_{hd} + A_2 I_{hc} + A_3 I_{hdc} + A_4 I_{md} + A_5 I_{mc} + A_6 I_{mdc} + B_1 u_1^2
$$
$$
+ B_2 u_2^2 + B_3 u_3^2 + B_4 u_4^2 + B_5 u_5^2)dt
\tag{3}
$$

subject to the system of ODE (2). The objective here is to find the 5-tuples $(u_1^*, u_2^*, u_3^*, u_4^*, u_5^*)$ such that

$$
J(u_1^*, u_2^*, u_3^*, u_4^*, u_5^*) = \min_{u_1,u_2,u_3,u_4,u_5 \in \Theta} J(u_1, u_2, u_3, u_4, u_5).
\tag{4}
$$

where $\Theta = \{u : \text{measurable and } 0 \le u_i(t) \le 1 \text{ for } t \in [0, T], i = 1, 2, \cdots, 5\}$ is the set of controls. Here, the time dependent control variables $u_1(t)$, $u_2(t)$, $u_3(t)$ are the treatment (medication) on dengue, chikungunya and dengue-chikungunya co-infected human population respectively, $u_4(t)$ pesticide control on mosquito population and $u_5(t)$ is the preventive measure control (bed-net, mosquito repellant, etc.) on human population. Also, $A_1, A_2, A_3, A_4, A_5, A_6$ are the constant weights corresponding to $I_{hd}, I_{hc}, I_{hdc}, I_{md}, I_{mc}$ and I_{mdc} respectively. The squares of the controls are taken to consider the side effect and over doses of the controls and B_1, B_2, B_3, B_4, B_5 are respectively, the positive constants associated with the squares of the control to balance the size of the terms.

Now the optimal control problem (3) is solved, subject to the condition (2). The Lagrangian of the problem is given by

$$
\begin{aligned}
L = & A_1 I_{hd} + A_2 I_{hc} + A_3 I_{hdc} + A_4 I_{md} + A_5 I_{mc} + A_6 I_{mdc} \\
& + B_1 u_1^2 + B_2 u_2^2 + B_3 u_3^2 + B_4 u_4^2 + B_5 u_5^2
\end{aligned}
$$

To minimize L, the Hamiltonian of the problem is formulated as

$$
\begin{aligned}
H = & L + \lambda_1(t)\frac{dS_h}{dt} + \lambda_2(t)\frac{dI_{hd}}{dt} + \lambda_3(t)\frac{dI_{hc}}{dt} + \lambda_4(t)\frac{dI_{hdc}}{dt} + \lambda_5(t)\frac{dR_{hd}}{dt} + \lambda_6(t)\frac{dR_{hc}}{dt} \\
& + \lambda_7(t)\frac{dR_{hdc}}{dt}\lambda_8(t)\frac{dS_m}{dt} + \lambda_9(t)\frac{dI_{md}}{dt} + \lambda_{10}(t)\frac{dI_{mc}}{dt} + \lambda_{11}(t)\frac{dI_{mdc}}{dt} \quad (5)
\end{aligned}
$$

where $\lambda_i(t), i = 1, 2, \cdots 11$ are adjoint variables to be determined by the following system of differential equations

$$
\begin{aligned}
& -\frac{d\lambda_1}{dt} = \frac{\partial H}{\partial S_h}, \quad -\frac{d\lambda_2}{dt} = \frac{\partial H}{\partial I_{hd}}, \quad -\frac{d\lambda_3}{dt} = \frac{\partial H}{\partial I_{hc}}, \quad -\frac{d\lambda_4}{dt} = \frac{\partial H}{\partial I_{hdc}}, \quad -\frac{d\lambda_5}{dt} = \frac{\partial H}{\partial R_{hd}} \\
& -\frac{d\lambda_6}{dt} = \frac{\partial H}{\partial R_{hc}}, \quad -\frac{d\lambda_7}{dt} = \frac{\partial H}{\partial R_{hdc}}, \quad -\frac{d\lambda_8}{dt} = \frac{\partial H}{\partial S_m}, \quad -\frac{d\lambda_9}{dt} = \frac{\partial H}{\partial I_{md}}, \quad -\frac{d\lambda_{10}}{dt} = \frac{\partial H}{\partial I_{mc}} \\
& -\frac{d\lambda_{11}}{dt} = \frac{\partial H}{\partial I_{mdc}} \quad (6)
\end{aligned}
$$

satisfying the conditions

$$
\lambda_i(T) = 0, \quad i = 1, 2, \cdots, 11. \quad (7)
$$

We have considered that $\bar{S}_h, \bar{I}_{hd}, \bar{I}_{hc}, \bar{I}_{hdc}, \bar{S}_m, \bar{I}_{md}, \bar{I}_{mc}, \bar{I}_{mdc}$ are the optimum values of $\bar{S}_h, \bar{I}_{hd}, \bar{I}_{hc}, \bar{I}_{hdc}, \bar{S}_m, \bar{I}_{md}, \bar{I}_{mc}, \bar{I}_{mdc}$. Also, let $\{\bar{\lambda}_1, \bar{\lambda}_2, \bar{\lambda}_3, \bar{\lambda}_4, \bar{\lambda}_5, \bar{\lambda}_6, \bar{\lambda}_7, \bar{\lambda}_8, \bar{\lambda}_9, \bar{\lambda}_{10}, \bar{\lambda}_{11}\}$ be the solution of the system (6).

Following Pontryagin et al. [25], the following theorems can be stated.

Theorem 1. *There exists optimal controls $u_1^*, u_2^*, u_3^*, u_4^*, u_5^*$ for $t \in [0, T]$ such that*

$$J(I_{hd}, I_{hc}, I_{hdc}, I_{md}, I_{mc}, I_{mdc}, u_1^*, u_2^*, u_3^*, u_4^*, u_5^*) =$$
$$\min_{u_1, u_2, u_3, u_4, u_5} J(I_{hd}, I_{hc}, I_{hdc}, I_{md}, I_{mc}, I_{mdc}, u_1, u_2, u_3, u_4, u_5)$$

subject to the system (2).

Theorem 2. *The optimal control 5-tuple $(u_1^*, u_2^*, u_3^*, u_4^*, u_5^*)$ which minimizes J over Θ is given by*

$$u_1^* = \max\{0, \min(\bar{u}_1, 1)\}, \ u_2^* = \max\{0, \min(\bar{u}_2, 1)\}, \ u_3^* = \max\{0, \min(\bar{u}_3, 1)\}$$
$$u_4^* = \max\{0, \min(\bar{u}_4, 1)\}, \ u_5^* = \max\{0, \min(\bar{u}_5, 1)\}$$

where, $\bar{u}_1 = \dfrac{(\bar{\lambda}_2 - \bar{\lambda}_5)\tau_1 \bar{I}_{hd}}{2B_1}, \ \bar{u}_2 = \dfrac{(\bar{\lambda}_3 - \bar{\lambda}_6)\tau_2 \bar{I}_{hc}}{2B_2},$

$$\bar{u}_3 = \frac{(\bar{\lambda}_4 - \bar{\lambda}_2 c_2(1 - d_2) - \bar{\lambda}_3 c_2 d_2)\tau_3 \bar{I}_{hdc}}{2B_3},$$

$$\bar{u}_4 = \frac{(\bar{\lambda}_8 \bar{S}_m + \bar{\lambda}_9 \bar{I}_{md} + \bar{\lambda}_{10} \bar{I}_{mc} + \bar{\lambda}_{11} \bar{I}_{mdc})}{2B_4},$$

$$\bar{u}_5 = \frac{1}{2B_5} \left[\frac{B}{N_h} \left[\bar{S}_h(\alpha_1 \bar{I}_{md}(\bar{\lambda}_2 - \bar{\lambda}_1) + \alpha_2 \bar{I}_{mc}(\bar{\lambda}_3 - \bar{\lambda}_1) + \alpha_3 \bar{I}_{mdc}(\bar{\lambda}_4 - \bar{\lambda}_1)) \right. \right.$$
$$+ \bar{I}_{hd}(\bar{\lambda}_4 - \bar{\lambda}_2)(\alpha_2 \bar{I}_{mc} + \alpha_3 \bar{I}_{mdc}) + \bar{I}_{hc}(\bar{\lambda}_4 - \bar{\lambda}_3)(\alpha_1 \bar{I}_{md} + \alpha_3 \bar{I}_{mdc}) \big]$$
$$+ \frac{B}{N_m} \big[\bar{S}_m(\delta_1 \bar{I}_{hd}(\bar{\lambda}_9 - \bar{\lambda}_8) + \delta_2 \bar{I}_{hc}(\bar{\lambda}_{10} - \bar{\lambda}_8) + \delta_3 \bar{I}_{hdc}(\bar{\lambda}_{11} - \bar{\lambda}_8))$$
$$\left. \left. + \bar{I}_{md}(\bar{\lambda}_{11} - \bar{\lambda}_9)(\delta_2 \bar{I}_{hc} + \delta_3 \bar{I}_{hdc}) + \bar{I}_{mc}(\bar{\lambda}_{11} - \bar{\lambda}_{10})(\delta_1 \bar{I}_{hd} + \delta_3 \bar{I}_{hdc}) \big] \right] \right]$$

4   Numerical Experiment

The following eight possible control strategies can be applied to control the disease. Namely, (i) medicine control only, (ii) insecticide control only, (iii)bed-net or mosquito repellent control only, (iv) medicine and insecticide controls, (v) medicine and bed-net or mosquito repellent controls, (vi) insecticide and bed-net or mosquito repellent controls, (vii) all the three controls and ($viii$) no control.

To graphically analyze the effect of various control strategies, the data set under consideration is shown in Table 1. All the input and output data are in appropriate units. The results are shown graphically in Figs. 3, 4, 5, 6, 7 and 8.

Table 1. Input Data for Optimal control problem

A_1	A_2	A_3	A_4	A_5	A_6	B_1	B_2	B_3	B_4	B_5
α_1	α_2	α_3	B	c_1	c_2	d_1	d_2	η_d	η_c	η_{dc}
δ_1	δ_2	δ_3	γ_1	γ_2	γ_3	μ_h	μ_m	N_h	N_m	
τ_1	τ_2	τ_3	τ_4	ρ_1	ρ_2	ρ_3				
$S_h(0)$	$I_{hd}(0)$	$I_{hc}(0)$	$I_{hdc}(0)$	$R_{hd}(0)$	$R_{hc}(0)$	$R_{hdc}(0)$	$S_m(0)$	$I_{md}(0)$	$I_{mc}(0)$	$S_{mdc}(0)$
1.0	1.0	1.0	1.0	1.0	1.0	0.1	0.1	0.1	0.1	0.1
0.001	0.001	0.001	.1	0.5	0.5	0.5	0.5	0.01	0.01	0.01
0.1	0.1	0.1	0.01	0.01	0.01	0.00004	0.02	500000	1500000	
0.5	0.5	0.5	0.5	0.5	0.5	0.5				
450000	15000	15000	7500	5000	5000	2500	1200000	100000	100000	10000

5   Discussion

The Optimal control problem (3) is solved in Sect. 4, Numerical Experiment using fourth order Runge-Kutta method in the time interval $[0, T]$. Using the transversality condition given by (3) and the optimality condition (cf. [9,12,13]), the state variables and the adjoint variables are solved simultaneously using Runge-Kutta forward and backward methods respectively. The set of parameters under consideration is presented in Table1 as input data. Here we consider the parameters $A_1 = A_2 = A_3 = A_4 = A_5 = A_6 = 1$ in order to minimize the total number of the infected population. The solution curves of different types of population for different control strategies are presented in Figs. 3 and 4. Time series plot for the bed-net/mosquito repellent control is given in Fig. 7. From Figs. 3, we can relate the effect of different control strategies on Co-Infected human population (I_{hdc}). Similarly, from Figs. 4 the effect of different controls on Co-Infected mosquito population(I_{mdc}) can be compared.

It is observed from Fig. 3 that the effect of all possible control is same as effect of medicine and bed-net control on human population. Also, it can be noticed that if we do not apply any control strategy, then the infected populations increase with an exponential nature from the initial point of consideration. In Fig. 4, it is found that the effect of all possible control is same as effect of pesticide and bed-net control on mosquito population and like human population infected mosquito population also grows exponentially if no control is applied to the system.

Further, it can be observed from the solution graph of the co-infected human population that only treatment control on infected human can control only human infection but not the co-infected mosquito population (cf. Fig. 3). Again, the pesticide control on co-infected mosquito population can control co-infected mosquito only, but not infected human population (cf. Fig. 4).

The role of the control strategies with both treatment and pesticide control is quite significant.

Only the bed-net control shows not much significant effect in controlling or removing the disease from the system, though when ever the bed-net control is combined with either medicine or treatment or both, it plays a significant role on the system.

Therefore, for simultaneous eradication or control of the co-infection of dengue and chikungunya in a system with human and mosquito population, all possible controls have to be applied.

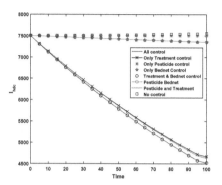

Fig. 3. Co-infected human population

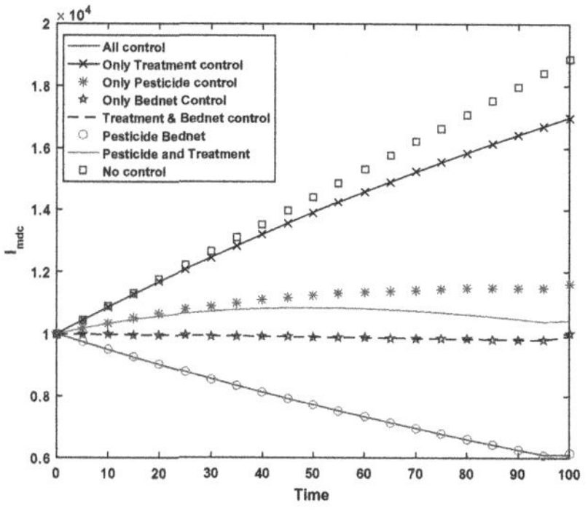

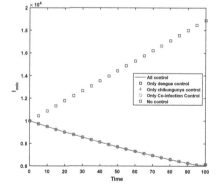

Fig. 4. Co-infected mosquito population

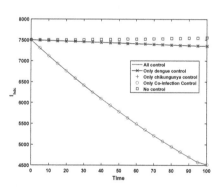

Fig. 5. Impact of different control on Co-infected human population

Fig. 6. Impact of different control on Co-infected mosquito population

In Fig. 7 it can be observed that the control variables u_5 is bounded between [0,1]. Similar type of figures can be obtained to u_1, u_2, u_3 and u_4. Figures 5 and 6 analyze the impact of no-control, single disease controls (only dengue control or only chikungunya control), only co-infection control and all control on co-infected human and co-infected mosquito population only.

Here, it can be noticed that single disease control measure fails to control the growth of co-infection on human and effectiveness of co-infection control is same as that of applying all control. So, in order to control co-infection, controlling co-infection is very important.

6 Sensitivity Analysis

The contact rate between human and mosquito is represented by the parameter B. From the optimum result of numerical experiment in Sect. 4, it is observed that, to control the infected mosquito insecticide control is more dominating in the control of the dengue chikungunya co-infection. If the contact rate between human and mosquito be regulated, then the disease can be removed completely from the system. From Fig. 8, it can be observed that the infected population for each species is monotonically increasing with the increase of B.

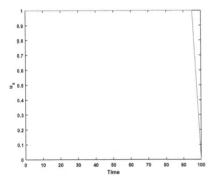

Fig. 7. Mosquito repellant and/or bed-net control for mosquito

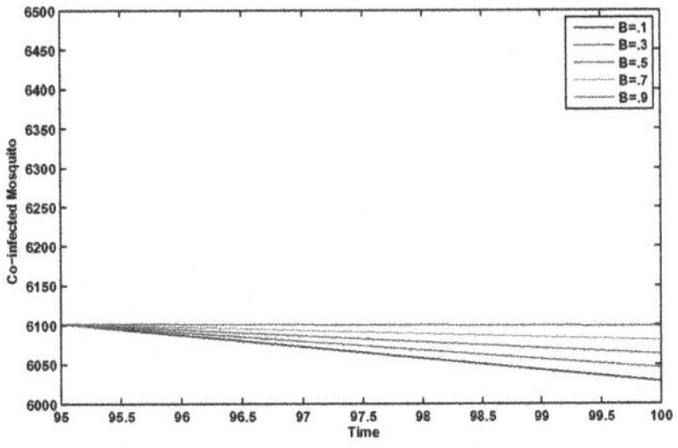

Fig. 8. Sensitivity analysis of B on co-infected mosquito population

7 Conclusion

In this paper, a dengue-chikungunya co-infection model is formulated mathematically and analyzed with three different types of control measures, namely treatment control on the human population, pesticide control on the mosquito population and bed-net control of the human population. Initially, for single infection, the basic reproduction numbers in each of dengue and chikungunya disease for the fixed control are obtained and analyzed. In the optimal control part, bed-net or mosquito repellent is introduced as new control and analyzed the effect of time-dependent control on the co-infected human mosquito system. The graphical solutions are obtained and compared the solutions with different control strategies. It is depicted by the figures that if there is no control, then the disease spreads exponentially with time from the initial state. It can be observed that a single control measure can not eradicate or control the infection completely from the system. Also, it can be noticed that the effect of all the three different controls is the same as the effect of medication and bed-net control of the human population. Again, the effect of all the three different controls is the same as the effect of pesticide and bed-net control on mosquito populations. Medication control only does not have any significant impact on the infected mosquito population. Similarly, pesticide control does not have any impact on the infected human population. So, for simultaneous controlling of co-infection with both human and mosquito population, simultaneous application of all the three types of control measures are needed. It is obtained that all the control measures are bounded. From the sensitivity analysis of the model it is observed that the contact rate between the human and the mosquito population plays significant role in amplification of infection.

This model can suggest the policymakers to take appropriate strategies to control when both the disease strikes at once to a population.

References

1. Afreen, N., et al.: Molecular characterization of dengue and chikungunya virus strains circulating in New Delhi, India. Microbiol. Immunol. **58**(12), 688–696 (2014)
2. Baba, M., et al.: Evidence of arbovirus co-infection in suspected febrile malaria and typhoid patients in Nigeria. J. Infect. Dev. Ctries. **7**(01), 051–059 (2013)
3. Bakare, E., Nwozo, C.: Bifurcation and sensitivity analysis of malaria-schistosomiasis co-infection model. Int. J. Appl. Comput. Math. **3**(1), 971–1000 (2017)
4. Bolarin, G., Omatola, I.: A mathematical analysis of HIV/TB co-infection model. Appl. Math. **6**(4), 65–72 (2016)
5. Carey, D.E., Myers, R.M., DeRanitz, C., Jadhav, M., Reuben, R.: The 1964 chikungunya epidemic at Vellore, South India, including observations on concurrent dengue. Trans. R. Soc. Trop. Med. Hyg. **63**(4), 434–445 (1969)
6. Caron, M., et al.: Recent introduction and rapid dissemination of chikungunya virus and dengue virus serotype 2 associated with human and mosquito coinfections in Gabon, Central Africa. Clin. Infect. Dis. **55**(6), e45–e53 (2012)
7. Chahar, H.S., Bharaj, P., Dar, L., Guleria, R., Kabra, S.K., Broor, S.: Co-infections with chikungunya virus and dengue virus in Delhi, India. Emerg. Infect. Dis. **15**(7), 1077 (2009)
8. Chang, S.-F., et al.: Concurrent isolation of chikungunya virus and dengue virus from a patient with coinfection resulting from a trip to Singapore. J. Clin. Microbiol. **48**(12), 4586–4589 (2010)
9. De, A., Maity, K., Jana, S., Maiti, M.: Application of various control strategies to Japanese encephalitic: a mathematical study with human, pig and mosquito. Math. Biosci. **282**, 46–60 (2016)
10. Furuya-Kanamori, L., et al.: Co-distribution and co-infection of chikungunya and dengue viruses. BMC Infect. Dis. **16**(1), 84 (2016)
11. Isea, R., Lonngren, K.E.: A preliminary mathematical model for the dynamic transmission of dengue, chikungunya and zika. arXiv preprint arXiv:1606.08233 (2016)
12. Kar, T., Batabyal, A.: Stability analysis and optimal control of an SIR epidemic model with vaccination. Biosystems **104**(2–3), 127–135 (2011)

13. Kar, T., Jana, S.: A theoretical study on mathematical modelling of an infectious disease with application of optimal control. Biosystems **111**(1), 37–50 (2013)

14. Khai, C.M., et al.: Clinical and laboratory studies on haemorrhagic fever in Burma, 1970–72. Bull. World Health Organ. **51**(3), 227–235 (1974)

15. Kularatne, S., Gihan, M., Weerasinghe, S., Gunasena, S.: Concurrent outbreaks of chikungunya and dengue fever in Kandy, Sri Lanka, 2006–07: a comparative analysis of clinical and laboratory features. Postgrad. Med. J. **85**(1005), 342–346 (2009)

16. Londhey, V., Agrawal, S., Vaidya, N., Kini, S., Shastri, J., Sunil, S.: Dengue and chikungunya virus co-infections: the inside story. J. Assoc. Physicians India **64**(3), 36–40 (2016)

17. Lukes, D.L.: Differential Equations: Classical to Controlled. Elsevier, Amsterdam (1982)

18. Madariaga, M., Ticona, E., Resurrecion, C.: Chikungunya: bending over the Americas and the rest of the world. Br. J. Infect. Dis. **20**(1), 91–98 (2016)

19. Mallela, A., Lenhart, S., Vaidya, N.K.: HIV-TB co-infection treatment: modeling and optimal control theory perspectives. J. Comput. Appl. Math. **307**, 143–161 (2016)

20. Mercado, M., et al.: Clinical and histopathological features of fatal cases with dengue and chikungunya virus co-infection in Colombia, 2014 to 2015. Eurosurveillance **21**, 22 (2016)

21. Murray, N.E.A., Quam, M.B., Wilder-Smith, A.: Epidemiology of dengue: past, present and future prospects. Clin. Epidemiol. **5**, 299 (2013)

22. Nayar, S., et al.: Co-infection of dengue virus and chikungunya virus in two patients with acute febrile illness. Med. J. Malaysia **62**(4), 335–336 (2007)

23. Neeraja, M., Lakshmi, V., Dash, P., Parida, M., Rao, P.: The clinical, serological and molecular diagnosis of emerging dengue infection at a tertiary care institute in Southern, India. J. Clin. Diagn. Res. JCDR **7**(3), 457 (2013)

24. Okosun, K., Makinde, O.: A co-infection model of malaria and cholera diseases with optimal control. Math. Biosci. **258**, 19–32 (2014)

25. Pontryagin, L., Boltyanskii, V., Gamkrelidze, R., Mishchenko, E.: The Mathematical Theory of Optimal Processes, translated from the Russian by Trirogoff, K.N.; edited by Neustadt, L.W. Interscience Publishers John Wiley & Sons. Inc., New York and London(1962)

26. Ratsitorahina, M., et al.: Outbreak of dengue and chikungunya fevers, Toamasina, Madagascar, 2006. Emerg. Infect. Dis. **14**(7), 1135 (2008)

27. Weaver, S.C., Reisen, W.K.: Present and future arboviral threats. Antiviral Res. **85**(2), 328–345 (2010)

28. Zaman, G., Kang, Y.H., Jung, I.H.: Stability analysis and optimal vaccination of an sir epidemic model. Biosystems **93**(3), 240–249 (2008)

Comparative Analysis of Machine Learning Algorithms for Categorizing Eye Diseases

Premaladha Jayaraman[1], R. Krishankumar[1], K. S. Ravichandran[2],
Ramakrishnan Sundaram[1], and Samarjit Kar[3]([✉])

[1] School of Computing, SASTRA Deemed University, Thanjavur 613 401, TN, India
[2] Rajiv Gandhi National Institute of Youth Development, Sriperumbudur, TN, India
[3] National Institute of Technology, Durgapur, WB, India

Abstracts. This paper presents a comparative study on different machine learning algorithms to classify retinal fundus images of glaucoma, diabetic retinopathy, and healthy eyes. This study will aid the researchers to know about the reflections of different algorithms on retinal images. We attempted to perform binary classification and multi-class classification on the images acquired from various public repositories. The quality of the input images is enhanced by using contrast stretching and histogram equalization. From the enhanced images, features extraction and selection are carried out using SURF descriptor and k-means clustering, respectively. The extracted features are fed into perceptron, linear discriminant analysis (LDA), and support vector machines (SVM) for classification. A pre-trained deep learning model, AlexNet is also used to classify the retinal fundus images. Among these models, SVM is trained with three different kernel functions and it does multi-class classification when it is modelled with Error Correcting Output Codes (ECOC). Comparative analysis shows that multi-class classification with ECOC-SVM has achieved high accuracy of 92%.

Keywords: Diabetic retinopathy · Glaucoma · BoF · Perceptron · SVM · LDA · CNN

1 Introduction

Machine Learning (ML) and Deep Learning (DL) have given remarkable results in various fields like bioinformatics, defence, agriculture, metrology, medicine, etc. Especially, in the field of medicine, the contribution of ML and DL is significant. Conventionally, ophthalmologists analyse the retinal fundus images and diagnose like the diseases like glaucoma, diabetic retinopathy (DR), age-related macular degeneration, macular edema, etc. This research focuses on analysing different machine learning algorithms to build models for classifying the retinal fundus images of Glaucoma, Diabetic retinopathy, and healthy eyes. There is an exponential increase in the number of people who are affected by eye diseases and there is a demand for the trained ophthalmologists to analyse & diagnose the images of the patients. In this scenario, it is better to have a computer aided

© IFIP International Federation for Information Processing 2021
Published by Springer Nature Switzerland AG 2021
Z. Shi et al. (Eds.): ICIS 2020, IFIP AICT 623, pp. 303–312, 2021.
https://doi.org/10.1007/978-3-030-74826-5_27

diagnosis system to analyse the images and diagnose the disease in a relatively faster manner. Computer aided diagnosis system comprises of the following phases: i) image acquisition, ii) image preprocessing, iii) feature extraction, iv) feature selection, and v) classification. In case of machine learning models are used for classification, then it is required to implement the feature extractor. In case of deep learning-based models, feature extraction is done by the model itself. Input images can be directly given to the model or it can be given as input after enhancement. Figure 1 sketches the flow of the research work.

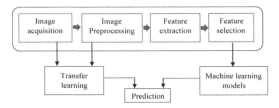

Fig. 1. Workflow of the proposed research work

1.1 Related Work

There are many profound research works carried out in the field of analysing the retinal fundus images for diagnosis and prognosis. Researchers prefer machine learning and deep learning algorithms to achieve higher accuracy in classification and prediction. Few research articles which are helpful in establishing the foundation for this research are discussed in this section. Tong et al., have done a detailed review on machine learning applications in ophthalmic imaging modalities [1]. They have emphasized the role of machine learning in automated detection and grading of pathological features in ocular diseases. These models can assist the ophthalmologists to provide high-quality diagnosis and personalized treatment plans. Hemelings et al., have used transfer learning to classify glaucoma images [2]. They have used color fundus images and 42% of the data were used for training. The model achieved 0.995 of Area Under Curve (AUC). Sengupta et al., presented a critical review on deep learning models to classify retinal diseases like age-related macular degeneration, glaucoma, and diabetic retinopathy [3]. They have presented a clear description about the databases and different algorithms to process and segment the retinal parts like optic cup, optic disc, macula, and fovea. Sarki et al., have trained a deep learning model to provide a overall classification of diseased and healthy retinal fundus images [4]. They have developed mild multi-class diabetic eye disease (DED) and multi-class DED. Fundus images annotated by the ophthalmologists were used for training the model. They have achieved an accuracy of 88.3% using VGG16 and 85.95% for mild-multiclass classification. Kamran et al., have proposed a convolutional neural network to distinguish between the different degeneration of retinal layers using optical coherence tomography (OCT) images [5]. Das et al., performed classification of retinal diseases using transfer learning [6]. They have used VGG19, pretrained model and trained it using retinal database which has glaucoma, diabetic retinopathy, and healthy eye images. Selvathi et al., have used thermal images to detect the diabetic retinopathy using support vector machines (SVM). From the thermography images, the authors extracted texture features from Gray level cooccurrence matrix (GLCM) and statistical features from RGB and HSI images [7]. The extracted texture and statistical features are classified using SVM and achieved 86.2% accuracy.

It is evident from the literature review that machine learning and deep learning algorithms provide a significant contribution in analysing the fundus images and diagnosing the diseases. In this research work, machine learning algorithms like perceptron, support vector machines, linear discriminant analysis are modelled with the key features of the images. Also, the results are compared with the pretrained model AlexNet with the retinal fundus images.

2   Materials and Methods

2.1   Image Acquisition

The retinal fundus images used for this research work are taken from High resolution fundus (HRF) images database [8]. As the name implies, HRF database provides images with resolution of 3504 × 2336. The database has 45 images, and it is categorized into glaucoma, diabetic retinopathy, healthy. Each category has 15 images. Sample images are shown in Figs. 2, 3 and 4.

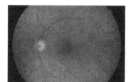
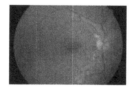
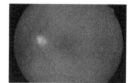

Fig. 2. Glaucoma   **Fig. 3.** Diabetic retinopathy   **Fig. 4.** Healthy eye

2.2   Image Enhancement

Image enhancement is the process which enriches the quality of the image. In this research work, contrast stretching is used for image enhancement. This technique finds the maximum and minimum pixel values of the image and normalizes the other pixels within the max-min range. The intensity values are stretched evenly between maximum and minimum. The contrast stretching is expressed mathematically as in Eq. (1).

$$I_{new} = \frac{I_{input} - I_{min}}{I_{max} - I_{min}} \times 255 \tag{1}$$

where I_{new} is the new normalized intensity value, I_{input} is the intensity value to be normalized, I_{max}, I_{min} are maximum and minimum values of the input image respectively, 255 is the global maximum value. The input image and the contrast stretched image are shown in the Figs. 5 and 6.

The contrast enhanced images are fed into feature extractor to get the interest points for classification.

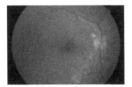

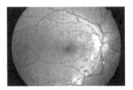

Fig. 5. Original fundus image of diabetic retinopathy

Fig. 6. Enhanced fundus image of diabetic retinopathy

2.3 Feature Extraction and Selection

The enhanced images are taken as input for this phase. Bag of words (BoW) approach is employed in this research work to carryout feature extraction and selection. BoW extracts the interest points/keypoints from the images using feature descriptors and selects the representative keypoints using k-means clustering algorithm. The cluster centers resulted from k-means algorithm are the bag of features or words that best represents the input image. These features are given as input for classification phase. k-means clustering algorithm resulted in 10,000 key points for each image and the matrix of 45 (samples) × 10,000 (interest points) is given as input to the classifier.

2.4 Classification

The extracted features are classified using perceptron, LDA, SVM, and deep learning algorithm (i.e.,) convolutional neural networks.

2.4.1 Machine Learning Algorithms

i) Perceptron: Perceptron is a commonly used machine learning algorithm for binary classification. It is a neural network with single layer. Perceptron takes the feature vector derived using BoF as input and it calculates weighted sum. Output from each neuron is triggered out using activation function in which the weighted sum is applied. The output of the network will be either $(0,1)$ or $(1,-1)$. HRF dataset has images of three classes. Three perceptron models are trained where the first model is inputted with the features of glaucoma and healthy images, the second model with features of diabetic retinopathy and healthy images, and the third model is trained with features of glaucoma and diabetic retinopathy images. The classification results are discussed in Sect. 3. The general architecture of perceptron nets is shown in Fig. 7

ii) Linear discriminant analysis (LDA): LDA can be used for classification when the data is categorical. Linear discriminant analysis tries to maximise the distance between inter class mean and minimise the intra class spread. This makes it an ideal candidate for classification problems especially binary classification. LDA projects the features in a plane making it linear in the projected plane. This projection of features leads to dimensionality reduction which is also one application of LDA. If the data follows a Gaussian mixture model, LDA yields the least classification error.

In LDA, empirical estimates of population covariance matrix and class mean are calculated from the training data. LDA uses probability to predict to which class the

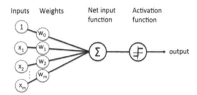

Fig. 7. Architecture of perceptron nets

data belongs to. LDA estimates the probability using Bayes' theorem which gives the probability estimate of the output class for the given input. LDA classifier is trained to classify DR & healthy images, Glaucoma & healthy images, and DR & glaucoma images. The results of these classifiers are discussed in Sect. 3.

iii) Support vector machines (SVM): Support vector machines (SVM) are used for binary classification. In this research work, it has been combined with Error Correcting Output Codes (ECOC) and it is used for multi-class classification. SVM-ECOC takes the input of all three classes and classifies it using OVO (One-versus-One) scheme. The structure of SVM-ECOC has three learners, where the first learner considers Glaucoma as positive class, DR as negative class and healthy eyes features are kept neutral. The second learner takes Glaucoma as positive class, Healthy eyes as negative class, and DR as neutral. The third learner considers DR as positive, Healthy as negative and Glaucoma as neutral. SVM-ECOC uses the Eq. (2) which makes use of coding matrix elements and predicted classification score to get the final output of the classifier [9].

$$\hat{k} = argmin\frac{\sum_{l=1}^{L}|cm_{kl}|g(cm_{kl}, score_l)}{\sum_{l=1}^{L}|cm_{kl}|} \tag{2}$$

where CM is the coding matrix with elements cm_{kl}, and $score_l$ is the predicted classification score for the positive class of learner l.

Arbitrary input is assigned to the class that minimizes the aggregation of losses for the L learners. The results of ECOC-SVM with the HRF dataset is discussed in the Sect. 3.

2.4.2 AlexNet

The original fundus images are trained with a pretrained model called AlexNet [10]. This model has 5 convolutional layers in which few of these layers are followed by max pooling layers, and three fully connected layers with softmax activation function. Dropout regularization is used to reduce overfitting. The architecture of AlexNet is depicted in Fig. 8.

This is a pretrained model which has been trained for nearly 22,000 categories. The trained model's knowledge is transferred to classify DR, glaucoma, and healthy eye images. This model has been trained for original images, enhanced images, enhanced and flipped images. The results of these model are discussed in Sect. 3.

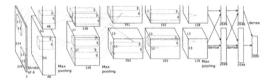

Fig. 8. AlexNet architecture [9]

3   Results and Discussion

Classification of retinal images are done by using perceptron, linear discriminant analysis, support vector machines with different kernel functions, and convolutional neural networks. Among these models, perceptron and LDA performs binary classification, ECOC-SVM and AlexNet performs multi-class classification. Binary classifiers are modelled with 10,000 features extracted and selected from the 45 input images (i.e.,) feature vector is of size 45×10000.

Perceptron: The results of perceptron algorithm are shown in Fig. 9a–9c. Figure 9a shows the error plot against the number of epochs for classifying Glaucoma & healthy images. By default, the algorithm executes 1000 epochs. "Early stopping" procedure is called to terminate the program if the algorithm fails to converge or results in oscillating pattern. Figure 9 shows the error plot against the number of epochs for classifying glaucoma & healthy images. Figure 10 shows the error plot against the number of epochs for classifying DR & healthy images. Figure 11 shows the error plot against the number of epochs for classifying DR & Glaucoma images.

LDA: In binary classification, LDA has been implemented with same feature set, whereas first model is trained with Dr and Glaucoma features, second model is trained with DR and healthy eye's features, and the third model is trained with Glaucoma and healthy eye's features. Three models of LDA have achieved an average accuracy of 84.5% and the results of individual classifier along with confusion matrix is tabulated in Table 1.

Table 1.   Classification accuracy for LDA

Input features	Accuracy
DR & glaucoma	96.7%
DR & healthy	80.0%
Glaucoma & healthy	76.7%

Figures 12, 13 and 14 show the scatter plots of the LDA models and Tables 2, 3 and 4 give the confusion matrices of the LDA models.

SVM: ECOC-SVM is trained with the features extracted from glaucoma, DR, and healthy eye images. Three learners are used in this model and it has achieved an average accuracy of 92%. These SVM learners are trained with linear, polynomial, and RBF

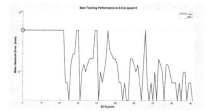

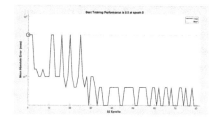

Fig. 9. Glaucoma vs Healthy

Fig. 10. DR vs Healthy

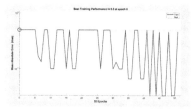

Fig. 11. DR vs Glaucoma

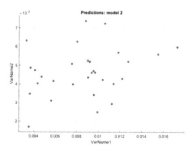

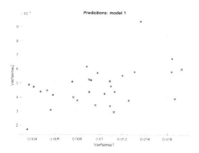

Fig. 12. LDA - Scatter plot for DR & Glaucoma

Fig. 13. LDA - Scatter plot for DR & Health

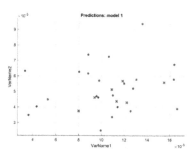

Fig. 14. LDA - Scatter plot for DR & Healthy

kernels, out of which linear kernel outperforms other kernels. ROC curves for the learners with linear kernel are given in the Figs. 15, 16 and 17.

Table 2. LDA – confusion matrix for DR & Glaucoma

True class	Predicted class	
	DR	G
DR	14	1
G	0	15

Table 3. LDA – confusion matrix for DR & Healthy

True class	Predicted class	
	DR	Healthy
DR	10	5
Healthy	1	14

Table 4. LDA – confusion matrix for Glaucoma & Healthy

True class	Predicted class	
	G	Healthy
G	11	4
Healthy	3	12

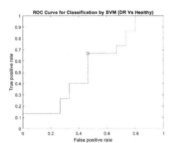

Fig. 15. DR & Healthy

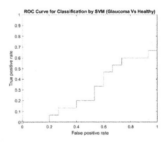

Fig. 16. Glaucoma & Healthy

AlexNet: AlexNet is trained with three sets of images (i.e.,) original images, enhanced images, enhanced flipped images. The accuracy achieved by these models are tabulated in Table 5. CNN extracts the features from images by itself and hence manual feature extraction, selection process is not required. The training progress and the loss incurred at each iteration is shown in Figs. 18, 19 and 20. From Table 5, it is inferred that the model trained with the original images provided better accuracy when compared to other two models.

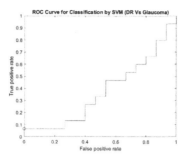

Fig. 17. DR & Glaucoma

Table 5. Results of AlexNet

	Original images	Enhanced images	Enhanced-flipped images
Accuracy	83.33	66.67%	75%

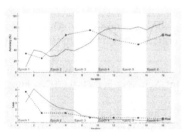

Fig. 18. Original images

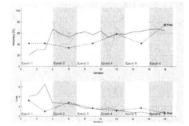

Fig. 19. Enhanced images

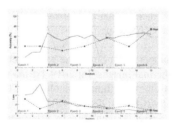

Fig. 20. Enhanced flipped images

4   Conclusion

This research work is carried out for classifying the retinal diseases, diabetic retinopathy, glaucoma, from healthy retina images. The dataset has 45 high resolution fundus images, with 15 images are there for each category (i.e.,) DR, Glaucoma, and healthy. These images are preprocessed for quality enhancement and significant interest points are extracted from each image to formulate the feature set to train different machine leaning models. These images are directly given as input for training the deep learning model because it has convolution layers to extract the significant features from the images. With respect to classification accuracy, SVM with linear kernel provided higher accuracy for multi-class classification. It is inferred that if the number of features is greater than the number of samples then the linear kernel will provide better accuracy. Huge drop in accuracy is found with the perceptron and LDA is because of the nature of the input vector which is linearly non-separable, and the dimensions of the features are high. Classification accuracy achieved by AlexNet is also less than the SVM model. If the model parameters are fine-tuned and the number of training samples are increased, then increase in accuracy can be expected.

References

1. Tong, Y., Lu, W., Yu, Y., et al.: Application of machine learning in ophthalmic imaging modalities. Eye Vis. **7**, 22 (2020). https://doi.org/10.1186/s40662-020-00183-6
2. Hemelings, R., et al.: Accurate prediction of glaucoma from colour fundus images with a convolutional neural network that relies on active and transfer learning. Acta Ophthalmol. **98**(1), e94-100 (2020)
3. Sengupta, S., Singh, A., Leopold, H.A., Gulati, T., Lakshminarayanan, V.: Ophthalmic diagnosis using deep learning with fundus images–a critical review. Artif. Intell. Med. **1**(102), 101758 (2020)
4. Sarki, R., Ahmed, K., Wang, H., Zhang, Y.: Automated detection of mild and multi-class diabetic eye diseases using deep learning. Health Inf. Sci. Syst. **8**(1), 1–9 (2020). https://doi.org/10.1007/s13755-020-00125-5
5. Kamran, S.A., Saha, S., Sabbir, A.S., Tavakkoli, A.: Optic-Net: a novel convolutional neural network for diagnosis of retinal diseases from optical tomography images. In: 2019 18th IEEE International Conference on Machine Learning and Applications (ICMLA), Boca Raton, FL, USA, pp. 964–971 (2019). https://doi.org/10.1109/ICMLA.2019.00165
6. Das, A., Giri, R., Chourasia, G., Bala, A.A.: Classification of retinal diseases using transfer learning approach. In: 2019 International Conference on Communication and Electronics Systems (ICCES), Coimbatore, India, pp. 2080–2084 (2019). https://doi.org/10.1109/ICCES45898.2019.9002415
7. Selvathi, D., Suganya, K.: Support vector machine based method for automatic detection of diabetic eye disease using thermal images. In: 2019 1st International Conference on Innovations in Information and Communication Technology (ICIICT), Chennai, India, pp. 1–6 (2019). https://doi.org/10.1109/ICIICT1.2019.8741450
8. Odstrcilik, J., et al.: Retinal vessel segmentation by improved matched filtering: evaluation on a new high-resolution fundus image database. IET Image Proc. **7**(4), 373–383 (2013)
9. Sundaram, R., Ravichandran, K.S.: An automated eye disease prediction system using bag of visual words and support vector machine. J. Intell. Fuzzy Syst. **36**(5), 4025–4036 (2019)
10. Krizhevsky, A., Sutskever, I., Hinton, G.E.: Imagenet classification with deep convolutional neural networks. Commun. ACM **60**(6), 84–90 (2017)

Extended Abstract

Extensions of Dynamic Programming for Combinatorial Optimization and Data Mining

Mikhail Moshkov

Computer, Electrical and Mathematical Sciences & Engineering Division,
King Abdullah University of Science and Technology (KAUST),
Thuwal 23955-6900, Saudi Arabia
mikhail.moshkov@kaust.edu.sa

Abstract. In contrast with conventional dynamic programming algorithms that return one solution, extensions of dynamic programming allows us to work with the whole set of solutions or its essential part, to perform multi-stage optimization relative to different criteria, to count the number of solutions, and to find the set of Pareto optimal points for bi-criteria optimization problems. The presentation is based on the results considered in three books published or accepted by Springer.

Keywords: Dynamic programming extensions · Bi-criteria optimization · Multi-stage optimization

The conventional dynamic programming algorithms for optimization problems include a structure of sub-problems of the initial problem, a way to construct a solution for a sub-problem from solutions of smaller sub-problems, and solutions for the smallest sub-problems. Such algorithms return only one solution.

In books [1–3] we consider extensions of dynamic programming that allow us (i) to make multi-stage optimization relative to a sequence of criteria, (ii) to describe the whole set of solutions or its essential part, (iii) to count the described solutions, and (iv) to construct the set of Pareto optimal points for a bi-criteria optimization problem. In this presentation, we discuss mainly results considered in the book [1] and mention some results presented in the books [2, 3].

In [1], we apply extensions of dynamic programming to the study of decision trees and rules for conventional decision tables with single-valued decisions, to the optimization of element partition trees for rectangular meshes which are used in finite element methods for solving PDEs, and to multi-stage optimization for such classic combinatorial optimization problems as matrix chain multiplication, binary search trees, global sequence alignment, and shortest paths.

In particular, for the decision trees, the applications of multi-stage optimization include the study of totally optimal (simultaneously optimal relative to a number of cost functions) decision trees for Boolean functions, improvements on the bounds on the depth of decision trees for diagnosis of constant faults in read-once circuits over monotone basis, computation of minimum average depth for a decision tree for sorting eight elements (a problem that was open since 1968), study of optimal reducts and

© IFIP International Federation for Information Processing 2021
Published by Springer Nature Switzerland AG 2021
Z. Shi et al. (Eds.): ICIS 2020, IFIP AICT 623, pp. 315–316, 2021.
https://doi.org/10.1007/978-3-030-74826-5

decision trees for modified majority problem, and designing an algorithm for the problem of reduct optimization.

The applications of bi-criteria optimization include the comparison of different greedy algorithms for construction of decision trees, analysis of trade-offs for decision trees for corner point detection (used in computer vision), study of derivation of decision rules from decision trees, and a new technique called multi-pruning of decision trees which is used for data mining, knowledge representation, and classification. The classifiers constructed by multi-pruning process often have better accuracy than the classifiers constructed by CART.

In [2], we generalize the considered tools to the case of decision tables with many-valued decisions. We also study inhibitory trees and rules, which instead of expressions "decision = value" use expressions "decision $\neq$ value". Inhibitory trees and rules can, sometimes, describe more information derivable from a decision table than decision trees and rules.

In [3], we apply the extensions of dynamic programming including multi-stage and bi-criteria optimization, and counting of optimal solutions to the following 11 known combinatorial optimization problems: matrix chain multiplication, global sequence alignment, optimal paths in directed graphs, binary search trees, convex polygon triangulation, line breaking (text justification), one-dimensional clustering, optimal bitonic tour, segmented least squares, matchings in trees, and 0/1 knapsack problem.

Acknowledgments. Research reported in this publication was supported by King Abdullah University of Science and Technology (KAUST).

References

1. AbouEisha, H., Amin, T., Chikalov, I., Hussain, S., Moshkov, M.: Extensions of Dynamic Programming for Combinatorial Optimization and Data Mining, Intelligent Systems Reference Library, vol. 146. Springer (2019). https://doi.org/10.1007/978-3-319-91839-6
2. Alsolami, F., Azad, M., Chikalov, I., Moshkov, M.: Decision and Inhibitory Trees and Rules for Decision Tables with Many-valued Decisions, Intelligent Systems Reference Library, vol. 156. Springer (2020). https://doi.org/10.1007/978-3-030-12854-8
3. Mankowski, M., Moshkov, M.: Dynamic Programming Multi-Objective Combinatorial Optimization, Studies in Systems, Decision and Control, vol. 331. Springer (2021). https://doi.org/10.1007/978-3-030-63920-4

Author Index

.

Printed in the United States
by Baker & Taylor Publisher Services